Cultural Development of Mathematical Ideas

Drawing upon field studies conducted in 1978, 1980, and 2001 with the Oksapmin, a remote Papua New Guinea group, Geoffrey B. Saxe traces the emergence of new forms of numerical representations and ideas in the social history of the community. In traditional life, the Oksapmin used a counting system that makes use of 27 parts of the body; there is no evidence that the group used arithmetic in prehistory. As practices of economic exchange and schooling have shifted, children and adults unwittingly reproduced and altered the system in order to solve new kinds of numerical and arithmetical problems, a process that has led to new forms of collective representations in the community. While Dr. Saxe's focus is on the Oksapmin, the insights and general framework he provides are useful for understanding shifting representational forms and emerging cognitive functions in any human community.

Dr. Geoffrey B. Saxe has conducted research on mathematical cognition and culture in a variety of settings, including remote parts of Papua New Guinea, urban and rural areas of northeastern Brazil, and elementary and middle school classrooms in the United States. His prior books include *Culture and Cognitive Development: Studies in Mathematical Understanding* (1991) and *Social Processes in Early Number Development* (with S. Guberman and M. Gearhart, 1987). He is currently a professor at the Graduate School of Education, University of California, Berkeley.

LEARNING IN DOING: SOCIAL, COGNITIVE AND COMPUTATIONAL PERSPECTIVES

SERIES EDITOR EMERITUS

John Seely Brown, *Xerox Palo Alto Research Center*

GENERAL EDITORS

Roy Pea, *Professor of Education and the Learning Sciences and Director, Stanford Center for Innovations in Learning, Stanford University*

Christian Heath, *The Management Centre, King's College, London*

Lucy A. Suchman, *Centre for Science Studies and Department of Sociology, Lancaster University, UK*

BOOKS IN THE SERIES

The Construction Zone: Working for Cognitive Change in School
Denis Newman, Peg Griffin, and Michael Cole

Situated Learning: Legitimate Peripheral Participation
Jean Lave and Etienne Wenger

Street Mathematics and School Mathematics
Terezinha Nunes, David William Carraher, and Analucia Dias Schliemann

Understanding Practice: Perspectives on Activity and Context
Seth Chaiklin and Jean Lave, Editors

Distributed Cognitions: Psychological and Educational Considerations
Gavriel Salomon, Editor

The Computer as Medium
Peter Bøgh Anderson, Berit Holmqvist, and Jens F. Jensen, Editors

Sociocultural Studies of Mind
James V. Wertsch, Pablo del Rio, and Amelia Alvarez, Editors

Sociocultural Psychology: Theory and Practice of Doing and Knowing
Laura Martin, Katherine Nelson, and Ethel Tobach, Editors

Mind and Social Practice: Selected Writings of Sylvia Scribner
Ethel Tobach et al., Editors

(Continued after the Index)

Cultural Development of Mathematical Ideas

Papua New Guinea Studies

GEOFFREY B. SAXE
University of California, Berkeley

with
INDIGO ESMONDE
University of Toronto

CAMBRIDGE
UNIVERSITY PRESS

CAMBRIDGE UNIVERSITY PRESS
Cambridge, New York, Melbourne, Madrid, Cape Town,
Singapore, São Paulo, Delhi, Mexico City

Cambridge University Press
32 Avenue of the Americas, New York, NY 10013-2473, USA

www.cambridge.org
Information on this title: www.cambridge.org/9780521761666

First published 2012

Printed in the United States of America

A catalog record for this publication is available from the British Library.

Library of Congress Cataloging in Publication Data
Saxe, Geoffrey B.
Cultural development of mathematical ideas : Papua New Guinea studies / Geoffrey B. Saxe ; with Indigo Esmonde.
p. cm. – (Learning in doing)
Includes bibliographical references and index.
ISBN 978-0-521-76166-6 (hardback)
1. Cognition and culture – Papua New Guinea. 2. Number concept – Case studies. 3. Constructivism (Education) 4. Social change – Papua New Guinea. I. Esmonde, Indigo. II. Title.
BF311.S3764 2012
155.8′49912–dc23 2011037488

ISBN 978-0-521-76166-6 Hardback

Additional resources for this publication at http://culturecognition.com and http://www.youtube.com/user/CultureAndNumber1

In memory of my parents, Carl and May

Contents

Figures

Series Foreword

This series for Cambridge University Press is widely known as an international forum for studies of situated learning and cognition. Innovative contributions are being made by anthropology; by cognitive, developmental, and cultural psychology; by computer science; by education; and by social theory. These contributions are providing the basis for new ways of understanding the social, historical, and contextual nature of learning, thinking, and practice that emerges from human activity. The empirical settings of these research inquiries range from the classroom to the workplace to the high-technology office, and on to learning in the streets and in other communities of practice. The situated nature of learning and remembering through activity is a central fact. It may appear obvious that human minds develop in social situations and extend their sphere of activity and communicative competencies. But cognitive theories of knowledge representation and learning alone have not provided sufficient insight into these relationships. This series was born of the conviction that new, exciting interdisciplinary syntheses are under way as scholars and practitioners from diverse fields seek to develop theory and empirical investigations adequate for characterizing the complex relations of social and mental life, and for understanding successful learning wherever it occurs. The series invites contributions that advance our understanding of these seminal issues.

Roy Pea
Christian Heath
Lucy A. Suchman

Preface

My aim in this book is to contribute to our understanding of relations between culture and thought. The context for key ideas and argumentation is my research in the Oksapmin region of Papua New Guinea, an area that encompasses two small valleys in the central highlands of the New Guinea island. The circumstances of the Oksapmin people have provided me with an extraordinary opportunity for research and reflection. The Oksapmin have historically used a 27-body-part counting system, and the change in their numerical practices since first contact by westerners in 1938 has been remarkable. In the pages that follow, I report my investigation of the history of quantification practices among the Oksapmin, with a particular focus on shifts that occurred over my three stints of fieldwork in 1978, 1980, and 2001. Having this historical perspective brings culture-cognition relations into unusual relief.

A Trajectory of Research – Tensions and Partial Resolutions

In a 2005 commentary on some of my Oksapmin research (Saxe & Esmonde, 2005), Anna Sfard identified a conceptual shift in my work over a period of 23 years (Sfard, 2005), which includes, at one end, my first visit to Oksapmin in 1978 and, at the other, my last visit in 2001 and the publication of the manuscript in 2005. She pointed to a movement away from a structural developmental perspective, building on the work of Piaget, to a different approach to problems of culture-cognition relations, one that focused on collective practices and the reproduction and alteration of mathematical ideas in communities.

My first reaction was to disagree with Anna. I saw my own development as having had no radical disjunctions, but rather as a continuous process that was firmly grounded in my early training. In preparing this book, however, I came to understand that what I had regarded as minor alterations constituted a marked conceptual shift in analytic approach. This preface provides a brief history of my disciplinary background and the trajectory of my work, providing an orientation to the perspectives that have shaped differently the waves of the Oksapmin fieldwork. I begin the short narrative with my undergraduate studies in psychology, and I conclude with the current treatment, which locates the reproduction and alteration of representations and ideas in community life as core conceptual and empirical problems that this volume will address. I believe knowing about my shifting orientation may be useful in understanding some nuanced methodological differences across the time frames.

Moral Reasoning in a Yu'pik Community

In 1969, as an undergraduate psychology major at U.C. Berkeley, I lived for a summer in an Eskimo village and conducted independent field research on moral development. My goal was to extend the structural-developmental treatment of moral reasoning elaborated by Lawrence Kohlberg (Kohlberg, 1969) to an investigation of the reasoning of Eskimo villagers. The methods that I used reflected a prominent paradigm of the time: A researcher, in garnering supportive evidence that a posited stage sequence was universal, sampled divergent communities and interviewed samples of people at different age levels, coding interviews by means of preestablished schemes. Following this paradigm, I adapted Kohlberg's notion of moral dilemmas into versions that I thought would be relevant to village life, and I interviewed village children and adults sampled from preplanned age groups. Although it was not my intention, I also became a participant observer in the community, simply by living among the Eskimos. I bathed daily with Yu'pik elders in an earthen sauna, a traditional practice; I attended ceremonies that blended indigenous and Russian Orthodox practices; I was an observer of interpersonal disputes inflamed by home brew and firearms; and I attempted (with only partial success) to thwart the rumors that I was a Russian spy. By the end of the summer, I left the region unclear whether I belonged in the Eskimo village as an odd-looking Yu'pik or in Berkeley as a student.

Back in Berkeley, I found myself struggling to make sense of the fieldwork. I learned Kohlberg's coding system, analyzed my interviews, and found that Kohlberg's codes reduced the complex moral issues in the villagers' lives to reasoning represented by the first two of Kohlberg's six stages. I was pleased that my findings converged with other cross-cultural studies from this era of research on moral development, but I was troubled as well, because my findings did not reflect the nuanced world of the Eskimo village, dense in moral complexities.

Upon reflection, I felt that my work failed to represent critical dynamics of culture-cognition relations in the moral domain, and I wondered if the same critique could be made of cognitive developmental research. The approach's emphasis on universals pushed the diversity and complexity of historical and cultural processes into a distantly backgrounded "context," and a question emerged that would later become the core of my Oksapmin research. How does one preserve the explanatory power of a developmentalist orientation to the progressive differentiation of forms of thought while creating techniques that yield greater insight into culture-cognition relations? The necessary focus on the individual's elaboration of cognitive structures in developmental treatments left culture as not much more than an appendage. But these issues were set aside after I entered graduate school in cognitive developmental psychology at U.C. Berkeley with Jonas Langer (1970–1975) and then held an 18-month postdoctoral fellowship in the cognitive neurosciences with Peter Wolff at Harvard University Medical School and Howard Gardner at the Boston Veterans Administration Hospital (1975–1977).

Graduate Study and Early Career

Throughout my graduate training and early post-Ph.D. years, I engaged with a thick reading of Piaget's and related developmental theories. Through a Piagetian lens, I examined, with graduate student colleagues and under the wise guidance of Jonas Langer, questions of the origins of human knowledge and corroborative evidence for the emergence of universal stages in logico-mathematical and physical understandings (Piaget, 1952). While I also read with great interest the work of Heinz Werner and Lev Vygotsky, their writings did not directly inform my dissertation research.

My dissertation engaged the long-standing discussion about relations between representation and understanding in cognitive development. In establishing a context for my work, I noted that Piaget, in his treatment of the development of numerical understandings, made offhand arguments about developmental relations between cultural systems of representing numbers, such as counting, and fundamental numerical understandings like number conservation. In my dissertation I endeavored to systematically explore counting-conservation relations, showing that counter to Piaget's argument, children used counting to mediate numerical evaluations prior to their understanding conservation. I developed arguments that showed how one could work with core elements of a constructivist framework like Piaget's to produce a more adequate treatment of counting-conservation relations, even though my findings were at odds with Piaget's claim.

My dissertation led me to an NIMH postdoctoral fellowship at Children's Hospital Medical Center in Boston and the Boston Veterans Administration Hospital (VA) in the developing field of cognitive neuroscience. My focus was on atypical populations – children with varied forms of learning disabilities related to number and adults who had suffered focal brain lesions. With these populations, I sought to explore dissociations between representation and understanding in the domain of numerical thought. The postdoctoral period opened my eyes to questions of physiological mechanisms and the organization of cognitive functioning. At the VA I regularly attended rounds and engaged in a number of case studies involving people who had suffered loss of cognitive functioning. At Children's Hospital I worked in a learning disabilities clinic and conducted related work on children with focal difficulties with number. As I worked to find ways to help children and patients to circumvent cognitive impairments, I became attuned to the character of atypicalities. It was during this period in Boston that I received a letter from David Lancy, at the time Principal Research Officer of the Ministry of Education in Papua New Guinea. David wrote to invite me to submit a grant proposal to study the development of numerical cognition in indigenous groups. He had learned of my dissertation research and was looking for recent Ph.D.s who might be up for an adventure into other worlds. For me, the glove fit well: I was excited about the possibility of extending my interest in comparative approaches to communities far removed from me as well as the opportunity to participate in another cultural world. I wrote a small grant proposal, which was funded. In the interim, I received my first academic appointment, at the Graduate Center of the City University of New York in a Ph.D. program in Educational Psychology.

I delayed the Papua New Guinea trip for a year, and during that academic year in New York, I met two doctoral students in anthropology, Tom Moylan and Virginia Guilford, who were bound for Papua New Guinea to do dissertation research in linguistic anthropology. We discussed our Papua New Guinea research plans, and several months after Tom was in the field, I received a letter from him. Tom wrote of his early experiences with the Oksapmin, a remote group, and their number system, and invited me to come for a stay and study. I worked out an arrangement with David

Lancy at the Ministry of Education to engineer the Oksapmin area as a target for my own work. After a lot of help from my friends and family, I took off for New Guinea at the end of my first year at the Graduate Center.

Oksapmin 1978

My first entry into the remote community was unnerving. After a couple of failed landing approaches trying to find holes in rapidly moving clouds, the pilot put down his small Cessna on a notoriously hard-to-land-on dirt strip adjacent to a Baptist mission station. The airstrip was about 12 kilometers from Kweptanap, the village where Tom and Virginia resided. After a quick greeting and questioning, I took up the resident missionary's offer for a ride on the back of his motored dirt bike as far as the trail permitted to Tom and Virginia's Kweptanap location. Of course, we were a spectacle to onlookers. What I saw on the bumpy and muddy trail appeared as a fantasy world. The terrain was alien, and the people were so different from me. We passed bare-breasted women in grass skirts carrying string bags of sweet potatoes as well as fierce-looking men with bows and arrows, their pierced noses displaying boars' tusks and beetles' horns, dressed in cane belts and penis sheaths. When the dirt bike could go no farther, the missionary pointed to a densely forested trail. I set off worried that only a few hours of daylight remained. After an hour's slog through deep clay mud, I encountered some boys hunting with their bows. Excited to find me, they helped guide me (patiently, at a pace that I could follow) to a mountain bluff, where they called out with a mountain yodel to signal our impending arrival at Kweptanap. I doubted that Tom expected me to take him up on his invitation, but there I was.

The 1978 studies represented my first effort to engage the tension between a developmental analysis of universals in numerical cognition and culture-specific cognitive developments. My resolution was to focus on a unique and important historical development in Oksapmin communities, the 27-body-part counting system, and Oksapmin children's developing understanding and use of it. With the orienting help of Tom and Virginia, I investigated developmental shifts in one-to-one correspondence operations as evidence of the ways that children turn their culture's counting system into forms for counting and doing simple arithmetic operations (Studies 8-1 and 8-2, described in Part III). While my findings preserved a Piagetian orientation, they represented my determination to investigate developmental processes that were tied to culture-specific systems. This work opened up new lines of inquiry on my subsequent visits.

Oksapmin 1980

Returning to the area with Tom Moylan and Maryl Gearhart (who later became my partner in life) in 1980, I conducted developmental studies with a clearer and better-articulated link to people's daily activities and their use of the 27-body-part system to solve emerging problems.[1] In comparison to my prior work, my approach was decidedly more functional though it retained its Piagetian-developmental orientation; I wanted to capture the way that people were adapting and extending the body-part system to solve problems in daily activities. In a departure from my earlier concentration on the assessment of cognitive structures as articulated by Piaget, I now focused on cognition in practice, in the form of local people solving local problems. In this shift in investigative posture, which at the time I hardly appreciated as a shift, I had been reading works in the sociocultural tradition. These included Luria's seminal studies in the postrevolutionary Soviet Union (Luria, 1976) and the work of Michael Cole (and his collaborators) on cross-cultural methods for investigating the representations of knowledge of non-Western peoples (Cole, Gay, Glick, & Sharp, 1971; Cole & Scribner, 1974; Scribner & Cole, 1981). I also revisited the developmental perspectives of Werner and Kaplan in their treatment of symbol formation (Werner & Kaplan, 1963). These efforts led to developmental studies on the Oksapmin's use of the body system to measure string bags (Study 8-3), to compute arithmetic problems in bush schools (Studies 9-1 and 9-2), and to solve emergent problems in economic transactions in trade stores (Study 4-1).

Oksapmin 2001

Twenty-one years later, I returned to the Oksapmin, this time with my 19-year-old son Josh and two graduate students, Indigo Esmonde and Cliff McIntosh. My analytic concerns had deepened during 20 years of research on collective activities and cognitive development in diverse settings, with projects that included the study of candy selling and straw weaving in Brazil, number games in the homes of Brooklyn preschoolers, educational games for elementary students in Los Angeles, and classroom practices and mathematics education. Across these projects, I had developed a commitment

[1] Tom pursued his own research on the Oksapmin language and consulted with me on studies of number. Maryl collected field observations for a Ministry of Education curriculum project directed by Randall Souviney.

to exploring the interplay between individuals' developing understandings and the recurring problems that they cognized and solved in collective activities. For example, in 1985, in my work in Brazil with young candy sellers, my concern was to explore the structure of the children's practice of purchasing from wholesale stores, marking up the candy for retail sale, selling it to customers, and then searching for a better buy at wholesale stores. In this complex practice, the numerical problems that emerged for the young entrepreneurs (some of whom had virtually no schooling) arose from an interplay between their own understandings and the conditions of their practice, which was steeped in norms, values, and social conventions. My research showed that sellers, through their engagement with selling, had developed specialized forms of numerical representations and understandings deeply interwoven with their practice, and that their use of these forms was both powerful and insightful (and quite different from those of their nonselling peers).

When I first considered returning to Oksapmin, I conceptualized the analytic focus in simple terms: I would ask, had people's approaches to quantification changed since my early visits? And if so, how? With these questions, I would extend my prior work on understanding cognitive development in collective practices to incorporate a historical perspective. But once in the field, I found that my initial "simple" questions quickly morphed into an engagement with foundational issues: ways of understanding social history as a central theoretical construct in the conceptualization of cognitive processes as well as ways of understanding cognitive processes in the historical development of mathematical ideas. Oksapmin in 2001 became an arena to flesh out and pursue new ideas as I framed new questions and empirical projects.

In the thinking that came to usefully organize my work in 2001, people's present-day cognitive activities were reproducing as well as altering the collective practices in which they were participants, including forms of numerical representations they were using and the functions that these forms were serving as they pursued their own goals. At the same time, the complex dynamics of historical processes made manifest in collective activities framed people's goal-directed actions. This dialectical relationship between individual and social history that emerged in people's engagement with collective activities became an important frame for me in 2001. Bringing strands of my earlier 1978 and 1980 Oksapmin work forward, I focused on two arenas for analysis in 2001. The first arena was people's engagement with collective practices of economic exchange (Studies 4-2, 4-3, 4-4, 5-1, 5-2, 6-1, 6-2, and 6-3), and the second was students' and teachers' participation with schooling (Studies 10-1, 10-2, 11-1, and 11-2). Though the shift in my own thinking about knowledge and its development seemed continuous and straightforward, the reader may detect a difference in analytic posture between the studies conducted between 1978 and 2001. To make the volume more coherent, I have organized it by making use of my current conceptual commitments related to understanding dynamic relations between individual, collective activity, and social history in the cultural development of mathematical ideas.

For readers interested in viewing video of scenes from Oksapmin communities in 1978, 1980, and 2001, please consult the following Web site: http://culturecognition.com/ or the YouTube channel: http://www.youtube.com/user/CultureAndNumber1. The videos include displays of the Oksapmin 27-body-part counting system and innovative uses of the system in arithmetical problem solving. The channel also includes video on collective practices, such as economic exchange and schooling; subsistence activities, such as hunting and gardening; and playing of (musical) percussion instruments.

Acknowledgments

The fieldwork on which this book is based could not have been accomplished without the help of many. The Oksapmin people are of course at the top of my list. Many Oksapmin extended their friendship and provided me with much-needed guidance, and I hope that this volume will contribute to an enduring record of their complex and remarkable community history. I am especially indebted to Bukura and his two sons, Aseng and Gasam (now known as Bonny and Henry), along with their extended family. They provided many opportunities to learn from them in 1978, 1980, and 2001, whether indulging me as I learned to shoot (poorly) with a bow and arrow, tolerating my efforts to keep up with them on difficult trails that they navigated with agility and grace, or instructing me in Oksapmin ways of number and other matters. I'm grateful for their help and support, and will always treasure my memories of them. Yangeh and his extended family were wonderfully supportive during my several visits. Many more elders, adults, and children tutored me in local practices, and I am grateful for their patient teaching. I am also indebted to the Oksapmin headmasters, teachers, translators, Mission Aviation Fellowship (MAF) bush pilots, Ministry of Education officials, and local missionaries who helped me during my research.

Funders and consultants from a number of organizations provided critical resources for this research. The studies I conducted in 1978 and 1980 were completed with the counsel of research officers from the Ministry of Education in Papua New Guinea; I am particularly thankful to David Lancy in 1978 and Randall Souviney in 1980. Funding for the 1978 work was provided by a grant from the Ministry of Education of Papua New Guinea (no. 534), a small award from the Wenner-Gren Foundation, and a postdoctoral fellowship from the National Institute of Mental Health (MH07361). The

National Institute of Education supported the write-up and analysis of the data I gathered on my 1980 trip (G-78–0076 and G-80–0119).

My studies in 2001 were completed with a grant from the Spencer Foundation (no. 200100026) and the organizational support of Wilfred Kaleva and the late (sadly) Rex Matang of the Glen Lean Ethnomathematics Centre in Goroka. Cliff McIntosh, a Ph.D. student, helped in my preparations for the trip and the initial phase of the research. I am indebted to Indigo Esmonde for her role in 2001. A second-year Ph.D. student at the time, Indigo's many contributions in the field included helping to pilot interview tasks, generating fieldnotes, and interviewing participants; in Berkeley, she helped to code and analyze volumes of interview records, assisted by a very adept undergraduate assistant, Serena Lehman. Indigo also took the lead in writing fieldnotes for Study 10-1 and drafting parts of the chapter that discusses it. I am also grateful to Michael Wesch, an anthropology doctoral student at the time, for his assistance in Tabubil. Finally, I thank my son Josh Saxe, who at age 19 in 2001 was a magnet for Oksapmin adolescents and children. He also completed his own independent historical study of the construction of the Bak airstrip. His presence in Oksapmin was special for me, and I am grateful to him for this pleasure.

Several graduate students at U.C. Berkeley have read and commented on earlier drafts of the manuscript, and I appreciate their help in clarifying points and editorial suggestions. They include Evra Baldinger, Kenton de Kirby, Nicole Louie, Nicole Leveille Buchanan, Katherine Lewis, Yasmin Sitabkhan, and Dragan Trninic. Roy Pea read a draft of the manuscript, and I'm grateful for his thoughtful comments. Maryl Gearhart, my partner in life (and with whom I worked on the 1980 research), has played the role of constructive critic in much of my thinking and scholarship; she has helped me turn turgid prose into readable narrative and shape loose thoughts into coherent arguments. My sister, Karen Westheimer, provided advice on several figures in the volume, and in preparation for my first 1978 field trip, she and her late husband, Dr. Eric Westheimer, provided critical support. I am grateful to Karen and still mourn the loss of Eric.

My preparatory training situated me well for the work in Oksapmin across my periods of fieldwork. I am indebted to Jonas Langer, with whom I studied as a graduate student. His influence on my own thinking about cognitive development has been tremendous, and over the course of my varied intellectual pursuits I've benefited in many ways from this early training. In the field in 1978, I drew on the wonderful support of, and good advice from, Tom Moylan and Virginia Guilford, two anthropologists living in Kweptanap. Tom and Virginia were extraordinarily generous, putting me up during my first extended stay. Marshall Lawrence, a linguist affiliated with the Southern Institute of Linguistics and a longtime resident of the Divanap area, provided good counsel in my early work.

Support for the preparation of this book was provided by the Center for Advanced Study in the Behavioral Sciences and the Rockefeller Foundation's Conference and Study Center fellowship in Bellagio, Italy. These two organizations provided extraordinary environments for me to work on this book and related manuscripts. My appreciation is also extended to the National Archives and Public Records services of Papua New Guinea for providing access to original patrol reports from the Telefomin[1] and Oksapmin Patrol Posts and granting permission to reproduce images in this volume. The National Library of Australia provided permission to publish photographs from 1938, the date of first Western contact with this remote region (from the papers of John Russell Black, ms. 8346, Series 3, Folder 13: nos. 124, 125, 128, 131, 133). Thanks go also to Denise Schmandt-Besserat and the University of Texas Press for permission to reproduce four images from her *Before Writing, Volume I: From Counting to Cuneiform* (Figure 5, p. 11; Figure 6, p. 11; Figure 568, p. 111; and Figure 89, p. 136). Finally, thanks are due to Oceania Publications for permission to reproduce maps from *Children of Afek* (Oceania Monograph, no. 40) drafted by Aub Chandica (nos. 1, 2, and 46).

This book draws on prior publications based on my fieldwork with the Oksapmin in 1978 (Saxe, 1981a, 1983), 1980 (Saxe, 1981c, 1982a, 1982b, 1985; Saxe & Moylan, 1982), and 2001 (Saxe & Esmonde, 2004, 2005). Parts of the text in this volume have been adapted from these earlier works, but much is new material. Further, the findings previously reported have been interpreted here in the context of a more elaborated, overarching framework for understanding the dynamic relations between historical and cognitive-developmental processes.

[1] Note that the spelling of the area is "Telefomin" and the spelling of the group is "Telefolmin."

Introduction

In daily life, we represent number in many ways: in specialized language forms like counting words, in written inscriptions like Hindu-Arabic or Roman numerals, and in representations of magnitude like number lines or coordinate graphs. There are also the many quantitative expressions that populate everyday speech, like "how much?" and "a lot," as well as temporal and spatial expressions like "before" and "after" or "in front of" and "behind." These varied representational forms – counting words, written inscriptions, graphical representations, everyday speech – afford ways to communicate and think about fundamental numerical ideas like order, magnitude, and arithmetical relations.

In this volume, I take up an issue core to the cognitive sciences: *Where do the numerical representations and ideas about number that we use in daily life come from?* A casual respondent might explain the origins of representations and ideas with statements such as "People learn them in school," or "We learn them from the languages we learn as young children," or "on the job," or "in conversation with others." But such answers beg the question, both historically and developmentally: Where do the representations that are learned in school, on the job, or in conversations with others come from? To pursue answers to this question, I will engage in an analysis of psychological, sociological, and historical processes.

One line of argumentation about developmental origins posits that individuals' representations directly reflect their social worlds – that is, the representations that we use are copies of representational systems used by others – and social history is not raised as a focus of inquiry in those accounts. Learning a community's number system is a matter of copying its representational forms, including (for some theorists) the logic of number embodied in the system. In this experiential view the study of the origins of numerical representations and ideas requires an analysis of the social environments in which individuals are reared. I will argue that an experiential account cannot in itself be adequate. People are exposed to instances of miscounting, errors in arithmetic, and varied speech acts that invoke quantities that can be taken up in many different ways. So, if we simply copy representations and ideas from a social world, how is it that we all end up with a relatively coherent elementary counting and arithmetical knowledge?

Another line of argument about developmental origins holds that our nervous systems are prewired with numerical understandings, or at least with strong predispositions set to support the acquisition of numerical representations. Cognitive neuroscientists use investigative techniques such as brain imaging, single cell recording, and analyses of cognitive performance of patients with focal brain lesions to study relations between brain structures and the processing of numerical information. Their findings reveal that as people and animals cognize number, specific areas of the brain are activated. These investigations and models of brain function can shed light on cortical mechanisms and, more generally, the material bases of cognitive processes. But biological approaches can never be entirely satisfactory, as they fail to deal with such issues as how and why collective representations like number words take form differently over the social histories of different human communities. Moreover, they cannot well address how diverse representations and ideas that are taught in school, that others use on the job, or that speakers use in conversation originate.

These questions about origins of representations and ideas and their varied nuances have been the subject of scholarly inquiry for millennia. In Plato's *Meno*, Socrates attributed numerical ideas to our immortal souls, arguing that ideas are recollected in our mortal lives. Contemporary discussions pursue some of the same issues, and the approaches span modern disciplines and often reach across traditional boundaries. Some of the key contributors to the discussions include those whose principal disciplinary affiliations are in developmental psychology (Piaget, Vygotsky, Werner), cultural anthropology (Barth, Bateson, Boas, Kluckhohn, Kroeber, Sapir, Sperber), linguistics (Chomsky, Croft, Keller, Lakoff, Whorf), epistemology (Fodor, Kant), and the cognitive neurosciences (Dehaene, Gallistel, Gelman, Spelke). My intent in this book is to inject into these discussions both new observations and a cognitive-historical framework that has some potential to integrate diverse strands of inquiry.

Part I of the book situates the overarching problem about origins in contemporary research in the cognitive sciences. I consider (and then reject) traditional dichotomies that treat culture and cognition as separable and often treat cognition as something determined by culture. I put forward my view that such approaches have handicapped productive research on the origins of representations and ideas. Drawing upon previous scholarly work, I sketch an alternative in which cognition and culture are treated as processes with intrinsic relations mutually constituted in activity. Focusing on numerical cognition as an illustration, I argue that *collective practices* – recurring socially organized activities in which cognitive problems emerge – are important arenas for analyses of relations between culture and cognition, and I support my argument with evidence from my research in the Oksapmin region of Papua New Guinea. I show that, in the dynamics of collective practices involving number, people unwittingly reproduce and alter cultural forms of representation to serve numerical functions in public displays of communication and problem solving. "Culture" and "cognition" are not static, nor can they be isolated in analyses of Oksapmin collective practices, either within the brief span of an exchange between two individuals, or over historical time. To understand the origins of number and social history of number, one must examine shifting relations between cultural representational forms and the functions they serve in relation to collective practices of daily life. A treatment of the reproduction and alteration of form-function relations in relation to shifting collective practices over historical time becomes the crux of the analysis in subsequent parts of the book.

In Parts II and III, I report findings on two kinds of collective practices, each occurring at multiple locations in the Oksapmin region and each engaging a shifting network of participants. Part II focuses on analyses of collective practices of economic exchange, tracing shifts in the character of exchanges and numeric forms and functions from the time of initial Western contact in 1938 through the present. I report the distribution of representational forms for number and currency through the Oksapmin community, and their reproduction and alteration over the targeted period of study. Part III focuses on collective practices associated with the introduction of Western schooling in the late 1960s, and the shifts in schooling practices. Again, I report the varied forms of representation used, their distribution, and their reproduction and alteration in classrooms over the period of study.

Part IV integrates the corpus of findings within a comprehensive treatment of the origins of numerical representations and ideas. A central thesis is that collective systems of representations emerge over time as the unintended consequences of local goal-directed actions of individuals. I conclude with a consideration of regulative mechanisms in the generation of form-function relations, and I point to ways that the framework and findings open up new territory for empirical investigation and conceptual analysis in any community.

Part I

The Origins of Number: Enduring Questions

1 Culture-Cognition Relations

My purpose in this first chapter is to review theoretical formulations and empirical work on culture-cognition relations, giving particular attention to accounts of the origins and development of numerical thought. I begin with investigations that treat culture and cognition as separate entities – cognition treated as a skill, intellectual ability, or cognitive structure, and culture treated as a part of a surrounding context. Investigators in this tradition study culture-cognition relations by correlating aspects of the surround with aspects of cognitive functioning; in this process, they explore the way variations in cognition may (or may not) be systematically associated with variations in culture or features of culture. The approach appears reasonable from the perspective of the cognitive and behavioral sciences, and studies have yielded some important findings. Nonetheless, I argue that the approach leads to a dead end, particularly when the questions focus on the origins of numerical representations and ideas. I will argue for a different conceptualization, one that treats both culture and cognition as processes jointly rooted in activity. In the latter half of the chapter, I introduce a new framework and argue for its potential to open up new and more productive paths for inquiry and analysis.

Culture and Cognition: Separate Variables with External Links

Empirical research that treats culture and cognition as separate and independent entities often carries with it two tacit assumptions.[1] One

[1] In this analysis, I draw on similar critiques by others, including Lewin (1931) and Vygotsky (1986) as well as more contemporary authors, like Berger and Luckman (1966), Cole (1996), and Scribner and Cole (Cole & Scribner, 1974; Scribner & Cole, 1981). I also draw on contemporary activity theory (e.g., Engestrom, 1987; Hutchins, 2006).

assumption is that culture is a property of populations, and cultural groups are relatively homogeneous and internally coherent – thus, the French, English, Samoans, or Navaho are distinct cultural groups that are distinguishable in their patterns of daily life, languages, belief systems, economic organizations, and other things cultural, including number systems. A second assumption is that cognition is a capacity of individuals, reducible to cortical structures that are the basis for psychological constructs like memory, motor abilities, perceptual systems, and cognitive structures; such individual capacities can be measured through assessments using cognitive tasks. Conceptualizing culture and cognition as separable domains frames quite naturally questions about their relations: Are variations in culture associated with variations in the cognitive capacities of individuals, and, if so, what is the evidence that cultural variation determines cognitive variations?

The study of culture and number fits quite well into methodological traditions that conceive of culture and cognition as separate domains of inquiry to be related with one another through correlational analyses. From a cultural perspective, the diversity in forms of number representation is astonishing. The Incas, for example, used the *quipu*, a complex system of knotted cords that served to record quantitative information such as gold storage in mines (Ascher & Ascher, 1981). The Mayans developed a place value system, principally for the recording of dates (Closs, 1986b). The Babylonians developed a sexagesimal system, aspects of which have descendants in today's clocks (Menninger, 1969). There are myriad lesser-known cultural forms for number representation that have emerged within and across the continents of Africa (Zaslavsky, 1973), the Americas (Closs, 1986b), and Asia (Menninger, 1969). Some of the variation still exists, despite increasing globalization. Indeed, on the island of New Guinea alone, there are more than 800 cultural and/or language communities using a wide range of culture-specific counting systems. Some of these groups use body parts, others use specialized words for number, and still others use physical tokens. Within each type there are many variations, and some groups use multiple types (Lancy, 1983; Laycock, 1975; Lean, 1992; Mimica, 1988; Saxe, 1981d; Saxe & Posner, 1982; Wassmann & Dasen, 1994).[2]

Scholars have made efforts to abstract the dimensions of cultural variation in numerical forms of representation. Analyses have come from varied disciplines, including anthropology (Closs, 1986a; Crump, 1990; Lean, 1992), linguistics (Hurford, 1987), psychology (Gelman & Gallistel, 1978;

[2] For compendia, see Lean (1992), Menninger (1969), Smith (1925), and Zaslavsky (1973).

Lancy, 1983), and economics (Harper, 2010). Proposed dimensions include the following:

1. *Material/nonmaterial properties of number signs.* In contemporary Western life, we use a specialized register of words to refer to numbers – count words like "one, two, three," et cetera in English. But other communities use forms that are not strictly verbal: In the Mt. Hagen area of Papua New Guinea, for example, one of the Melpa's systems consists of physical tallies (Saxe, 1983), and many central highlands groups in Papua New Guinea make use of body parts (Lean, 1992). Over the course of world history, remarkable forms of number representation have emerged with varied material embodiments, whether the Inca's knotted cords, Neolithic notched bones, or the cuneiform records of ancient Mesopotamia.
2. *Number of number signs in systems.* Representational systems use different numbers of elements in counting. Some, as in the Western use of Hindu-Arabic orthographic numerals (0, 1, 2, ..., 9), include a potentially infinite number of representations made possible by base-structure principles (for example, 1001). Others make use of a restricted number of number names, and indeed it has been reported that cultural groups in Brazil, including the Pirahã (Everett, 2005) and the Munduruku (Pica, Lemer, Izard, & Dehaene, 2004), have minimal if any representations for exact quantities. In the case of the Pirahã, Everett reports that the group's language lacks a singular-plural morphology and their counting numbers are restricted to not more than two (Frank, Everett, Fedorenko, & Gibson, 2008).
3. *Cyclic structure of signifiers.* A feature of most contemporary number systems is the use of elementary signs to generate composite signs. Lean (1992) refers to the simplest systems as two-cycle, that is, the use of two elementary signs, with additional number signs being composites of these. Thus the first and unique elements are signs for one and two, with consecutive numbers taking the form of $3 = 2 + 1$, $4 = 2 + 2$, $5 = 2 + 2 + 1$, and so on (Hurford, 1987; Lean, 1992). In his documentation of South Pacific groups, Lean describes systems that employed a range of such structures, including two-, four-, five-, six-, and ten-cycle organizations (Lean, 1992). As I will discuss in Chapter 2, one of the Oksapmin's two number systems is a modified two-cycle "conversational" number word system not linked to the body. The system is used to represent five numbers (one to five), in which the word for two is *yota* and the word for four is *yota*

yota. (The Oksapmin also use a body-part counting system, making use of 27 points on the body as signs for numbers.) Languages for which Ancient Chinese is the ancestral language have a ten-cycle organization, as does English. In the case of English (and in many other Romance and Germanic languages), the expressions "eleven" and "twelve" are irregular in the ten-cycle structure.

4. *Numerical functions of signifier use*. Signifiers serve different numerical functions within a group and across groups. For example, in the English-speaking world, we use our own register of number words to serve nominal functions (bus routes), ordinal functions (positions in a sequence), and cardinal functions (the number of objects in a group), to name a few. Forms may take on function-specific properties. In English, for example, word forms for cardinal and ordinal numbers differ, as in "four" and "fourth."

The widespread use of forms of number representation, coupled with striking cultural variation in the forms, has been the focus of varied positions regarding culture-cognition relations. Let's consider four prominent ones.

1. *Cultural Forms of Number Representation Determine the Numerical Thought of Individuals*

The first position identifies cultural forms of number representation as the principal determinant of quantitative thought. The logic of number in language becomes the logic underpinning the numerical thought of the individual. This position, of course, is a variation of the Whorfian position applied to numerical thought (Whorf, 1956). This focus leads to identifying cultural variation in representational forms for number and then associated variations in numerical capabilities of individuals.

In research on number with the Pirahã, a remote Amazonian group, Daniel Everett (2005, 2007) and Peter Gordon (2004) initially took up this position, arguing that language differences in number (along with differences in the structure of the indigenous language) determine thinking differences among individuals and across cultures. Everett, a linguistic anthropologist, found representations that the Pirahã used for number to be very limited. Gordon, a cognitive psychologist, followed up the linguistic findings by administering to Pirahã adults tasks that required numerical comparisons and reproductions of arrays of varying set sizes, and he found that these adults rarely produced successful solutions beyond very small sets (Gordon, 2004).[3] Based upon these findings, Gordon argued that the Pirahã's fundamental numerical competence was limited and that these limitations were a result of the Pirahã's limited number system.

In subsequent work with the Pirahã, Frank, Everett, and additional collaborators (Frank et al., 2008) offered an alternate hypothesis: The Pirahã merely lacked the "technology" of a cultural system of number words, not the knowledge of number.[4] Gordon's number tasks required counting (or possession of an extraordinary eidetic memory), and Pirahã people may have possessed an understanding of the logic of one-to-one correspondences but lacked the number word resources to produce representations that supported adequate solutions. Indeed, Frank and colleagues replicated the findings on Gordon's original tasks, but they found that their Pirahã participants had no difficulties with a one-to-one correspondence task, regardless of set size. Frank and colleagues interpreted these findings as evidence that language and number systems are *not* the source of fundamental numerical ideas like one-to-one correspondence. This is a position at odds with the strong claim that cultural forms of representation determine the numerical thought of individuals.

2. *Cognitive Structures for Number Are Determined by Biological Evolution*

A second position answers the question about origins of correspondences (or more generally, cognitive structures of numerical thought) by appealing to evolutionary processes. Over evolutionary history, the ability to extract and process quantitative information provided the ancestral species with selective survival advantages leading to dedicated brain circuitries; over the millennia, this neurobiological capacity provided a substratum on which numerical thought was formed (Butterworth, 1999; Gallistel & Gelman, 2005).

Noam Chomsky (1980), the renowned linguist who posited native structures specialized for language, suggests that his argument for language is also applicable for number. As in his remarks on language acquisition (Chomsky, 1967), he asked how people could come to an understanding of number so early in life based upon experience.

[3] The findings from the Pirahã studies that showed abilities to discriminate small sets were echoed in research with the Munduruku, another South American group with a number system having limited count words (Pica et al., 2004).

[4] These tasks were similar to some that Piaget used in his studies of children's developing understandings of one-to-one correspondence relations (Piaget, 1952).

> To gain further understanding of the general nature of the human mind, we should ask in what domain humans seem to develop complex intellectual structures in a more or less uniform way on the basis of restricted data. Wherever this is the case, we can reasonably suppose that a highly structured genetic program is responsible for the achievement.... Language is an obvious area.... Are there other systems ... ? Perhaps so.... [O]ne curious property of the human mind is our ability to develop certain forms of mathematical understanding – specifically concerning the number system. (Chomsky, 1980, pp. 248–249)

Support for native capabilities has been corroborated in a variety of programs of research. One research strand focuses on infants' numerical discriminations of very small quantities; infants successfully discriminate two versus three (Antell & Keating, 1983) and make more or less exact and reliable discriminations with larger magnitudes (Xu & Spelke, 2000; Xu, Spelke, & Goddard, 2005). Whether infants are capable of anticipating elementary arithmetical compositions with small quantities is a matter of dispute (Wakeley, Rivera, & Langer, 2000a; Wakeley, Rivera, & Langer, 2000b; Wynn, 2000), but evidence for an arithmetical competence has been produced more consistently with toddlers (Klein & Starkey, 1988).[5]

Additional support for the idea that native neural circuitry is dedicated to processing numerical information comes from studies of cognition in patients who have suffered cortical lesions. Patients with lesions in the parietal cortex show abnormalities processing numerical information, even as their other cognitive capacities appear intact (Cohen, Dehaene, Chochon, Lehéricy, & Naccache, 2000; Saxe, 1981b). Studies using functional magnetic resonance imaging (fMRI) have revealed that the parietal regions show increased activation when processing numerical information (Chochon, Cohen, van de Moortele, & Dehaene, 1999; Dehaene, Piazza, Pinel, & Cohen, 2003). More exacting studies conducted with nonhuman primates (monkeys) show specialized activation of particular kinds of neurons in the parietal regions in response to stimuli that vary in magnitude (Dehaene, 2003; Nieder, Diester, & Tudusciuc, 2006; Nieder & Merten, 2007).

[5] Evidence of numerical competence in nonhuman species further bolsters a biological argument (Gallistel, 1992; Gelman & Cordes, 2001). Lions, for example, apparently decide whether to attack an intruding group by discriminating the number of unfamiliar roaring individuals heard in relation to the number of their own pride (McComb, Packer, & Pusey, 1994). Rhesus monkeys appear to select the greater quantity of food when presented with a choice (Hauser & Carey, 2003), a finding also documented with salamanders (Uller, Jaeger, Guidry, & Martin, 2003).

Empirical research supporting a biological argument for elementary numerical cognition includes studies about cross-species competence for discriminating number, numerical circuits dedicated to processing information relevant to number, and early numerical competence in humans. In my view, these findings do not adequately inform our knowledge of the development of complex numerical understandings in human ontogeny. The nativist stance denies that underlying logical structures undergo any significant development, with investigators arguing that changes in memory, confidence, or motivation inhibit native structures from showing themselves in early development (Bever, Mehler, & Epstein, 1968; Mehler & Bever, 1967). But explanations like Mehler and Bever's seem implausible, especially given findings that show that children construct transitional stages of numerical understanding that show both continuities and discontinuities with prior stages (e.g., Piaget, 1968; Steffe, von Glasersfeld, Richards, & Cobb, 1983).[6] I will discuss the constructivist position shortly.

3. *Native Cognitive Structures for Number Are Amplified by Cultural Forms of Number Representation*

A third position is a union of the first two: that numerical thought has its roots in the number systems of culture as well as native structures of individual thought. This line of argument highlights the way that cultural forms of representation serve to amplify native capabilities.

In his early book *Studies in Cognitive Growth*, Jerome Bruner offered the amplification metaphor, in part to explain what he argued were cultural differences in cognitive capacities (Bruner, 1966). The basic idea is that, when native capacities like knowledge of elementary number become hooked to powerful cultural systems of representation, the power of thought is magnified; further, cultures differ in the amplifiers they provide. With this position, cultures provide differing number systems (amplifiers of thought), and individuals provide the biologically rooted cognitive structures for number.

In the 1970s, David Lancy undertook research that built in part on Bruner's proposal. Lancy investigated the effects of variation in forms of

[6] The documented cross-species numerical discriminations also are suspect as building blocks for higher mathematical functions. Despite surface similarities, higher functions involving number may not grow out of perceptual discriminations at all. Some, for example, have argued that the roots of number lie in transformative actions on the physical world (e.g., Langer, 1986; Piaget, 1952; Piaget, 1963) or develop through other kinds of constructive activities, like metaphor (e.g., Lakoff & Nuñez, 2000).

number representation in Papua New Guinea groups and its effects on the quantitative thought of individuals. In his project, Lancy classified 11 groups according to his analysis of the cognitive sophistication of the counting systems based on four generic types, from body counting (Type I) to a Western-like 10-cycle system (Type IV). He and his associates then administered standard cognitive/quantitative tasks to children in each cultural group. Though his goal was to demonstrate a relationship between the complexity of number systems (the amplifiers) and complexity of numerical thought, Lancy reported no clear association. Lancy indicated that missing in his analyses were systematic accounts of children's activities and the ways children used number in their everyday lives.[7]

There is no doubt that the use of a number system, just like the use of a computer, amplifies human capabilities. A computer allows us, for example, to run complex statistical programs that would be difficult to accomplish by hand. Similarly, without a counting system, our capacity to compare numbers of great magnitude or reproduce a set of objects numerically or engage in more complex arithmetical operations distant in time or space would be severely limited. But while artifacts like computers and number systems may amplify the products of human activity, the amplification metaphor provides little insight into cognitive processes. Indeed, using a software computer program as opposed to paper and pencil leads to a different organization of the cognitive activity of statistical analysis, and using a number system to compare number leads to a new organization of arithmetic activity, each with different challenges (Cole & Griffin, 1980; Pea, 1985). Thus, like others, I take the amplification metaphor to provide limited utility for understanding how posited elementary and biologically rooted capacities are magnified into powerful ways of solving mathematical problems.

4. *Cognitive Structures for Number Are Constructed in Individual Development*

A fourth position is the genetic epistemology of Jean Piaget. As a part of his seminal studies of the origins of number in human ontogeny, Piaget (1952, 1968, 1970) made two arguments. First, the development of young

[7] Without an analysis of children's activities, as Lancy became aware, one has little basis for making strong inferences about relationships between cultural number systems and individuals' cognition, regardless of any association (or lack thereof) between cultural number systems and individuals' performances on cognitive tasks.

children's numerical understandings such as one-to-one correspondences is not reducible to an imitation of cultural forms of representation (as Whorf and others had argued); second, neither are these structures of understanding reducible to preformed knowledge in the human genome. Let's consider some of the findings that Piaget used to support these arguments.

In one well-known study, Piaget and his collaborators presented Genevan children with a model array of discrete objects and asked them to produce a matching array with the same number; the task was similar to the much more recent tasks used with the Pirahã (Frank et al., 2008). Piaget used his results to argue that, early in development, young children do not differentiate the space occupied by a discontinuous quantity from its number and, as a result, that the children based their evaluations on endpoints of the arrays, even if they were fluent with number words (Piaget, 1952, 1968). Somewhat older children began to differentiate space and number, creating a one-to-one correspondence between groups, but they did not conceptualize the one-to-one correspondence as lasting. When one array was spread apart or crowded together, the children judged one group to contain more, again revealing a difficulty in maintaining a distinction between number and space occupied; at other times, they vacillated in their judgments, showing conflict as they tried to coordinate compensatory relations between changes in lengths and densities of arrays. Only the oldest children in his sample (approximately age seven) argued that one-to-one correspondence is conserved over spatial transformations; their justifications focused on inverse relations between lengths or areas of arrays and densities of the objects.[8]

Piaget's constructivist argument offers a conceptualization of cognition as a developmental process. Many cross-cultural studies have used adaptations of Piaget's methods to evaluate his claims for the universality of the developmental sequence of stages, and their conclusions have largely corroborated the original findings, though rate of progress through the sequence has varied across studies (Dasen, 1972, 1975, 1977; Hallpike, 1979; Posner & Baroody, 1979). However, Piaget's treatment offers little analytic leverage to enable us to understand the interplay between cultural forms of number representation and children's developing use of correspondences to serve numerical functions. In Piaget's account, cultural forms for number are external markers, used appropriately only when

[8] However, various researchers have argued that Piaget's methods tend to underestimate the numerical capabilities of younger children and infants (Gelman & Gallistel, 1978; Siegal, 1991).

the developing child has constructed adequate cognitive structures with precursory roots in sensorimotor action.

The four positions on origins that I have just reviewed share a common property. Each position treats culture and cognition independently. In so doing, they do not well illuminate the interplay of culture and cognition in both development and social history. While Piaget's constructivism does provide a basis for understanding the ontogeny of numerical structures of thought as they develop in interaction with a physical and social world, it yields little insight into the interplay between cognitive and sociohistorical processes. In my view, inquiry grounded in any of these positions leaves the complex interrelations between cultural-historical and cognitive-developmental processes understudied and unexplainable. What is needed is a theoretical framework appropriate for conceptual and empirical analyses of intrinsic and mutually constitutive relations between sociohistorical and cognitive developmental processes – the phenomena that should be the focus of study and analysis.

Culture and Cognition: Jointly Rooted in Activity with Intrinsic Relations

In this section, I frame an alternative methodological approach.[9] The approach is rooted in the idea that both culture and cognition should be understood as processes that are reciprocally related, each participating in the constitution of the other. This is an idea with roots in prior treatments in sociology (Berger & Luckmann, 1966; Bourdieu, 1977; Garfinkel, 1967; Giddens, 1986), psychology (Cole, 1996; Vygotsky, 1986; Wertsch, 1998), and historical linguistics (Croft, 2000, 2008; Keller, 1994). I extend and build on this idea in the remainder of this chapter, and conclude with an approach to empirical inquiry. I begin with considerations relating to the treatment of culture and cognition as processes. I then consider units of analysis that coordinate the two in a single analytic frame.

Culture as Process

Culture has been treated in countless ways across the disciplines of anthropology, psychology, and sociology. In their seminal review of treatments

[9] By methodology, I do not mean merely a set of empirical techniques like cognitive interview tasks or observational protocols and associated coding schemes. More fundamentally, I mean a way of conceptualizing problems as well as constructing empirical techniques to address them.

of culture in psychological anthropology, Kroeber and Kluckhohn (1952) note that, across many treatments, cultural groups tend to be conceptualized as if they were homogeneous, with little regard for variation in the activity of individuals.

> Not only is every individual different, but concretely, the cultural forms differ too with the individuals who color them with their own needs and presses.... [E]ven the cultural heritage of each individual is unique, even though abstractly the total cultural heritage is available to all.... The same form can be used for an almost endless variety of purposes and for expressing an almost infinite shading of meanings. (pp. 363–364)

Claims like "the French like snails," "Italians are expressive," "Irish are drunks," or "Asians are smart in math" are abstractions. They reduce a multitude of variations in people's activities to a single pattern and turn that pattern into a simplified notion of culture. Abstractions of behaviors may be useful for some purposes, but the reduction obscures activity at the individual level and the way that the activities of individuals are constitutive of cultural processes.

In an incisive critique, Sperber and Hirschfield (2001) further problematize the received view of culture. They point to the importance of a process-oriented treatment, and question the notion that cultures exist in the traditional sense.

> The standard anthropological characterization of cultures as relatively bounded, homogeneous, and coherent entities has repeatedly been challenged (e.g., Fried, 1975; Leach, 1954). The idea of discrete tribes each with its own culture was a colonial administrator's dream – a dream they forced on people – before being an anthropologist's presupposition. In fact, different flows of cultural information – linguistic, religious, technological – have different boundaries, or, quite often, do not even have proper boundaries, just zones of greater or lesser intensities.... From such a point of view, the notion of a culture should not have more of a theoretical status than that of a region in geography. Culture is best seen not as a thing, but as a property that representations, practices, and artifacts possess to the extent that they are caused by population-wide distribution processes. (p. cxxvii)

Sperber and Hirschfeld's concern with "population-wide distribution processes" is a resource for the approach I elaborate here. Their focus on "information flow" draws attention to individuals as "conduits" of information, and thus to activities of individuals as constitutive of cultural phenomena. But, as stated, Sperber and Hirschfeld's process-oriented treatment of culture is problematic. How does information "flow" from one individual to another through historical time? What does the metaphor of flow hide in analysis of the processes of reproduction and alteration of representations in communities? A complementary approach is needed that captures cognition as process.

Cognition as Process

Just as Sperber and Hirschfeld question a static and monolithic conception of culture, Gregory Bateson (1972) questions cognition as self-contained cognitive structures, or information storage separable from an environmental surround. In his apocryphal story of the blind man and the stick, Bateson argues that what is sometimes regarded as external to an individual or "context," like the red and white stick used by the blind, is really a part of a working cognitive system.

In his story, Bateson describes a blind man tapping his stick as he progresses down a street, avoiding obstacles as he makes his way. Bateson points out that when the blind man is using his stick to serve the function of navigating, the "out-of-body" stick (a cultural form) is an ingredient of his navigational cognition, part of how he construes space, obstacles, and routes. Bateson argues that to understand the man's construction of a route, it makes little sense to draw boundaries between what's enclosed by the skin and the way cognitive activity is extended using the stick. Further, Bateson points out that the stick shifts in cognitive status as activity changes. For example, when the man is eating lunch, the stick is then no more part of the man's cognitive processes than any other distant material artifact would be. Bateson's account motivates a focus on cognition as process, with attention to cultural forms (like red and white sticks) and the cognitive functions they serve in individual activity (like navigating).[10]

[10] Bateson's account also has implications for the analysis of cognitive processes as they are constituted in collective activities. When we consider the blind man as engaged in a collective practice, the man's emergent navigational problems and goals are shaped not only by the stick and obstacles; they also take on properties relative to how the man positions himself and is positioned by others (Davies & Harré, 1990; Harré, Moghaddam, Cairnie, Rothbart, & Sabat, 2009; van Langernhove & Harré, 1999). For example, the blind man's use of the stick positions pedestrians as potential "helpers" and the blind man himself as in potential need of assistance. Thus if the blind man is moving toward a pothole in the sidewalk, a passerby may become a "good Samaritan," and the blind man will never confront the cognitive challenge of moving around the pothole (or stumbling). Further, the material world is also adjusted for the blind with sounded stoplights and Braille inscriptions adjacent to elevator buttons, artifacts that alter the organization of cognitive activities and cognitive challenges people generate in activity.

Understanding the blind man's cognitive constructions as they take form in his navigation of the street requires a different kind of framework for analysis than the received approaches reviewed in the prior section. The framework that I develop focuses on cultural forms, such as the blind man's red and white stick, and the functions they serve in collective activities, such as navigation through city streets. Of particular concern is to understand form-function relations as processes in motion – as they are constituted and shift in processes of microgenesis, sociogenesis, and ontogenesis. I begin by considering how these genetic processes might apply to the blind man and his stick and then extend the discussion to a more general framework.

Microgenesis. How does the blind man transform the cultural form of the stick into a means of finding his way in the moment-to-moment activity of getting from point A to point B? I will refer to this process as the *microgenesis* of the man's way-finding activity.[11] The stick contains no inherent navigational function, and the route itself must be created in activity. As the man moves the stick in the direction of his possible route, he draws upon his knowledge of the terrain from prior walks, his knowledge of space and geometry (one can create different paths to arrive at the same point), and effective ways of holding the stick to explore. As he navigates, the man's coordination of these cognitive resources in the haptic construction of obstacles and pathways constitutes a microgenesis of activity that results in his route. Insofar as the navigated route results from the man's constructive effort, his activity defies an explanation of development as a simple transmission of information from individual to individual.

Sociogenesis. The second strand concerns the emergence, reproduction, and alteration of the collective practice of using white and red sticks for navigation in the social history of the man's community. Such sticks, constructed of particular materials and with standard cross sections and lengths, came to be manufactured in the social history of way finding in the United

[11] Some researchers have used the term "microgenesis" to refer to a methodological approach involving the intensive study of shifts in children's strategies and/or cognitive structures over short periods of cognitive development (see, for example, Siegler & Crowley, 1991, 1992). My use of the term is more consistent with earlier treatments of the construct (Vygotsky, 1986; Werner & Kaplan, 1963), in which the process of schematization of a phenomenon, perceptually or conceptually, is understood as a short-term developmental process. As conceptualized in the present discussion, microgenesis is neither a methodological approach nor a small-scale version of ontogenetic change. Rather, microgenesis is used to refer to a process by which forms – with the cognitive functions they afford – are transformed into means for accomplishing emerging goals in activity.

States. How did the industry arise? How did populations of pedestrians and motorists come to orient to red and white stick users? What was the social history of the population-wide processes constituted in the local actions of individuals whereby the forms of navigation and socially recognized functions of stick use occurred? These are questions about the *sociogenesis* of navigation in a special population. Not only are sociogenetic processes of import in their own right, but they also have implications for the character of microgenetic processes in activities like navigation of blind people (which in turn have implications for sociogenetic processes). Sociogenetic products, like red and white sticks (and associated norms for how to treat people with such sticks), enable and constrain the microgenesis of the blind man's navigational activity as he walks from one city block to another, and, from a sociogenetic perspective, in this microgenetic process, the man is unwittingly reproducing and altering the use of red and white sticks in the social history of a community. To understand cognition as a process involves a treatment of sociogenetic processes in the reproduction and alteration of collective activities and associated form-function relations in communities through time.

Ontogenesis. The third strand focuses on the blind man's stick use in his own development from infancy to adulthood. The man's first efforts to use a stick as a child would have differed from how he used it later as a skilled navigator in his adult life, and these differences cannot be reduced to shifts in the sociogenesis of practices of the community during the man's lifetime. Indeed, as a child, obstacles may have initially stymied the boy before he was able to explore routes around them systematically. Qualitative shifts over time in the boy's spatial thinking constitute changing resources for his navigation, and the boy's shifting use of the stick to serve differentiated navigational functions is referred to as the *ontogenesis* of form-function relations of red and white stick use. Indeed, the boy will progressively build more adaptive capabilities to cognize routes and detours as well as haptic techniques to explore objects with sticks and construct useful representations to serve his navigational goals leading to new functions for the stick. To understand cognition as process involves an analysis of the ways that individuals appropriate new forms in activity (such as the red and white stick) and the shifting functions of these forms over individuals' developments (Werner & Kaplan, 1963).

Units of Analysis

We cannot isolate culture and cognition as if each were a monolith if we are to develop adequate understandings of either. The analysis of culture without regard to its roots in individual activity leaves unexplained how collective practices and representations are reproduced and altered across different locations and through historical time. The analysis of individual development without regard for collective life leaves unexplained how individuals' cognitive constructions of ideas and representations take on properties of the collective and are interwoven in fundamental ways with the material, semiotic, and institutional artifacts of community life. Clearly there is a need for units of analysis that leverage efforts to understand their mutually constitutive relations.

In his treatment of the ontogeny of language and thought, Vygotsky (1986) presents a compelling case for ways of conceptualizing useful units of analysis. In Vygotsky's treatment, his focus is on relations between language and thought (or speaking and thinking). He takes issue with approaches that sought to break each down separately into their most elemental forms and then to analyze connections between them. For Vygotsky, a key analytic is to consider core emergent properties of the intersection and then to identify units that would support an analysis of their relations. He argues, for example, that if we were studying the properties of water, we could reduce water into its parts, hydrogen and oxygen. Yet, studied separately, both the elements of hydrogen and oxygen burn, but as molecules, H_2O, they put out fire. In his treatment, Vygotsky argues for word meaning as a unit of analysis, a unit, which, he argues, captures intrinsic relations between the historically emergent and communicative functions of language (words) and the generalizing properties of human thought (meanings).

Scholars in the Vygotskian tradition have extended and revised Vygotsky's semiotic (word meaning) focus to one or more molar units of human activity. Leontiev, for example, has argued for activity as a unit of analysis (Leontiev, 1981), because activity affords a perspective on the constituting conditions and operations that people use to accomplish goal-directed actions (energized by motives). Luria investigated psychological resources for human action – attention, involuntary memory, voluntary memory, higher cognitive functions – and the way these varied resources were organized into "functional systems" (Cole, Levitin, & Luria, 2006; see also Cole & Scribner, 1974; Luria, 1973). Contemporary scholarship has proposed other units, building upon treatments of activity. They include a focus on activity as situated in a matrix of artifacts, divisions of efforts, and rules (Cole, 1996; Engestrom, 1987; Nardi, 1996; Roth & Radford, 2011; Wartofsky, 1973); legitimate peripheral participation as individuals shift from being learners on the periphery of a socially organized activity to central participants (Lave & Wenger, 1991); distributed cognition, or ways that collective work is distributed across the social and material world

(Goodwin & Duranti, 1992; Guberman & Saxe, 2000; Hutchins, 1995; Pea, 1993; Saxe & Guberman, 1998; Taylor, 2009); and "commognition," a treatment of human thought as a communicative act with deep roots in human interaction (Sfard, 2008).

Like other treatments that build upon Vygotsky's seminal analyses, I make use of activity as a key orienting construct. My focus is on the origins of numerical representations and ideas in culture-cognition relations, and I have argued that appropriate units should include collective practices on the one hand and numerical form-function relations on the other, with particular regard for their microgenesis, sociogenesis, and ontogenesis over the social history of communities.

Toward a Genetic Treatment of Intrinsic Relations

In the sections to follow, I build on the ideas of collective practices and genetic strands to frame the work on number in Oksapmin communities. I take collective practices to be recurring structures of social activity that are constituted as people construct, communicate about, and accomplish recurrent problems over historical time. Just as navigation by blind people in the United States can be regarded as a kind of collective practice, I regard economic exchange and schooling in Oksapmin communities as collective practices. But in contrast to way finding, in economic exchange and schooling in Oksapmin, recurrent problems, like the recurrence of making change in face-to-face economic transactions, involve numerical representations and culturally elaborated forms – like the Oksapmin body-part counting system, currency tokens, and commodities – to serve varied cognitive and social functions.

To introduce ideas about collective practices in the Oksapmin world, consider an observation of an economic exchange in 1978 at a trade store near the Baptist mission station in an Oksapmin valley (Tekin). The observation is captured in the series of photographs in Figure 1, in which an Oksapmin clerk can be seen standing behind a counter facing customers.[12] Moments prior to the scene captured in the photos, the customers had requested cans of fish from the clerk, the clerk had retrieved them, and the customers presented payment. In the sequence depicted in Figures 1a through Figures 1i, one customer retrieves payment from a small can that

[12] Stores were commonplace in 1978, but the store in this description was unusual. It was built of prefabricated materials imported by the Baptist mission and was located on the mission grounds.

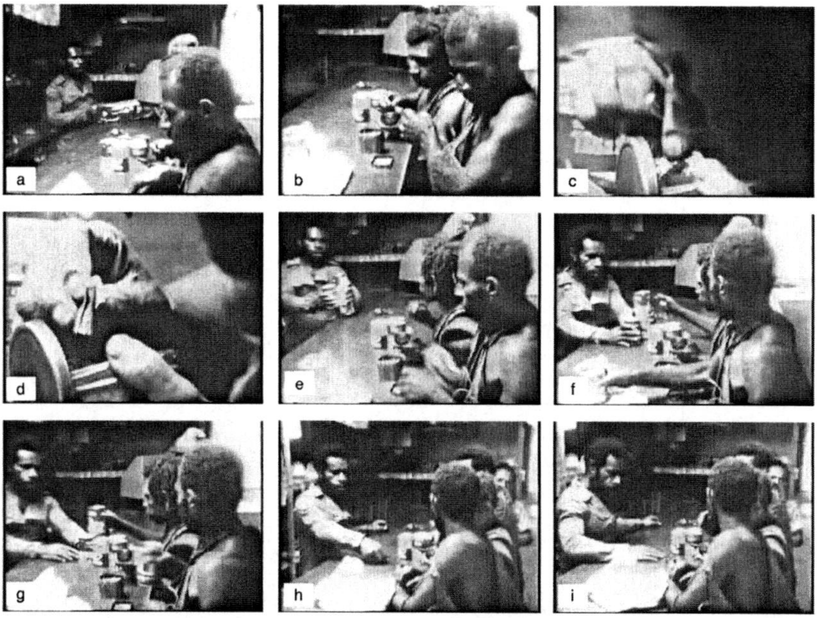

Figure 1. Two men making a purchase of tins of fish in the mission trade store from an Oksapmin trade store clerk, 1978.

he keeps in his string bag and requests additional cans (Figures 1a–d). In Figure 1e the clerk returns with additional cans, and in Figures 1f–1i payment is made and change is returned.

The structure of the collective practice portrayed in the images of Figure 1 has no permanence, other than its re-creation in the activity of the participants as they work to complete an exchange of goods for cash. Similarly, the artifacts that are constitutive of emergent numerical problems in exchanges, the cash and commodities, have no inherent numerical or exchange value. Artifacts take on value as they are used and interpreted in the recurrent collective problems that constitute the exchange. But the re-created structure and meanings attributed to artifacts are more or less predictable.[13] Interlocutors' actions are rooted in precedent, rendering them interpretable by one another. Further, alterations in the practice's structure (for instance, to accommodate a customer's request to pay later) or the properties of the collective problems (for instance, the government's

[13] The predictable quality of collective practices like economic exchange in Oksapmin is captured in constructs of cultural scripts (Nelson, 1981, 1986), cultural schemas (D'Andrade, 1990), and cultural models (Quinn & Holland, 1987).

issue of a new currency token) may lead to a back-and-forth that renders new ways of coordinating actions and new precedents that support communication in subsequent interactions.

In trade store exchanges in 1978 like the one depicted in Figure 1, I observed different kinds of recurrent collective problems in exchanges. For example, recurrent problems were created as a customer anticipated (or did not anticipate) the cost of an item and evaluated the amount of currency in his possession, or when clerk and customer negotiated a cash-for-commodities transaction. As people structured and accomplished collective problems, they generated local goals in their activities, drawing upon cultural forms like the body system, currency tokens, or names for tokens to serve various cognitive functions. These functions included using tokens to pay for goods or using the body system to quantify tokens. The character of these collective problems and emergent goals that take shape in activity is complex, and tensions and coordination problems across interlocutors necessarily emerge. The process is illuminated if we return to Sperber and Hirschfeld's tripartite critique of culture and the way their critique bears on collective problems that take form in the conduct of exchanges.

Recall that Sperber and Hirschfeld questioned received views of culture, views in which a "culture" is conceptualized as a homogeneous group that is bounded by a particular geography and engages in practices that display an underlying coherence. In their critique of the notions of cultural homogeneity, boundedness, and coherence, Sperber and Hirschfeld refer to such approaches as "a colonial administrator's dream" that became "an anthropologist's presupposition." But when we apply their descriptors to collective practices like economic exchange in a trade store, we find elements of heterogeneity and homogeneity, boundedness and unboundedness, and coherence and incoherence. These pairings enable me to frame an analysis of the communicative tensions that emerge for interlocutors in collective practices of economic exchange and the varied resources that interlocutors draw upon for their resolution.

Collective Practices of Economic Exchange: The Emergence and Resolution of Numerical Problems

In the trade store exchange depicted in Figure 1, there is evidence of population *homogeneity*, which is a resource for people who need to establish common communicative ground in practices of economic exchange. The men depicted in Figure 1 participate in material worlds of similar artifacts (dress, weaponry, tools), use a common language (Oksapmin), and use the same body-part counting system. At the same time, consistent with Sperber and Hirschfeld's argument, there is considerable *heterogeneity* in the clerk's and customers' backgrounds and experiences, and some of these differences can lead to communicative challenges in exchanges. For example, there is heterogeneity in the way the body is used to represent currency in the Oksapmin community. As I will discuss in Chapter 4, many store clerks have developed complex body-part counting strategies to solve arithmetical problems, but traditional individuals are typically unsuccessful in extending the counting system for this purpose. There is also heterogeneity in the dialects and speech registers used; the clerk, in addition to the Oksapmin language, speaks Tok Pisin (learned when he worked outside the area) and can use Tok Pisin number words. There is heterogeneity as well in their knowledge of devices that originate in the West; the clerk in Figure 1 has some knowledge of the mechanical cash register, whose keys are labeled with Hindu-Arabic numerals (somewhat visible in Figures 1a, b, e, f, and g). These variations in knowledge of numeration, in language use, and in literacy can lead to breaches of communication in the collective practices of economic exchange. To repair them, interlocutors must draw on common resources that support getting across one another's communicative intents.

Using Sperber and Hirschfeld's term, the community has "*hard boundaries*" – there is limited inflow of foreigners into the Mountain-Ok area.[14] But the boundaries also have some *permeability*. Oksapmin is one of a number of Mountain-Ok communities, and there is some movement of Mountain-Ok people among communities, often for purposes of trade. Prior to Western contact, such trading relations may have contributed to the common use of the 27-body-part counting system by Mountain-Ok groups who otherwise speak different languages. Permeability is also reflected in the influence of Western missionaries, patrol officers, and people like me. While hard boundaries contribute to the homogeneity of people participating in collective practices of exchange, soft boundaries contribute to heterogeneity and communicative challenges across interlocutors.

Finally, the various representational forms for number used in Oksapmin economic exchanges have unique histories leading to patterns of both *coherence* and *incoherence* across the use of representational forms and functions. For example, the 27-body-part counting system is used across a range of practices; traditionally it is used as a way to produce cardinal and ordinal

[14] The geography is rugged, and warfare with neighboring groups has tended to insulate highland communities from one another.

values (to count and order) but does not serve arithmetical functions (as I will report in Chapter 4). It is used across a range of practices outside of economic exchanges in the trade store. But a juxtaposition of this traditional practice and some of the cultural forms used in the trade store has occurred that creates difficult-to-reconcile systems of representation. For example, the cash register, patented in the latter part of the nineteenth century, with its Hindu-Arabic forms inscribed on the keys, was invented by an Ohio saloon owner, who rigged the machine with a bell to alert him to swift-handed employees who might be pocketing sales. In contrast, there is the indigenous number system, with its 27 constitutive and ordered body parts and no base structure.

Forms and Functions in Trade Store Exchanges (1978): Developmental Processes

In a process akin to Bateson's blind man's use of the stick for navigation, the clerk and customer in Oksapmin stores draw upon cultural forms – like the body-part counting system or Tok Pisin number words – as they work to accomplish their own emerging goals in relation to collective (and often recurrent) problems. But in contrast to the blind man, who uses the stick as a *tool* to detect obstacles, the customer and clerk are using forms for *symbolic* purposes – to represent number and currency. Symbol-using activities require a differentiation between a symbolic vehicle, in this case body parts, and referential objects, in this case material coins or idealized numerical values or numbers. To introduce additional ideas specific to the development of symbol-using as distinct from tool-using, I build upon Werner and Kaplan's (1963) developmental treatment of symbol formation. Their account treats symbol formation as rooted in communicative situations similar to problems that emerge in trade store transactions.

Werner and Kaplan offer an account of symbolic acts in which interlocutors construct and interpret symbols, referents, and their relations. In their analysis, the production and interpretation of symbols is accomplished through a process of "dynamic schematization." In the semiotic act, an interlocutor coextensively schematizes a symbolic vehicle (word forms, number signs), a referential object (denotative or connotative referents), and a relation between the two such that the vehicle comes to stand for the object, and the denoted or conceptual object comes to be represented by the vehicle. It is in this moment-to-moment activity that we find an interplay between the construction of the symbolic vehicle and the referential object. Symbols enter into the process of constructing objects of reference (for example, by assigning unique body parts in one-to-one correspondence to coins), and objects of reference have implications for the character of their symbolic representation (for example, cans of fish have physical properties that constrain the use of body parts, because they are discontinuous quantities of particular sizes). Thus, symbol and referent take on meaning in the symbolic act of production and/or interpretation. In Werner and Kaplan's words,

> symbolizing enters directly into the construction of the "cognitive objects," determining how events are organized and what they mean. Our thesis is thus opposed to the widespread view which treats symbolic vehicles and referents as two fully formed entities that are externally linked to each other through contiguous pairing (and reinforcement). If one accepts this latter view, one implicitly denies to symbolization (including language) any creative role in the cognitive organization of experience and thought: symbolic vehicles then become reduced to a complex system of markers, useful merely for routine indication of referents and for communication about reformed judgments and concepts. (p. 15)

Werner and Kaplan formulated their treatment largely in the context of early language development. In their analysis, symbolic forms are schematized in relation to two dimensions of the communicative situation, a symbol-referent dimension and, additionally, an addressor-addressee dimension. In their treatment of ontogenetic development, both dimensions are undifferentiated in communicative acts at early developmental levels and their principal developmental concerns are to trace a progressive distancing between the schematization of symbol-referent and addressor-addressee relations. At early developmental levels, reference might be accomplished through pointing; the object of reference simply stands for itself, and the infant or toddler does not coordinate the audience's perspective with his or her own. Over development, in the symbolic act, both dimensions undergo progressive differentiation (see Figure 2 for a simplified sketch of Werner and Kaplan's model). Symbolic forms become more autonomous from the objects to which they refer, and referential objects undergo a progressive differentiation and integration.

Werner and Kaplan's treatment of symbol formation provides a useful initial frame for understanding the interplay between symbolic forms and the functions they serve in the Oksapmin case. To illustrate, imagine two Oksapmin people engaged in communication about a one-pound currency note after introduction of the Australian currency to the community.

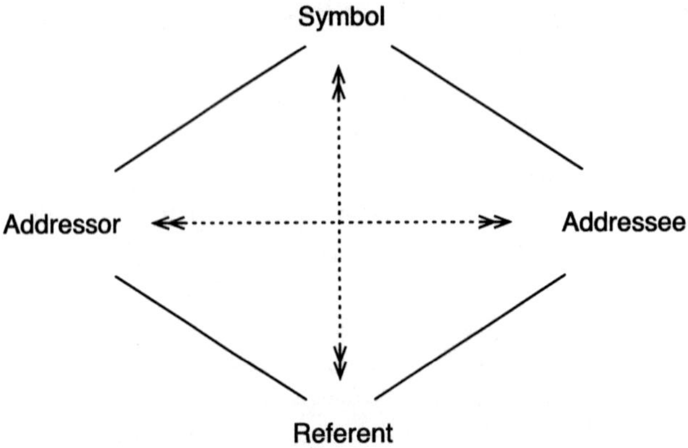

Figure 2. A sketch of Werner & Kaplan's (1963) representation of a progressive distancing between symbol-referent and addressor-addressee relations.

Neither interlocutor yet knows a common name for the note. How do the two communicate their intentions to reference the note to one another?[15]

One means of referring to an object is by pointing (deixis). Referring to the banknote with a pointing gesture makes little differentiation (or distance) between symbol and referent; the pointing gesture is directly dependent upon its object (the banknote) to get across reference. But in many circumstances pointing alone would be inadequate, and we may well find a shift to speech forms, for instance, to a *trope*, a figure of speech, such as "leaf," the referent for which has some physical similarity to the one-pound note (both leaf and note are thin and about the same size and texture). From the Werner and Kaplan perspective, the shift from deictic pointing to the use of the figure of speech "leaf" is a progressive differentiation in symbol-referent and addressor-addressee relations. In the symbol-referent relation, the speaker differentiates what was a deictic act in which there was little differentiation between symbol and referent from an act in which "leaf" (symbol) stands for the note (referent). In the addressor-addressee relation, the speaker takes into account the perspective of the addressee as distinct from his own, anticipating that his interlocutors know the expression "leaf," and this knowledge will support the addressee's identification of the leaf-like note as the object of intended reference.

15 Lewis, in his treatment of conventions, frames this as a coordination problem – or a problem in how interlocutors coordinate their actions in relation to one another (Lewis, 1969). More recently, Clark (1996, 2006) frames coordination problems and their solutions in interlocutors' efforts to establish a "common ground."

As I argued with respect to the blind man's use of his stick, the interplay between cultural forms and functions occurs through three genetic processes, but here form-function relations are involved in symbolic acts of representation and quantification, not tool use. I focus on two broad classes of form-function relations related to representation. One is naming practices for currency tokens. The other is representational systems for number. To illustrate the analytic approach to microgenetic, sociogenetic, and ontogenetic processes in representational activity, for now let's consider people's use of the body system.

The first strand of form-function relations in the use of body parts to serve a numerical function is the process of *microgenesis*, the transformation of the body-part form into a vehicle to represent the coins numerically. Consider an elder's microgenetic construction as he uses the body-part form in an economic transaction in a count of a payment of seven 10t coins. Using the cultural form of body counting, the elder counts a payment of seven coins, touching the coins as each is counted and treating each coin as a body part in an ordinal sequence. This is at once a representational and transformational process – coins are tracked such that they are included once and only once in a count, and the body system is interwoven with the tracking. In this process, the elder transforms the body-part form into a vehicle of numerical representation (forearm [7] as the representation for seven coins), and the coins into a referential object (the coins as a numbered group). From the perspective of the microgenesis of quantification, nothing inherent in the coins or in the body part contains the number seven. Forearm (7), taken as seven coins, is a microgenetic construction in which body parts and coins become sequenced and summed in a representation such that one can stand for the other. There can be rapid shifts in the microgenesis of vehicles and referents. For instance, sometimes a speaker will shift from using coins as referential objects of body parts, as in the counting example, to using them as vehicles of representation, as when they are used to reflect the worth of a bag of rice.

The second strand, *sociogenesis*, involves the microgenetic activities of multiple individuals at multiple sites that, collectively, constitute a process in which representational forms and functions are reproduced and altered in a community over time. Sociogenesis is a population-wide process, distributed across microgenetic constructions in local interactions. For example, at multiple trade stores in the Oksapmin community, people use the body system to accomplish similar problems, presupposing that others will understand their numerical meanings as they communicate with interlocutors. But occasionally, because of variations in personal histories as well as communicative adjustments to make intentions clear to one another, variations in forms emerge across sites in how or whether the body is used, especially as collective problems shift with changing economies and social conditions. Some people may start using body parts to serve different numerical functions, while others may use counting words in Tok Pisin, a language that has become prominent in the Oksapmin world in recent years. Such variations can lead to miscommunications in the back-and-forth of communication, and these communicative acts are sources of both the reproduction and alterations in representational systems through time. It is here we find that the "flow of information" alluded to by Sperber and Hirschfeld is not a "flow" at all. Rather than flow, these reproductions and alterations are solutions to problems that interlocutors generate as they work through communications with one another involving complex cognitive acts in the generation of symbols.[16]

The third strand, *ontogenesis*, involves shifts in form-function relations in activity over the course of an individual's development. The process of ontogenesis itself is constituted in microgenetic constructions of form-function relations. In the microgenetic act, people interpret and produce relations between forms and functions, and in this process adapt forms and functions to new problems, providing continuity with prior knowledge but also introducing discontinuity with novel coordinations. Consider that early in a store owner's development, he likely used counting to assign body parts to objects in order to construct cardinal representations of groups. But in Oksapmin trade stores, prices for items are not listed to support arithmetical compositions, and difficulties of producing arithmetical compositions of nonpresent objects may emerge. As I document in Chapter 4, some people whose participation in the money economy (in 1980) was limited had not created keeping-track functions for body parts in arithmetical compositions. In their effort to add five coins to seven coins when the coins were not present to count, they began with the forearm (7), enumerating the elbow (8), biceps (9) and were unclear where to stop (see Figure 3a). Others who dealt frequently with currency turned the correspondences that they typically established between body parts and objects inward, creating correspondences internal to the body system (Figure 3b),

[16] Sociogenetic processes are thus synchronic, involving activity at various sites in a period as interlocutors address similar problems. They are also diachronic, as actors over time draw upon prior interactions to communicate their intents and work to solve problems, leading to the reproduction and alteration of representational systems over the social history of communities.

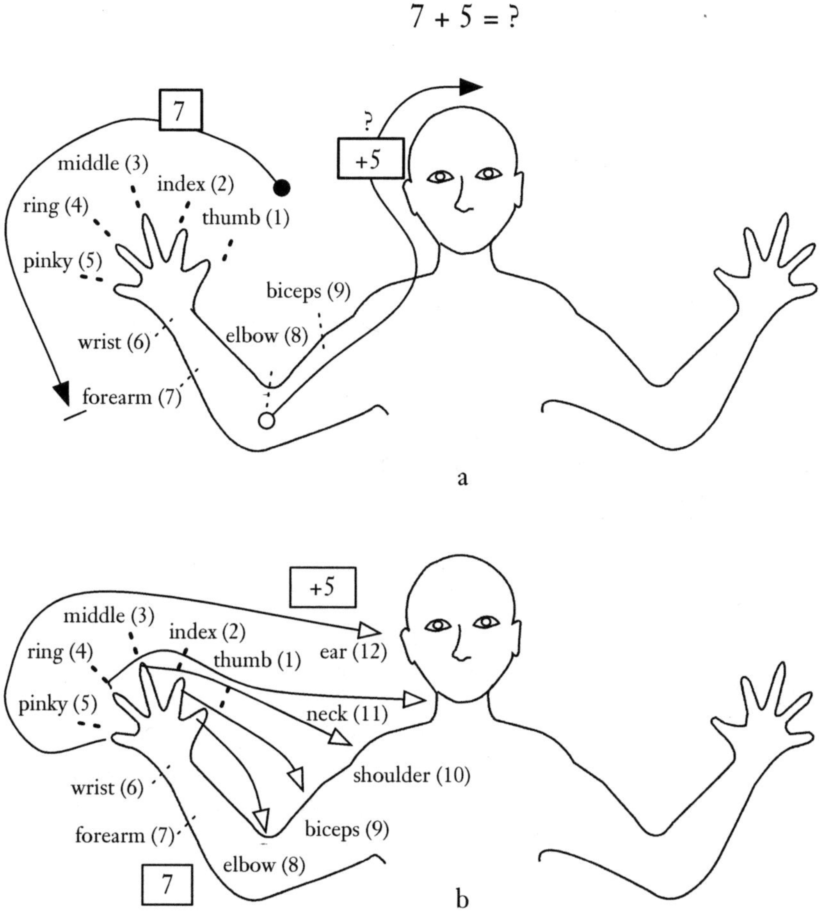

Figure 3. Two body-part strategies to solve 5 + 7 = ? coins. (a) Direct extension of the body system; (b) internal correspondences established to keep track of the addition of 5 to 7.

an ontogenetic development that constitutes a shift in relations between body parts and their numerical functions. These individuals solved problems like 7 coins + 5 coins by producing a double enumeration with a count of the thumb (1) through the forearm (7), and then proceeding five more body parts by counting, but now using the body parts thumb (1) to pinky (5) to keep track of the addition. Thus, the microgenetic construction of solving a particular addition problem is at once a point in sociogenetic processes in the reproduction and alteration of representational forms and functions through the community, and also an ontogenetic development of an individual.

Summary: Collective Practices of Economic Exchange and Genetic Strands of Form-Function Relations

In sum, collective practices like economic exchange in Oksapmin communities are complex, semidurable structures reproduced and altered in exchanges. Over historical time, even as actors change, emergent collective problems recur, like the exchange of cash for goods. In any act, the strands of microgenesis, sociogenesis, and ontogenesis converge as people use representational forms to accomplish problem-linked goals in exchanges. Indeed, constructive acts of communication or problem solving can be understood as at once a process of using representational forms to serve particular functions (microgenesis), which are moments in an individual's own development (ontogenesis) and, at the same time, reproducing and altering forms and functions in the social history of a community (sociogenesis).

Figure 4 contains a schematic of the interplay between form and function in relation to the three genetic processes. The schematic focuses on three individuals (I_1, I_2, I_3), in past, present, and future activities in collective practices of exchange. At the center, in the present, is an individual engaged in a microgenesis of a representation using the body system (I_{2b}), an act that occurs in the back-and-forth with the constructions of others (I_{1b}, I_{3b}). In this process, the actors are reproducing forms and introducing alterations (new variants of forms and functions). Thus, the very same microgenetic acts are moments in sociogenetic processes involving the reproduction and alteration of forms of representation in the community. Moreover, the figure also reflects that these acts are moments in each individual's own ontogenetic trajectory (I_{2a}, I_{2b}, I_{2c}) as he or she draws upon sociogenetic developments in relation to his or her own prior understandings, thereby creating adaptations of form-function relations to local circumstances.

A focus on collective practices as a key arena for analysis addresses the problem of the bifurcation of the individual and the collective with which I began the chapter. In studies of collective practice, the individual and the collective become part of the same frame for analysis. On the one hand, collective activities are constituted by the concerted work of individuals at moments in the microgenesis, sociogenesis, and ontogenesis of

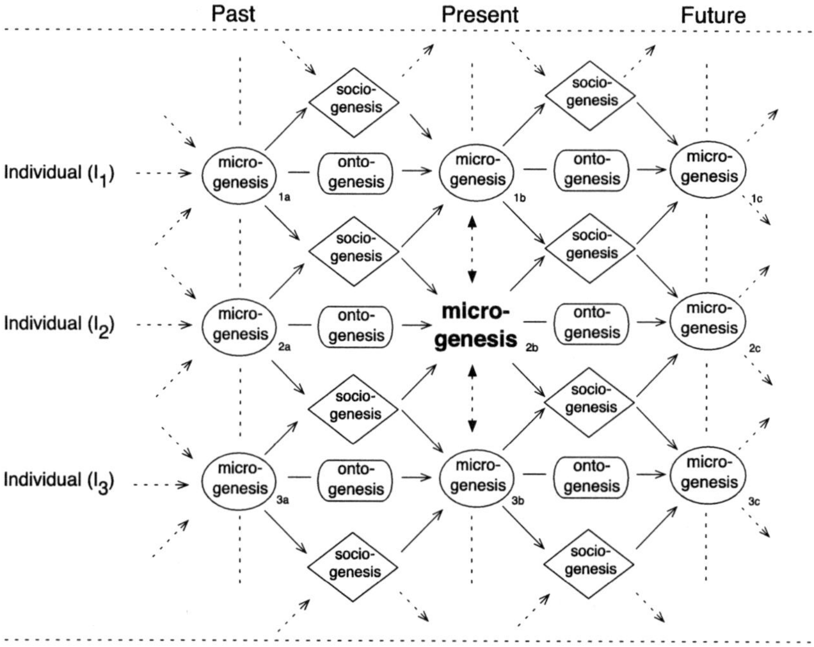

Figure 4. The interplay between genetic processes.

form-function relations. On the other hand, processes of microgenesis, sociogenesis, and ontogenesis take on social meaning in relation to social norms, conventions, artifacts, and institutions themselves as they are reproduced and altered in collective life. The complex and reflexive process in which individual and collective activity take form in relation to one another leads to both continuities and discontinuities in form-function relations as well as collective practices over historical time.

Corroborative Empirical Support for the Utility of the Approach: The Micro- and Ontogenesis of Form-Function Relations in Collective Practices

To my knowledge, no prior empirical research offers a historical treatment of cognitive processes of the sort sketched here, but pockets of prior research do point to the potential utility of the approach. These studies document collective practices in which numerical problems emerge and the cognitive constructions that individuals produce as they solve recurrent problems.

One relevant body of research focuses on collective practices of economic exchange, such as buying and selling commodities. The studies are set in a wide range of sites, with exchange practices differing in the character of the forms used, the conventions for selling and pricing, and the backgrounds of the participants. Most bear on the participants' microgenetic constructions, and some include an additional concern with ontogenetic analyses using cross-sectional methods. Studies include children in urban Brazil who sell candy (Saxe, 1991) and other commodities (Nunes, Schliemann, & Carraher, 1993) or purchase commodities when running family errands (Guberman, 1996); urban Indian children who sell assorted commodities on trains (Sitabkhan, 2008); adults who work in small businesses in Lebanon (Zebian, 2004); and U.S. children who purchase commodities in Hawaii (Brenner, 1998) and purchase snacks at liquor stores in poor urban neighborhoods in the San Francisco Bay area (Taylor, 2009). Investigations have documented the ways that cultural forms such as currency, Hindu-Arabic numeration, and physical commodities enable and constrain the efforts of children and adults to frame and accomplish mathematical problems and goals.

A second arena of research concerns the playing of games. A common focus is on the mathematical problems that emerge in play and the way participants make use of forms to serve varied cognitive functions interwoven with the emerging structure of play. Some of this work focuses on games that are played in people's everyday lives. Nasir's work on dominoes (Nasir, 2000a, 2008) and basketball (Nasir, 2000b) illustrates this focus well. In her work on dominoes, for example, she notes that the game has long been popular in some African American communities, with similarities and variation across communities. She documents the structure of game play and engages in a careful analysis of the emergent problems in play with attention to the way that students structure material forms, with some attention to issues of identity and broader institutional relations in which collective practices take form (Nasir & Saxe, 2003).

Another body of work focuses on much younger children in their everyday play and in teaching-learning situations that emerge in families. In a study conducted in New York in the early 1980s with working and middle-class communities (Saxe, Guberman, & Gearhart, 1987b), our focus was the numerical activities that were occurring in the everyday lives of 2½- and 4½-year-old children. Our analyses included children's use of representational forms (like number words, written numerals) and the functions the forms served (counting to produce cardinal values, counting to compare groups, counting to produce arithmetical compositions). A principal focus was on shifting form-function relations as children constructed and accomplished numerical goals in everyday numerical activities with their parents. The results provided insight into differing trajectories of development among children growing up in different households.

Other work on game play focuses on the educational design of games that make use of forms designed to support children's developing understanding of ideas. For example, in a project I conducted in collaboration with my graduate students, we designed a board game called Treasure Hunt. In the game, third and fourth graders purchased and sold supplies using a currency of gold doubloons (base-10 blocks) and then reported to one another the worth of their doubloons using Hindu-Arabic numerals on gold registers. The numerical values (in the hundreds and thousands) were designed to engage students in solving problems of place value in the context of additions and subtractions. The game led students to create new functions for representational forms related to place value, multidigit arithmetic, and arithmetic with base-10 blocks (Saxe & Bermudez, 1996; Saxe & Guberman, 1998).

A third arena for the study of emergent goals in collective practices involves the analysis of classrooms. In research on the "travel of ideas" in classroom communities, I have investigated the ways in which mathematical ideas are reproduced and altered over the course of lessons (Saxe et al., 2009). Related investigations in discussion-rich classrooms have focused on the "snowballing" of recurrent analogical reasoning strategies (Dong, Anderson, Kim, & Li, 2008; Lin et al., 2010) or the diffusion of tool use in a science class (Roth, 1996). Still other studies of emergent goals in classroom communities focus on the interplay between software used in classroom communities and students' prior understandings (Lagrange & Erdogan, 2009). Building on analyses of social positioning and power relations, ethnographic studies (for example, Davies & Harré, 1990; van Langernhove & Harré, 1999) have explored in classroom interactions the interplay between positional identities and the elaboration of mathematical goals (Esmonde, 2006; Langer-Osuna, 2010).

Corroborative Empirical Support for the Utility of the Approach: The Sociogenesis of Form-Function Relations in Collective Practices

To date, research that takes a sociohistorical perspective on collective practices and genetic processes is limited to a small corpus of studies, though for some time researchers have pointed to the potential utility of a historical approach in which a focus is on the sociogenesis of knowledge forms

(Scribner, 1985; Vygotsky, 1986). One relevant line of research was conducted by Greenfield (Greenfield, Maynard, & Childs, 1997; Greenfield, 1993, 2004) and more recently by Maynard (Maynard, 2009; Maynard & Greenfield, 2006; Maynard, Greenfield, & Childs, 1999), documenting the shifting collective practices of weaving in a Mayan community in Chiapas, Mexico. The investigators show how historical changes from an agricultural to a commercial economy penetrate the collective practice of weaving, and they examine the shift in cognition of individuals linked to weaving. Luria (1976) has also championed a historical approach using cross-cohort methods to study shifts in reasoning linked to a movement toward collectivism in the postrevolutionary Soviet Union.

Historical research based on archival sources provides additional evidence of form-function shifts in relation to historical changes in collective practices. For example, Swetz (1987), in his analysis of one of the first arithmetic texts – the *Treviso* – investigates the gradual shift in the early Renaissance from the use of counting boards and Roman numerals to the ascendancy of Hindu-Arabic numeration and bookkeeping (see Figure 5). The shift was gradual, and Swetz's analysis points to the entrenchment of cultural forms of representation like Roman numerals and counting boards in the economics of a particular era as well as shifts in cultural forms associated with shifts in collective practices of economic exchange and schooling.

Denise Schmandt-Besserat, whose focus has been on the emergence of cuneiform number representation in ancient Mesopotamia, presents a fascinating case of sociogenesis of form-function relations based upon archeological records (Schmandt-Besserat, 1992, 2004). Schmandt-Besserat asks about the origins of cuneiform numeracy, a cultural form found inscribed on many clay tablets at sites throughout ancient Mesopotamia.

According to Schmandt-Besserat and other scholars of ancient Mesopotamia (e.g., Nissen, Damerow, & Englund, 1993), farming settlements existed in the Near East around the Jordan and upper Euphrates rivers in about 9000 BC, and people engaged in exchanges of commodities. About a thousand years later, small clay objects of distinctive shapes appeared – tetrahedrons, cones, spheres – that Schmandt-Besserat argues were used to serve numerical record-keeping functions in collective practice of exchange (see Figure 6a). Sometimes these tokens were found inside hollow clay spheres called *bullae*, with impressions on the outside that resembled the tokens within, and other times the tokens were found outside of the bullae. Schmandt-Besserat suggests that bullae were used to contain a record of an exchange, analogous to a modern-day receipt, and

Figure 5. Calculating-table by Gregor Reisch: Margarita Philosophica, 1508. The woodcut shows Arithmetica instructing an algorist and an abacist. (Located at the Bibliothèque de la Faculté de Médecine, Paris, France.)

that their outer markings were iconic inscriptions of the bullae's contents (see Figure 6b).

How did cuneiform representations come to take the form that they did in tablets like that shown in Figure 6c (and in tablets more complex than shown here)? Schmandt-Besserat argues that, over time, people who used inscribed bullae came to treat the tokens within as superfluous for a transaction record; all that one needed was the set of exterior marks inscribed in clay to record the transaction, and thus the small clay tablet (Figure 6c), the progenitor of cuneiform writing, was born. In later periods,

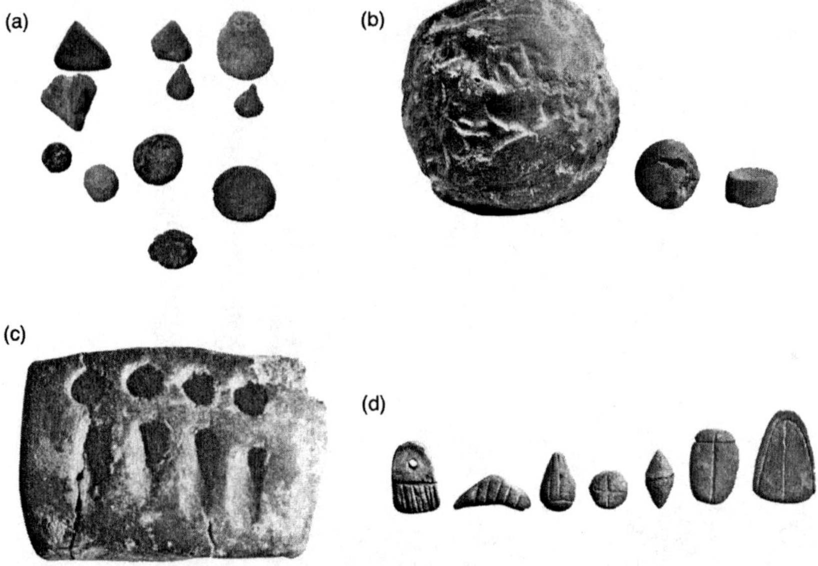

Figure 6. (a) Plain tokens, Jarmo, Iraq. (b) Envelope and its contents, a sphere and a disk, Susa, Iran. (c) Tablet showing a line of four circular signs followed by a line of four short wedges, Godin Tepe, Iran. (d) Complex tokens, Susa, Iran. (Images reproduced from *Before Writing, Volume I: From Counting to Cuneiform* by Denise Schmandt-Besserat, Copyright 1992. Courtesy of the author and the University of Texas Press.)

archeological strata show the appearance of more complex tokens that may have served new functions associated with an urban economy (Figure 6d). These later tokens appeared to represent commodities like bread, oil, perfume, and wool as well as artifacts made in workshops like bracelets, types of cloth, and tools. Schmandt-Besserat's research shows how, through collective practices of exchange, sign forms and their functions for quantity might have shifted.

Schmandt-Besserat's argument about the emergence of cuneiform numerical records is supported by her analysis of ancient artifacts; of course, she could not investigate directly the ancient peoples' activities and their uses of forms to serve numerical functions. As a result, many of her claims about activity lack the kind of empirical corroboration one would want. In a critique of Schmandt-Besserat's evidence and argumentation, Zimansky (1993) questions the continuity of tokens, bullae, and cuneiform writing across this 5,000-year period of study, and the paucity of evidence that the same clay forms were used for only a single function over this stretch of Mesopotamian history. He also argues that Schmandt-Besserat is overattributing numerical functions to excavated objects that may not have been tokens at all and that the numerical renderings on the bullae may not have been pictographs of their interiors. Zimansky's critiques notwithstanding, Schmandt-Besserat's historical treatment of human activity in the emergence of writing and number representation is a resource for the work I report in this volume. Her diachronic perspective on form and function, and her allusions to broad-scale sociogenetic processes rooted in collective practices, point to the importance of social history as a central construct in culture-cognition relations.

The Promise of the Oksapmin Case

The bifurcation of cultural and cognitive phenomena in studies of their relations has led to some important contributions in the cognitive sciences, but research in this tradition cannot address questions that lie at the crux of culture-cognition relations, like the origins of numeric representations and the shifting functions of these representational forms in collective practices of daily life. The studies that I report in this volume serve two broad purposes. The first is to document how numeric representations and ideas have been reproduced and altered over time in Oksapmin communities. As such, the research contributes to the social history of a little-known group whose forms of representation, like those used by many other non-Western communities, may soon be lost if this history remains undocumented (Evans, 2010). My second purpose is to elaborate the skeletal framework that I have outlined in this chapter – a general methodological framework in which sociocultural history is central to cognitive analyses and vice versa.

2 Cultural Forms of Number Representation Used in Oksapmin Communities

Located just north of Australia and just south of the equator, New Guinea is the second largest island on the globe. The first human migration is estimated to have been 50,000 years ago by way of Indonesia. Subsequent migratory patterns, the island's rugged geography, and local warfare have contributed to the isolation of communities, leading to remarkable linguistic and cultural diversity: The island is the site of more than a thousand cultural groups that speak more than eight hundred languages. The Oksapmin reside in the Mountain-Ok area of central New Guinea, a region in which distinct groups share a material culture such as dress, tools, and weapons (Craig, 1988). The Mountain-Ok geographical region is home to seven language groups, as depicted in Figure 7.[1] The Oksapmin speak a language that has for some time been considered an isolate, with unknown relations to the other Mountain-Ok languages (Healey, 1964; Wurm, 1983), though its isolate status has been recently questioned. The argument is that similarities previously thought to be due to borrowing may well be rooted in a prior proto-language (Loughnane, 2009; Loughnane & Fedden, 2006). As of 2005, Oksapmin language speakers numbered about 8,000 people (Loughnane, 2009).

Traditional subsistence practices in Oksapmin, as in other Mountain-Ok communities, include cultivating taro and sweet potato, keeping pigs, and hunting with bow and arrow for birds and small mammals. Other sources of protein in the diet include varied types of insects, worms, and tadpoles. As in the other Mountain-Ok communities, traditional dress is a grass skirt

[1] The related Mountain-Ok languages include Ngalum, Tifal, Miyanmin, Telefol, Bimin-Kuskusmin, Faiwol, Fetik, and perhaps Oksapmin, though this is a matter of dispute (Healey, 1964; Loughnane, 2009; Loughnane & Fedden, 2006; Wurm, 1983).

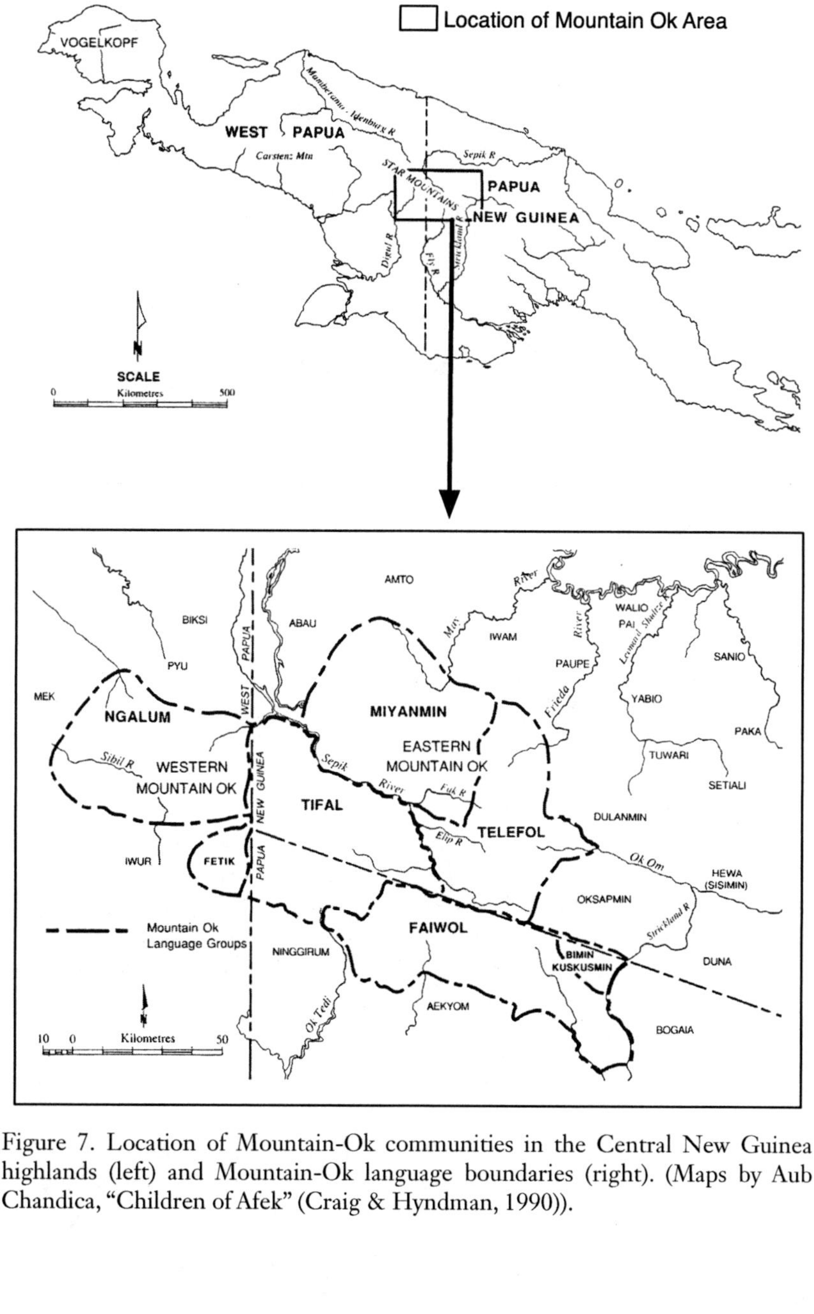

Figure 7. Location of Mountain-Ok communities in the Central New Guinea highlands (left) and Mountain-Ok language boundaries (right). (Maps by Aub Chandica, "Children of Afek" (Craig & Hyndman, 1990)).

Figure 8. Man in traditional dress – cane belt and penis sheath – smoking (indigenous) pipe in his taro garden.

for women and a penis gourd and cane belt for men. Figure 8 shows an Oksapmin man in his taro garden wearing traditional dress (photographed in 1978).

The cosmology of all the Mountain-Ok groups is rooted in an origin myth of a magical woman (known as Uanku in the Oksapmin area but referred to as Afek by the Telefolmin and other Min groups), though versions of the myth vary across groups. In a prototypical version, Uanku is said to have traveled long ago through the Oksapmin area from the east toward the western Mountain-Ok, and she brought with her culture, including language (perhaps number), horticulture, and gendered roles for men and women (Brumbaugh, 1990; Brutti, 2000). Some believe that Uanku shifted languages as she encountered groups on her travels, a belief that accounts for the language differences among neighboring groups. All Mountain-Ok groups know some version of the myth, though accounts of Uanku's travels differ (Barth, 1987). Initiation in some Mountain-Ok groups involves difficult challenges, and, as these challenges are met, more secret knowledge is revealed, such as progressively deeper meanings of Uanku's activities (Brumbaugh, 1990).

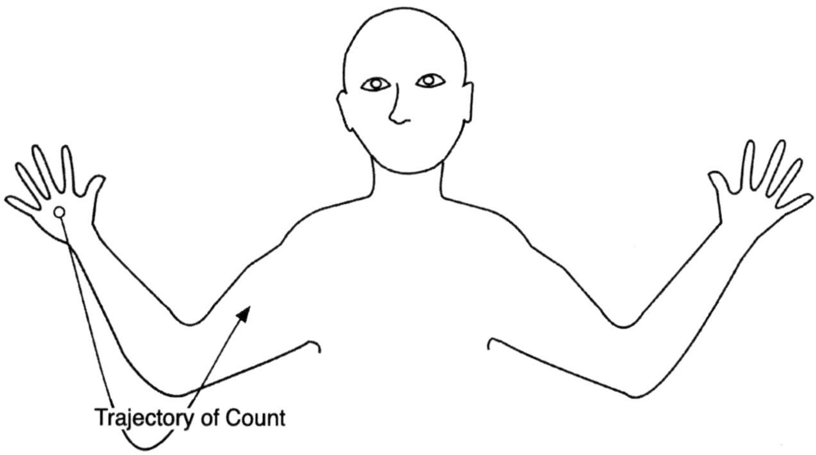

Figure 9. Generic trajectory of body parts used in highlands counting systems.

Body Counting in the New Guinea Highlands

The counting system of the Oksapmin is one of many body-count systems used among Papua New Guinea highland groups. In a comparative analysis, Laycock (1975) notes that in almost all of the highlands body systems, counting proceeds *up* one side of the body to a midpoint like the nose or the breastbone and then *down* the other side, duplicating positions in the reverse direction. Often the naming of the duplicate positions is accompanied by a linguistic marker indicating "again." Because of the midpoint, most body systems end the series on an odd number. Laycock also notes that systems differ in their symmetry: some systems are asymmetrical, with fingers on both right and left hands counted in the same order, while others are symmetrical, beginning and ending, for example, on the little finger of each hand. The trajectory of a count across systems is depicted in Figure 9.

Figure 10 contains Laycock's analysis of what he regarded as the better-documented systems traditionally used in highlands areas. In the table, columns indicate cultural group, and rows indicate the ordinal position of a body part in a group's count sequence. The final row indicates the number of ordinal positions in each counting system, respectively. All of the groups included in the table use the five fingers for the first five positions, though some groups start with the thumb and others start with the little finger. For subsequent positions, there is variation. Only the Kewa use the heel of the palm for the sixth position and use the palm as the seventh. Almost all other groups represented (except the Enga) use the wrist for the sixth position.

Body Part	Language Group									
	Kewa	Gende	Orokolo, Telefol, Hewa	Sibil	Yuri, Kalam	Anggor	Namau	Enga	Gidra	Bine, Awyi
Little finger (thumb)	1	1	1	1	1	1	1	1	1	1
Ring (index) finger	2	2	2	2	2	2	2	2	2	2
Middle finger	3	3	3	3	3	3	3	3	3	3
Index (ring) finger	4	4	4	4	4	4	4	4	4	4
Thumb (little finger)	5	5	5	5	5	5	5	5	5	5
'Heel of palm'	6	-	-	-	-	-	-	-	-	-
Palm	7	-	-	-	-	-	-	-	-	-
Wrist	8	6	6	6	6	6	6	-	6	6
Forearm	9	7	7	7	7	7	7	6	-	-
Radius	10	-	-	-	-	-	-	-	-	-
Ulna	11	-	-	-	-	-	-	-	-	-
Elbow (inner)	12	-	8	8	8	8	8	7	7	7
Biceps	-	-	9	9	9	9	-	-	-	-
Upper arm	14	9	10	-	-	-	-	-	-	-
Armpit	-	-	-	-	-	-	-	-	8	8
Edge of shoulder	15	-	-	-	-	-	-	-	-	-
Top of shoulder	16	10	11	10	10	10	9	8	-	-
Neck muscle/clavicle	17	11	-	-	11	-	-	9	-	-
Side of neck	18	12	-	11	-	-	10	-	9	-
Nipple	-	-	-	-	-	11	11	-	-	9
Side of jaw	19	-	-	-	-	-	-	-	-	-
Ear	20	13	12	12	-	-	-	10	-	-
Cheek/temple	21	14	-	-	-	-	-	-	-	-
Eye	22	15	13	13	-	-	-	11	-	-
Inside of corner of eye	23	-	-	-	-	-	-	-	-	-
*top of breastbone	-	-	-	-	12	12	12	-	10	10
*nose-bridge	24	16	14	14	-	-	-	12	-	-
Total	47	31	27	27	23	23	23	23	19	19

*Denotes central point

Figure 10. Body-part counting in the New Guinea area (from Laycock, 1975, p. 223).

The Oksapmin Counting System

The Oksapmin body system includes 27 body parts, the same 27 body parts used in the Telefol language represented in Figure 10.[2] The body words used in the Oksapmin count system are listed in Figure 11. In conversations with other visitors to Mountain-Ok regions, I learned that other Mountain-Ok groups use an identical set of 27 body parts, but with some using a symmetrical order of fingers at the beginning and ending of the 27 body parts (e.g., beginning a count with one thumb (1) and ending with the other (27)). Some groups west of Oksapmin do not start with the thumb (1) as the Oksapmin do but instead with the pinky (1) – this is true of the Sibilmin (as noted in Laycock's Table 1), who reside in the westernmost part of the Mountain-Ok area (Brongersma & Venema, 1962), the Nekalimin,

[2] Note that Sibil is a dialect of one of the Mountain-Ok languages.

Number	Body Parts (English)	Body Parts (Oksapmin)	Specialized Lexemes (not body parts)	Number	Body Parts (English)	Body Parts (Oksapmin)
1	Thumb	*Tip^na*	Tit	15	Eye	*Tan-kina*
2	Index finger	*Tipnarip*	*Yota*	16	Ear	*Tan-nata*
3	Middle finger	*Bumrip*	*Yetir*	17	Neck	*Tan-gwer*
4	Ring finger	*H^tdip*	*Yota yota*	18	Shoulder	*Tan-kata*
5	Little finger	*H^th^ta*	*Hanen*	19	Biceps	*Tan-tow^t*
6	Wrist	*Dopa*	-	20	Inner elbow	*Tan-kir*
7	Forearm	*Besa*	-	21	Forearm	*Tan-besa*
8	Inner elbow	*Kir*	-	22	Wrist	*Tan-dopa*
9	Biceps	*Tow^t*	-	23	Thumb	*Tan-tip^na*
10	Shoulder	*Kata*	-	24	Index finger	*Tan-tipnarip*
11	Neck	*Gwer*	-	25	Middle finger	*Tan-bumrip*
12	Ear	*Nata*	-	26	Ring finger	*Tan-h^tdip*
13	Eye	*Kina*	-	27	Little finger	*Tan-t^th^ta*
14	Nose	*Aruma*	-			

Note. The trajectory for body parts from 1 to 13 ascends up one side of the body and, for body parts from 15 to 27, down the other side. To indicate a cardinal value for body parts, a counter uses the suffix *–hai* to a body part name. To count beyond 27, the counter loops up and around the body to the wrist (28) on the same side as the 27th body part, continuing with the forearm (29), elbow (30)...

Figure 11. Oksapmin number systems: the body-count system and the specialized lexemes for number.

in the region bordering West Papua (Wesch, 2004), and the Telefolmin people (Healey, 1965). However, the Asabano, allies and trading partners of the Oksapmin, use the same sequence of body parts (Lohmann, 2004), and, like the Oksapmin, loop back around the body to count numbers greater than 27. The reports on the Sibilmin, Nekalimin, and Telefolmin counting systems do not describe the looping procedure. Further, all groups distinguish the first 13 body parts from the "mirror image" last 13 by the use of a morpheme. The groups more to the west cited in the figure use a suffix to distinguish symmetrical body parts, whereas in Oksapmin people use the prefix *tan* (meaning, roughly, "other side").

To count using the Oksapmin body system, one begins with the thumb of one hand and enumerates the 27 places around the upper periphery of the body (with accompanying body-part names), ending on the little finger of the opposite hand accompanied by respective body-part words (Moylan, 1982; Saxe, 1981a). To count past the 27th body part, as depicted in Figure 12, the counter loops from the little finger (27) to the wrist, then proceeds to the forearm, and on up and around the body.[3] There is no distinction between the name for the 21st body part and the name for the 29th body part; both are called *tan besa* ("other forearm," indicating the forearm on the opposite side of the body from where the count was initiated). To

[3] As I report in Chapter 4, in 2001 there was variation in this looping procedure and the treatment of the last body parts in the count.

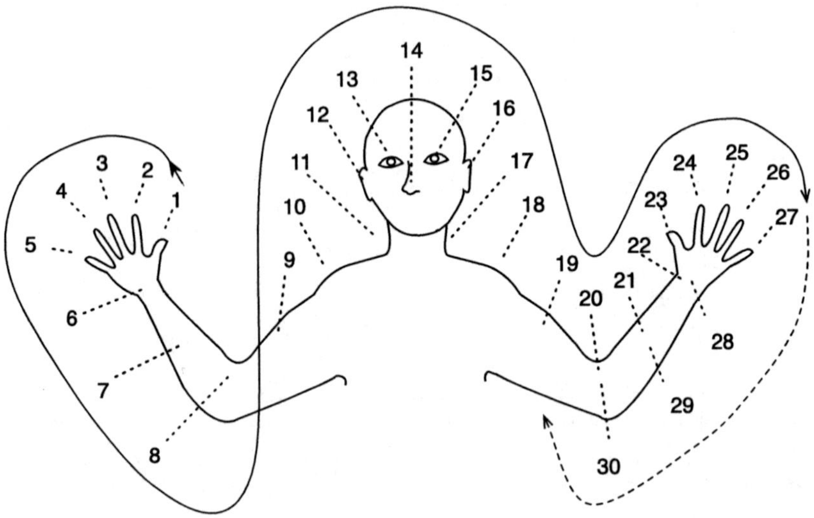

Figure 12. The trajectory of a count using the Oksapmin body-part system (Saxe, 1981a).

indicate a particular number, the counter points to the appropriate body part and says the body-part name. For example, to indicate the number 12, he or she points to the ear, which is the 12th body part in the sequence, and says the word for ear, *nata*. To create a cardinal representation, he or she uses the morpheme "*hai*" after the body-part word.

Fu! *and Other Ancillary Expressions in Oksapmin Counting*

In Oksapmin, upon completion of the little finger (27th body part, see Figure 12), it is customary to produce the stereotypic gesture of raising one's fists and exclaiming "*fu!*" or "*tit fu!*" or "*tit fu gon a*" (see Figure 13). This expression signals the completion of a count of all of the body parts. Note that if one is counting further than 27, one loops back to the wrist of the second hand and then names the body parts, proceeding in the reverse direction.

Auxiliary words and morphemes also serve specialized counting functions. As noted previously, the suffix *hai* is added to a body-part word to mark the body part as an expression of a cardinal value; without this marking, the body part could refer to either a count word or an ordinal value (or, more rarely, a cardinal value). Direction words may be used to indicate directional orientation in the trajectory of a count: Examples are *wasa* (indicating "going down"), as in proceeding down the second side of the

Figure 13. Woman reaching 27th body part and making stereotypic gesture of raised fists accompanied by the exclamation "*tit fu!*"

body, and *wuris* (or *rusi*), indicating "going up," as in going back up the body after having circled around the 27th body part.

In many respects, the body-counting system used by the Mountain-Ok groups appears well tailored to solving numerical problems in traditional life, akin to number systems across other human communities (Crump, 1990; Menninger, 1969). The system was (and to some extent still is) used in daily activities that involved counting valuables, measuring string bags (a common artifact), and indicating ordinal position (such as ordinal relations along a path). Arithmetic, as westerners know it – the composition of cardinal values into sums and differences – appears to have been nonexistent in traditional life.[4]

In his review of groups that use body systems in the New Guinea highlands, Laycock noted that many of these groups use additional specialized words for number that are unrelated to body-part names. Typically there is a "conversational" number-word register that includes no more than seven

[4] I base this assertion on my observations, discussions with Oksapmin people, local westerners living in the Oksapmin area, and the findings of a 1980 study reported in Chapter 4 (Study 4-1).

number names. As in the groups Laycock described, the Oksapmin use a limited two-cycle set of five words specific to representing the integers one through five: *tit* (1), *yota* (2), *yetir* (3), *yota yota* (4), and *hanen* (5).[5]

Reproduction and Alteration of Numerical Representations and Ideas

This volume, which contains the results of research on the social history of number in Oksapmin communities that cut across a significant span of time in the Oksapmin world, provides an extraordinary vantage point from which to study processes of reproduction and change. I have partitioned the empirical studies into two parts, each focused on a set of collective practices. Part II is concerned with economic exchange and Part III with schooling. Across Parts II and III, I use similar methodological approaches to investigate forms of representation and the functions that these forms serve in relation to collective practices of exchange and schooling.

For historical and analytic purposes, there is good reason to separate the practices of economic exchange and schooling. Negotiations involving economic exchanges with currency were inherently local: no central Oksapmin authority set down conventions, save the exchange value of currency as determined by a mission store. Western schooling, in contrast, brought with it the authority of the missionaries and the government teachers. What was taught came from the central authority of the school with the backing of the weight of the penetrating Western world. In the early years of schooling in Oksapmin, the teachers themselves were expatriates, mostly from other parts of Papua New Guinea, who had been schooled in teachers' colleges and knew little, if anything about Oksapmin ways. Though I treat collective practices of economic exchange and schooling in different parts of the volume, I do investigate the ways in which participation in one collective practice may lead to alterations in the other. For example, sometimes Oksapmin children come to school with a prior history of buying items at a trade store, and sometimes teachers use lessons that apply Western algorithms to solve problems with currency. In the field, I was careful to design empirical studies that could reveal how the developments that individuals created in one practice may also have been used to frame and solve problems in the other.

[5] I observed dialect variations in the Oksapmin community that others have previously documented (Lawrence & Snyder, 1993; Moylan, 1981). I have included dialect variations for number-related words in Appendix A, Figure 140.

Part II

Economic Exchange

Collective practices of economic exchange have shifted dramatically over the past 60 years in the Oksapmin world. In Chapter 3, I provide a historical sketch; I draw upon archival records and fieldwork to examine the changing social, political, and economic conditions interwoven with these shifts with a focus on the emergent numerical problems with which people have been engaged and the changing character of these problems related to collective practice of economic exchange. Then Chapters 4 through 6 report on empirical investigations of form-function shifts in the numerical representations and currency token representations that people have used in shifting collective practices. Chapter 4 focuses on numerical representation, and Chapter 5 on the representation of currency tokens and the quantification of tokens. Chapter 6 concludes with studies of form-function shifts in the use of one expression, *fu!* The expression serves as a microcosm for an analysis of the interplay between shifting collective practices of economic exchange and the microgenesis and sociogenesis of form-function relations in numerical representations.

Figure 14 provides an overview of the methods used across the studies on economic exchange. For each study, the figure provides study number, name, year, empirical techniques, research designs, and the principal forms and functions that were the study's focus. The data-gathering techniques and research designs I used are eclectic: some techniques are qualitative in character, and others yield quantitative information. Together, these studies reveal, with some level of detail, the interplay between individual and collective activity in the reproduction and alteration of representational forms and the functions they serve in collective practices of economic exchange.

Study/period number, (Date), Title	Empirical techniques	Data sources/ designs	Principal forms & functions of interest: Forms	Functions
Chapter 3. Collective Practices of Economic Exchange: A Brief Social History				
Period 1. Pre-Western contact (before 1938)	Review of archival records	Archival	Body parts, quantifiers, *bonang*, commodities	Counting, communication
Period 2. Early colonial period (1938-1960)	Review of archival records	Archival	Body parts, quantifiers, *bonang*, commodities, currency tokens	Counting, communication
Period 3. Late colonial period, early independence, and the burgeoning cash economy (1961–1980)	Review of archival records; ethnography	Archival, ethnographic	Body parts, quantifiers, *bonang*, commodities, currency tokens	Counting, arithmetic, communication
Period 4. Post independence and the cash economy (1981–2001)	Review of archival records; ethnography	Archival, ethnographic	Body parts, quantifiers, commodities, currency tokens	Counting, arithmetic, communication
Chapter 4. Reproduction and Alteration of Numerical Representations				
Study 4-1 (1980). Body arithmetic	Task-based interviews	Cross-cohort	Body parts, currency tokens	Counting, arithmetic
Study 4-2 (2001). The body and Tok Pisin: Forms used for number representation	Task-based interviews	Cross-cohort	Body parts, Tok Pisin count words	Displays of Oksapmin/ Tok Pisin counting
Study 4-3 (2001). Languages of quantification used in stores and emergent arithmetical problems	Observational records: Exchanges, speech	Cross-cohort	Oksapmin/Tok Pisin language use	Representation and quantification of values
Study 4-4 (2001). Emergent arithmetic in store transactions	Observational and interviews	Cross-cohort	Tok Pisin & Oksapmin quantifiers, goods, tokens	Exchange of commodities for currency
Chapter 5. Reproduction and Alteration of Currecny Token Representations				
Study 5-1 (2001). Shifting ways of naming tokens	Task-based currency token identifications	Cross-cohort	Oksapmin/Tok Pisin language use	Token reference
Study 5-2 (2001). Shifting quantification practices with currency	Task-based interviews	Cross-cohort	Body parts, currency tokens	Addition of multi-denominational tokens
Chapter 6. Representational Forms, Functions, Collective Practices, and 'Fu': A Microcosm				
Study 6-1 (2001). Shifting meanings, locations, and uses for *fu*	Interviews & task-based interviews	Cross-cohort	*Fu*, body parts, quantifiers, tokens	Varied quantitative functions
Study 6-2 (2001). Distribution of *fu* use in representing values of currency and number of stones	Task-based interviews	Cross-cohort	*Fu*, currency tokens	Communicating summations of currency
Study 6-3 (2001). *Fu* use in the reading of prices written with Hindu-Arabic numerals	Task-based interviews	Cross-cohort	*Fu*, Hindu-Arabic inscriptions for number	Reading multidigit numbers, communicating read prices

Figure 14. Research designs and empirical techniques used in chapters on economic exchange.

3 Collective Practices of Economic Exchange

A Brief Social History

Collective practices of economic exchange have a complex history in Oksapmin communities. In the following pages, I partition this history into the period just prior to Western contact and then delineate three more recent periods extending over approximately six decades. During this 60-year stretch, trade practices shifted from a traditional subsistence and barter economy to an economy rooted in exchange of commodities for currency.

Figure 15 shows the four periods on a timeline. The timeline portrays a growing differentiation between a subsistence and a Western cash-based commercial economy. In the figure the vertical axis is an estimate of commercial economic activity, roughly indexed by export of vegetables, cash in circulation, number of commodities sold in Oksapmin trade posts, and proportion of people who participate in the cash economy. I estimated that this differentiation began in about 1960, although processes before the 1960s seeded the differentiation. Also represented on the timeline are significant events related to the economy, such as the issuance of new currencies, national independence, the completion of airstrips, the opening of the Ok Tedi mine, and the introduction of schooling. Though the line shows an upward trajectory for the amount of currency in circulation, I expect that in the future the trajectory will not always be upward. Indeed, when the Ok Tedi mine in Tabubil closes in several years as the ore is exhausted, Oksapmin communities will lose a principal market for the export of vegetables. Further, remittances from Oksapmin people who have found employment in Tabubil will likely also decline.

Pre-contact

Archeology in the Telefomin area shows that people lived in the central New Guinea highlands at least 17,000 years ago, with varying periods of

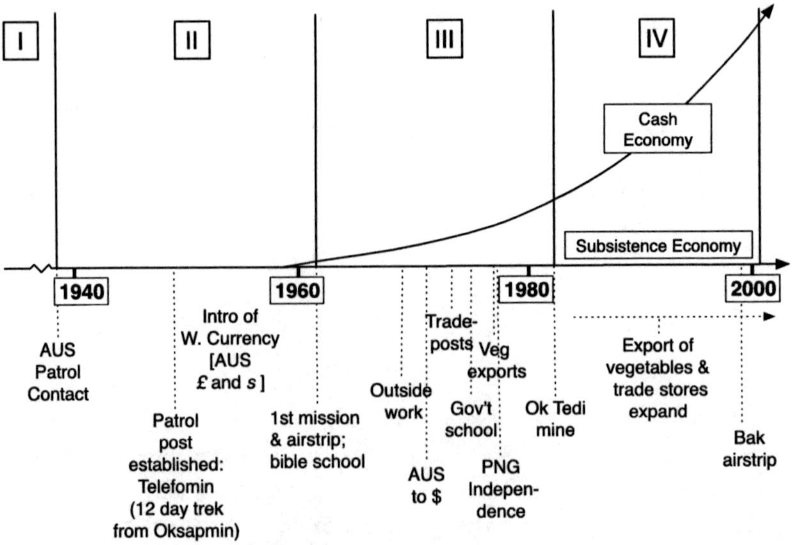

Figure 15. A growing differentiation between subsistence and cash economies in Oksapmin (1940–2001). Weeks (1981a) presents a more complete chronology of events related to contact in the Oksapmin area.

migration since that time (Swadling, Mawe, & Tomo, 1990). Hyndman and Morren (1990) argue that the introduction of the sweet potato in the Mountain-Ok area about three or four hundred years ago – a comparatively recent event – had a profound effect on life in the region, since previously the principal staple for people in central New Guinea was a type of taro that grew only at limited elevations. Unlike taro, Hyndman and Morren note, the sweet potato could be cultivated at a wide range of elevations. Thus, sweet potatoes became a staple both for human consumption and also as fodder for pigs, a source of protein. The increase of food supply (due to the rise of both the pig population and the sweet potato harvest) and the need for additional land to support it led to population increases as well as outward expansion from a central core into alternative biomes.

Trade was central to prehistoric life in Mountain-Ok groups (Cranstone, 1990). Figure 16 depicts known trajectories of commodities (from Cranstone, 1990). The Oksapmin valleys are located toward the lower right part of the map in Figure 16, and the principal trade routes that include the area are numbers 2, 4, 5, and 6. As indicated in the figure, Oksapmin obtained stone axes from the north; black palm, shells, and bows

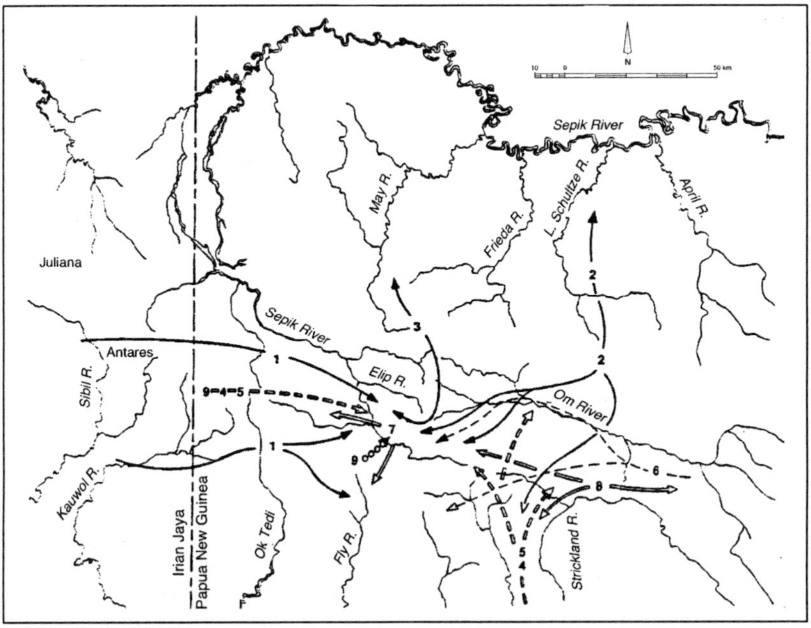

1. Stone adzes and nassa shells: stone adzes (fubi) from Mumyeme/Red Digul sources reach the Wopkeimin via Moyansil bridge, the Fegolmin via the Wopkeimin and the Ifitaman Telefolmin via both the Atbalmin and Wopkeimin. Nassa shells (bonang) from West Papua follow the same routes as the fubi adze; nassa shells are the primary shell valuable for the Mountain-Ok.
2. Stone adzes and cowrie shells: stone adzes (mok) from sources on the Tau and Mepu Rivers are traded both north to the Sepik and south to the Om River, thence to the Telefolmin, Falamin and Oksapmin, and from the Oksapmin to the eastern Faiwol groups and beyond; cowries (kol) are traded from the Sepik area via the Leonard Schultz and Om Rivers.
3. Stone adze: stone adzes (bangkele) from Mt. Stolle are traded north and south by the Miyanmin; they may reach the Telefolmin and Falamin via the Om River and the Fegolmin via the Oksapmin.
4. Bows: Black palm bows reach ifitaman Telefolmin via the Atbalmin, and the Oksapmin via eastern Faiwol groups; the eastern Faiwol groups also supply brown palm bows to the Ifitaman Telefolmin.
5. Handdrums: Handdrums (ot), associated with taro ritual, are made by the Atbalmin and traded to the Ifitaman Tlefolmin; drums of identical type are made by the Kwermin and other Faiwol groups, and traded to the Falamin, Bimin and Oksapmin; similar drums are made on the Leonard Schultz.
6. Salt: Salt made by the Oksapmin from ash was traded to the Faiwol, and to the Ifitaman Telefolmin via the Om River. Salt from the Enga was passed on by the Duna to the Oksapmin and became part of the other salt trade network.
7. Tobacco: Tobacco (suk) is traded from the Telefolmin west to the Atbalmin and south to the Fegolmin; the Wopkeimin obtain their tobacco from the Tifalmin.
8. Pandanus oil: Red pandanus oil was traded from the Bimin west to the Ifitaman Telefolmin, south to eastern Faiwol and east across the Strickland.
9. Dogs' teeth: Bands of dogs' teeth are highly valued by the Mountain-Ok but are more plentiful in the lowlands to the north and south; they reach the Ifitaman Telefolmin primarily via the Fegolmin to the south and the Atbalmin in the west.

Figure 16. Movement of commodities through successive trades with neighboring groups in Mountain-Ok prehistory (Figure 46 from Craig & Hyndman, 1990).

from the south; salt from the east; and kundu drums from the west (see also Cranstone, 1990; Weeks, 1981a).

In pre-contact times, people traded commodities, which included shell valuables, a form of currency (*bonang*) used across Mountain-Ok groups.

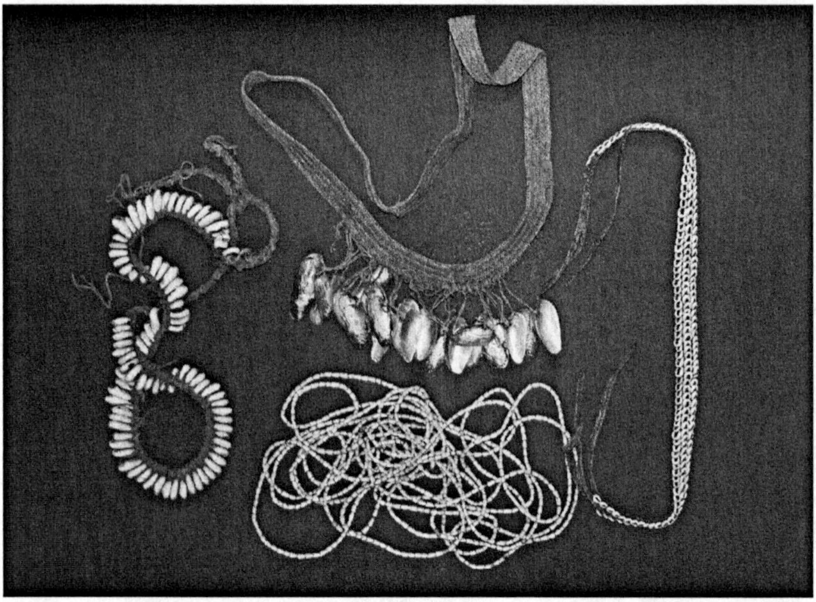

Figure 17. *Bonang*, used in traditional trading practices, and cassowary-bone dagger, used in warfare (from Oksapmin area).

Robbins (1999) provides a fascinating discussion of the function of such valuables in the case of the Urapmin, a group located near the Oksapmin. Robbins notes that some Urapmin men traveled to other groups that were their longstanding trading partners.

No one knows how body-count systems came to be used by New Guinea highland groups nor why 27-body-part systems became prevalent in the Mountain-Ok area. But a pattern of findings has led to some informed conjectures. Laycock points to pre-contact trade relations as an important factor, suggesting that body systems "are related to the measurement of rope-like objects such as shell-money or rattan, where one end is held in the left hand, while the other hand stretches the length to obvious tally-points such as (inner) elbow, shoulder, top of breastbone, and finally to the extended arm-span" (Laycock, 1975, p. 220).

The trade of shells, woven onto long narrow weaves (Figure 17), to which Laycock refers, made its way through Oksapmin and other Mountain-Ok communities; these were valued objects used as *bonang* (Robbins, 1999). Measuring such objects in public displays may well have seeded the emergence of body-count systems through sociogenetic processes leading to the use of the body at various locations in a community of trading partners, with a process of conventionalization taking place within networks of trading partners. But whether such "rope-like objects" were measured in this way, and, if they were, how such practices may have emerged and spread, is not accessible for study.

Period 1 (1938–1960)

Led by the Australians James Lewis and John Black early in 1938, 350 members of the Hagen-Sepik Patrol embarked on a 19-month trek west through central New Guinea. The patrol's purpose was to search for minerals and locate sites for stations. The patrol members included people from various parts of New Guinea who served as carriers and police. Because of the need to keep this enormous number of people fed, the patrol splintered into two groups early in its expedition. John Black led a party of about a hundred, which made first contact with the Oksapmin in June 1938 as it moved through an Oksapmin valley; it established a base camp near Tekin. The photographs in Figure 18 and Figure 19 were taken in June 1938 by John Black. They show a mix of Oksapmin men (with boars' tusks, cane belts, feathered string bags, penis sheaths, and bows and arrows) and patrol police (with hats, rifles, and machetes). Figure 20 is a photograph of John Black, with local people in the background.

Gammage (1998) provides a synthetic account of the first contact of this patrol with Oksapmin people, culled from extensive original patrol reports and follow-up interviews with both members of the patrol and surviving people from the contacted groups (done years after the patrol was completed). In a description of first contact, Gammage describes a scene at Kunanap, an Oksapmin area. A large group of Oksapmin men had assembled with bows and arrows. A member of the patrol, in an effort to appease them for setting up camp in their territory, offered a cowry shell as a friendly gesture. But the gesture was refused, and members of the patrol killed these Oksapmin men. A few days later, the patrol entered Tekin, an area that 40 years later would become a site of my own work. Gammage provides the following vivid portrayal based upon reports pieced together from Oksapmin peoples upon this entry:

> The Tekin knew of the Kunanap killings, and heard that witches were eating the dead from the inside out, as witches did. They thought they too would be eaten and were determined to resist, clearing shooting lanes and stacking stones and clubs against an invasion. Allies came from as far as the upper Om to help. Six miles west an easy track led into the valley: that way

Figure 18. Police meet Min men, Yendanap area, Bak Valley, 19 July 1938. Photo 124 from papers of John Russell Black (1908–1988), MS 8346 from the National Library of Australia. By permission of the National Library of Australia.

Figure 19. Police meet Min men, Yendanap area, Bak Valley, 19 July 1938. Photo 125 from papers of John Russell Black (1908–1988), MS 8346 from the National Library of Australia. By permission of the National Library of Australia.

Figure 20. John Black in foreground with men at Sembati camp (near Tekin), 3 August 1938. Photo 133 from papers of John Russell Black (1908–1988), MS 8346 from the National Library of Australia. By permission of the National Library of Australia.

> a friend would come. Instead, as at Kunanap, the line crossed a steep range onto their rear. Yet only four spirits advanced [members of the Patrol]. The Tekin thought them easy meat. They would come from the bamboo, take them by the hand, slash out their eyes with a knife, and lead them away to be eaten. That was the custom. (p. 102)

Despite these terrible first encounters, Black's patrol and the Oksapmin community managed to build a more peaceful relationship. The patrol needed food and other resources that only local people could provide, and the Oksapmin, in their turn, desired the commodities that the patrol could offer in trade.

Based upon Black's field reports, Gammage notes the unintended inflationary effects of very early trades with westerners.

> As at Hagen and Hoiyevia, [Oksapmin] people traded food for axes, salt, shell and beads, then at a profit bought food from neighbors to trade again. They were the "envy of people for miles," John noticed, "Travelers are already coming long distances to trade … no doubt paying toll to the local people for the privilege." (p. 108)

It was not until 1948 – ten years later – that the Australian government would establish an outpost in Telefomin, a multiday trek from Oksapmin. (Telefolmin people were traditional enemies of Oksapmin, and the hostility probably inhibited direct contact between Oksapmin and Western officers.) Patrol reports from that era describe initial contact with many different groups in the area, many of which were suspicious of the foreign element. Some reports make mention of early efforts to trade with local people for food. Some note the difficulties encountered and extant trade relations with other groups. Patrol Officer Rogers (1949) writes:

> From the experience gained during the patrol and from information gained by questioning natives on the station it was learnt that the best trade was tambu, salt, and large girigiri in that order. Unfortunately, the patrol did not have any tambu and the girigiri held by the patrol was of the small variety. There is little interest and no demand for goldlip shell.
>
> The natives in the area are of course hungry for steel in the shape of tomahawks and knives. A few steel tomahawks have found their way into the area via native trade routes leading down into the Fly and Sepik rivers. The majority of the stone tomahawks held by the natives have also come to the area via these trade routes. A few stone tomahawks were seen that resembled the Mount Hagen type in design, these come from a trade route up the Om river. (p. 14)

I have not found patrol or missionary reports that discuss cash payments in shillings and pounds during this early period. Perhaps some currency made its way to the Oksapmin region through successive trades via trading partners, but I have no evidence of this.

Period 2 (~1960–1980)

More sustained contact with members of Oksapmin communities was initiated in the early 1960s. A principal arena for contact came with the building of a patrol post in Oksapmin in June 1961 and then a Baptist mission in Tekin.[1] I surveyed early patrol officers' reports from the Papua New Guinea national archives with special concern for early exchange practices during this era. One of the first patrol officers posted at Oksapmin, John S. Hicks, made notes of special interest. His patrols were multiweek efforts that involved tasks like exploring sites for landing strips or gathering census

[1] Weeks (1981c) provides a detailed chronology of contact-related events in Oksapmin.

information. For each, Hicks needed to recruit and compensate carriers. In his narrative for Patrol Report no. 2, Hicks (1962a) writes: "No permanent carrier line was attached to the Patrol but little difficulty was experienced in gaining sufficient carriers in walks from group to group.... Carriers were paid in salt or matches depending upon the distance carried and the nature of terrain" (p. 10). Hicks (1962b) also describes a practice of paying people in salt for sweet potatoes and other vegetables:

> Except for a small group at UANMIN where extremely acidic podsol makes the growing of staple foods difficult, food was available at all times and much of that offered far exceeded the requirements of the Patrol. Sweet potato, ginger, pit pit and some leaf vegetables were all bought at different times, payment being effected in salt. People, especially from the Upper Valley, offered to follow the Patrol with food but were discouraged on the grounds that each group was made to feel, to some extent, responsible for the support of the Patrol. (Report no. 3, p. 10)

Later in the same report, Hicks wrote about the distribution of new kinds of seed to local people. Vegetables grown from seed were not only intended for local consumption but became food sources for patrols:

> During the first section of the Patrol lettuce, cabbage, bean and tomato seeds were distributed and it was gratifying to see, during early November, these growing vegetables attended and thriving in gardens previously given entirely to the production of sweet potato. Corn seeds given out by Mr. W. T. Brown's patrol have also been carefully nurtured and many gardens throughout the valley boast small plots of this welcome addition to the staple. (Report no. 3, p. 12)

Particularly noteworthy are a series of events that began to shift traditional trade practices to a cash economy, an economy in which pounds and shillings became a medium of exchange (see Figure 21). Hicks's practice of trading salt, matches, and other Western artifacts for food and work began to give way to cash with the opening of a mission trade store. In Patrol Report no. 2, Hicks (1962a) writes:

> The peoples' interest in trade items, in spite of the use of such items throughout the Patrol, is waning. There were many instances of people, although accepting salt in payment for food, preferring money. The cause of this elementary and somewhat historic change in the economy of the people of the TEKIN Valley can be attributed to the opening, at OKSAPMIN Patrol Post, of a Baptist Mission Trade Store which has been in operation

Figure 21. Australian shilling and pound (now obsolete). Twenty shillings was the equivalent of one pound.

> for three months. They have found, not unnaturally, that all types of useless and unnecessary items can be purchased for varying amounts of money and that such luxuries cannot be bought in exchange for what is all they have to exchange – food and their labour. Such is a normal and necessary development, and perhaps they are becoming fed up with a salt economy anyway. (Report no. 2, page 10)

In 1962 the first Baptist mission station and its accompanying landing strip were built in the Tekin area; they are depicted in the local map in Figure 22. The missionaries brought with them a host of concerns. Of course, a principal concern was the conversion of local people to the Baptist faith, which included change in traditional dress, elimination of warfare with neighboring groups, and alternative living practices. They also included an end to Oksapmin initiations, in which boys learned progressively deeper meanings and facets of the origin myths and other sacred knowledge, involving a series of sometimes painful rites of passage (Brumbaugh, 1990). The hamlets closest to the mission were perhaps most immediately affected by the mission's evangelical concerns.

The first substantial influx of currency to the area, in Australian shillings and pounds, came when plantations and mines in neighboring provinces began recruiting Oksapmin for stints of work. Recruitment began in the mid-1960s. The work sites were melting pots. People from different Papua New Guinea groups came, and many learned Tok Pisin, the lingua franca of the country, along with other practices foreign to their local communities.

After their stints away from home, many Oksapmin men returned to their hamlets, some investing their cash wealth in commodities such as

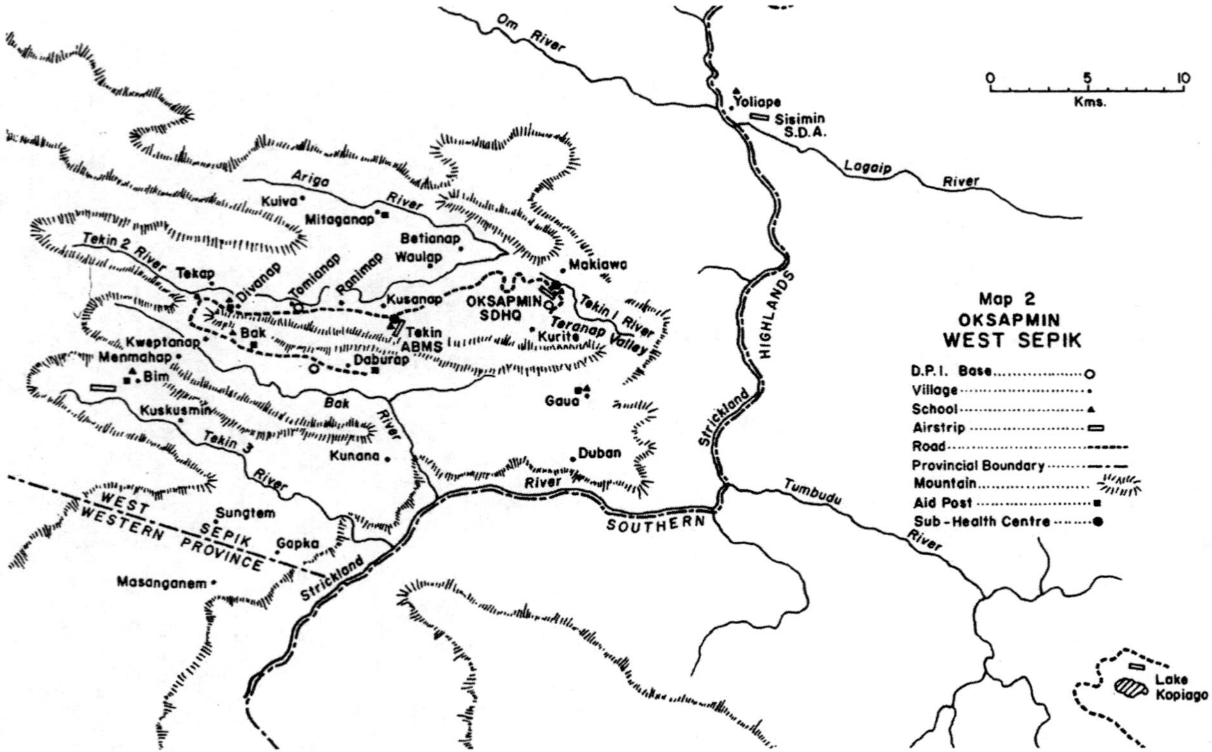

Figure 22. Oksapmin area. From Weeks, 1981d. The map depicts the area shortly after the building of the Oksapmin Station Headquarters (SDHQ), the Australian Baptist Mission Station (ABMS), and the two landing strips, one adjacent to the Headquarters (Oksapmin) and the other to the Baptist Mission Station (Tekin).

Figure 23. A trade store in 1978. The window and the door are one and the same; the interior functions as a storage place for goods.

bags of rice and cans of fish. Guilford (1994), who conducted fieldwork in Oksapmin from 1978 to 1979, provides an analysis of the trade store as a cache for storage of surplus wealth. She argues that the traditional practice of sharing surplus wealth was adaptive in a community in which surplus was best stored through a web of social obligations and allegiances produced through giving, a practice well documented throughout highland communities. Trade stores marked a break with this traditional practice, as instead of sharing surplus wealth, trade store owners kept their wealth enclosed in their shops.

The proliferation of trade stores supported currency as a medium of universal exchange in Oksapmin. Weeks (1981a) reports that in 1967, in the Oksapmin area, there were only two mission-run trade stores. Guilford (1994) reports that in 1978 there were more than one hundred licenses for trade stores. Figure 23 contains a photograph of one such store.

Currency System	Dates of Issue and Retraction	Key Equivalence Relations
Australian pounds and shillings	Pre-contact through 1966	20 shillings = pound note
Australian dollars and cents	1966–1975	20 10¢ coins = 2-dollar note
Papua New Guinea kina and toea	1975–present	20 10-toea coins = 2-kina note

Note. In the Oksapmin language, "pound" and "shilling" have become borrowed terms. They are pronounced "faun" and "siling," respectively.

Figure 24. History of Western currencies used in Oksapmin area, with dates of issue and retraction, and key equivalence relations.

Another source of currency resulted from a move on the part of some to sell their surplus vegetables. Missionaries, like the patrol officers before them, distributed new types of seed to Oksapmin people, and the exchange of these new crops for money contributed to an increase of cash in circulation. Later, in the 1970s, missionaries began to export vegetables on small mission aircraft, creating additional demand for cash crops and contributing to cash flow in the community.

The rise in the use of currency in the 1960–1980 period brought about a rise in new kinds of quantification problems for some segments of Oksapmin communities. These problems included identifying the various denominations of currency as elements of varying values, identifying numerical equivalences between values, and adding and subtracting the values. The problems emerged more regularly for some sectors, such as those engaged in trade store transactions.

The growing participation in a cash economy in Oksapmin coincided with three shifts in currency systems. Figure 24 shows, for each system, the dates of issue and retraction, as well as the key equivalence relations.

As I mentioned, the first Western currency to enter Oksapmin was Australian pounds and shillings, with 20 shillings to a pound. Pounds and shillings were in use through the time of the early trade stores.

In 1966, Australia shifted its system of currency to a decimal system of dollars and cents (1 dollar equaling 100 cents). In this new decimal system, the 10-cent Australian coin was equivalent to the old 1-shilling coin, the 1-pound note was equivalent to a 2-dollar note, and these tokens bore some similarities in appearance.

In 1975, the newly independent country of Papua New Guinea introduced its own currency, consisting of kina and toea. The kina replaced the Australian dollar and the toea replaced the Australian cent. At this time,

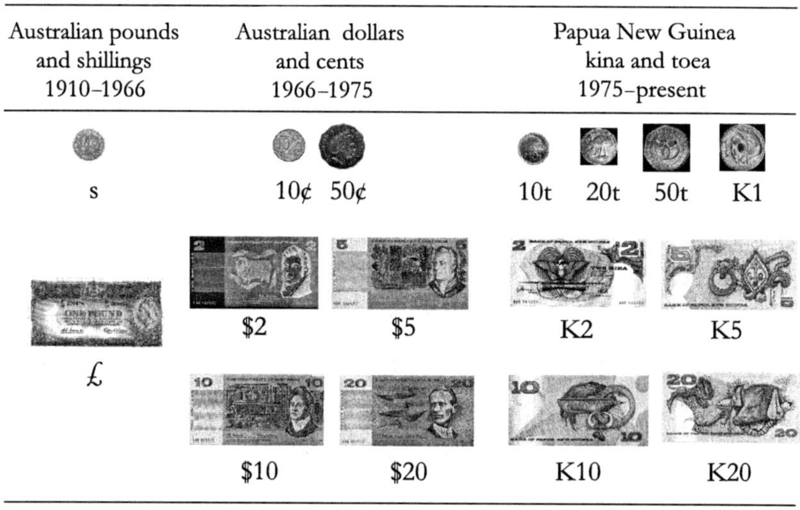

Figure 25. Selected currency tokens across three waves of changes in currency in Oksapmin social history.

both coins (in denominations of 1t, 2t, 5t, 10t, 20t, and K1) and notes (K2, K5, K10) were minted with an effort to match the appearance of the new currency tokens to features of the corresponding Australian system of dollars and cents. A K20 note was introduced in 1978, and a K50 note was introduced in 1989. Throughout the period of my own data-collection efforts, many people who engaged in economic exchange used the early pound/shilling names when referring to the new currency. They called the 10t coin a shilling (*siling*) and preserved the 20-to-1 relation between pounds and shillings by calling the K2 note a "pound" (*wan faun*).[2]

Exemplars of the currency tokens – pounds and shillings, dollars and cents, and kina and toea – are shown in Figure 25. In the figure, we see notable physical similarities between coins of approximately the same worth. The similar configuration of the 10¢ and 10-toea (10t) coins should be apparent; note also the corners on both the 50¢ coin and the 50t coin. The colors of the Australian dollar note and the Papua New Guinea kina note also correspond.

In addition to the introduction of a cash economy and Christianity during the 1960–1980 period, Western-style schooling was also introduced. The Baptist mission supported a Bible study group in the early 1960s.

[2] I suspect that during the short period in which Australian dollars and cents were used in the Oksapmin area, the 10-cent piece and the 2-dollar note were also referred to as "*silings*" and "*fauns*," respectively, but I have no hard data on this.

Then, in the late 1960s, the government opened up its first community school in the area. The children who attended school (only a small proportion of the population) were schooled in English and in Western-style arithmetic. Marshall Lawrence, a linguist with the Summer Institute of Linguistics, had worked in concert with a small group of Oksapmin people to translate the Bible (Lawrence, 2010). To this end, he produced a written form of the Oksapmin language, along with an Oksapmin-English dictionary (Lawrence & Snyder, 1993).

In sum, the 1960–1980 period gave birth to a number of strands – a new religion, new forms of economic exchange, new languages, and new forms of schooling. It also introduced new kinds of heterogeneity, since these varied strands were taken up differently by different segments of the population. Some of the population became participants in some of these practices, others in all of them, and others in none.

Period 3 (~1980–2001)

In the early 1980s, active work on the Ok Tedi copper and gold mine commenced. Tabubil, a small town constructed to support the workings of the mine, grew rapidly into a small city with many features to attract expatriate administrators and related industries.[3] Many Oksapmin men and women found work in the mine or in supporting industries, and many of them sent remittances back to family in the community. Many others who remained in Oksapmin sold vegetables for export to Tabubil.

By 2001 the export of vegetables had grown and the effort was no longer operated solely by the missionaries. Vegetable brokers were local Oksapmin men, who bought from local subsistence gardeners and sold to buyers for the mine and other large stores in Tabubil. A report of exports in 1990 and 1991 indicated that 3 tonnes[4] per week were being exported from the Oksapmin community to Tabubil (Bourke, 2002). In 2003, this figure had grown to an estimated 6 to 10 tonnes of vegetables weekly (Lawrence, 2003). The widespread increase in interest in vegetable exporting in the community, with the possibility of increased revenue from vegetable sales, led local people in the Bak Valley to complete a 15-year project of the leveling of land to support air transport of vegetables to distant markets. Figure 26a shows vegetables prepared for shipment to Tabubil. Figure 26b

[3] Tabubil was the town where I first heard the new use of the word *fu*. For an analysis of the effects of Tabubil and the Ok Tedi mine on local communities, see Hyndman (1990).

[4] One metric tonne is equivalent to 1,000 kilograms or 2,204 pounds.

Figure 26. (a – top) Vegetables to be loaded on a twin-engine plane at Tekin airstrip in 2001; (b – bottom) a common public display of money being paid to people for their vegetables sold for export to Oksapmin middlemen (photograph extracted from video of payment found at http://www.culturecognition.com/middle-men/).

Figure 27. Trade stores in Oksapmin, 2001.

Figure 28. Interior of Oksapmin trade store, 2001.

shows people being paid for vegetables that middlemen had sold to the Ok Tedi mining company in Tabubil. Other contexts for exchange of currency at that time included a local informal display and sale of vegetables, in which people sold produce to one another. On days on which a flight was scheduled to land on the new airstrip, some local people would bring extra produce like sugar cane and bananas to sell to one another.

The influx of cash contributed to an expansion in the size and stock of trade stores. Three typical stores are pictured in Figure 27. The first trade stores, built in the later part of the 1960s, were small structures, and their inventories were typically limited to just two items – tinned fish (mackerel) and bags of rice. In 2001, the stores were still made of materials indigenous to the community, save the roof, which was sometimes made of imported metal. They differed markedly from the trade stores of the 1970s in size, the variety of commodities sold, and the amount of cash exchanged.

Inside the stores of 2001, as we see in Figure 28, the store owner or clerk would stand behind a counter. The commodities for sale were displayed behind the counter, and often prices were posted with written numerals. Prices were always rounded to the nearest unit of 10 toea. Though trade stores varied in size and selection of commodities in 2001, most carried a wide variety of items, including nonperishable foods, clothing, school supplies, and household goods like soap and batteries. A few stores had generators that ran coolers, allowing them to stock perishable food items – most frequently, frozen chickens, a luxury item that most people could rarely afford.

In 2001, elements of the traditional economy had not entirely vanished. Robbins (1999) provides a fascinating analysis of Urapmin, a nearby Mountain-Ok group. Robbins describes the side-by-side use of traditional shell money (*bonang*) and cash. To varying degrees, people in Oksapmin combined subsistence practices with a growing involvement with the local cash economy. Most of the food in Oksapmin came from subsistence gardens, and most of the houses were still made almost entirely from traditional materials. But almost everyone participated in the cash economy in some way. The number of Oksapmin people who were employed outside of the region had increased dramatically, and the number of trade stores within the region also increased. Money had become a necessity of life in Oksapmin, used by different segments of the population to purchase clothing, to contribute to bride price, and to pay school fees.

4 Reproduction and Alteration of Numerical Representations

In June 1978, 40 years after the Hagen-Sepik patrol first marched into the Oksapmin area, I made my way with some difficulty (not comparable to that of the early patrol!) into the isolated world of Oksapmin. My destination was Kweptanap, an Oksapmin hamlet, where I would stay with my friends Tom Moylan and Virginia Guilford, graduate students who had already been in residence for some time working on their dissertations in linguistic anthropology. Soon after my arrival, Tom and Virginia introduced me to an Oksapmin man named Yangeh. Interactions with Yangeh and my discussions with Tom and Virginia seeded the studies conducted in 1980 and 2001, subsequent to my 1978 visit.

In 1978, Yangeh worked for Tom both as a key informant and as an all-around assistant. A few years older than me, perhaps in his thirties at the time, Yangeh was one of the men who years before had been recruited to work outside the Oksapmin community for a stint, and returned having learned Tok Pisin. Yangeh had built a small trade store in his hamlet, where he sold basic commodities like bags of rice and cans of fish. As I learned from him about Oksapmin number, Yangeh showed me how he counted, sometimes stopping at the elbow on the other side (20), because that was a pound (the equivalent of 20 shillings).[1] Yangeh also taught me other complicated ways that he composed values. At the time, I only had an inkling of the importance of Yangeh's alterations in the body system and the functions he used it to serve relative to other Oksapmin people whom I was coming to know (and who also became my teachers of Oksapmin number).

[1] Throughout the remainder of the text, I differentiate symmetrical body parts by a number in parentheses. Numbers greater than 14 indicate that the named body part is on the "other" side of the body. Thus, elbow (8) and elbow (20) refer to symmetrical body parts occurring at different positions in an enumeration: elbow (8) occurs first (in the 8th position) and elbow (20) occurs second (in the 20th position).

These 1978 interactions with Yangeh and many other Oksapmin people had some time to incubate; it was not until my next visit, in 1980, that I would follow up on these conversations with Yangeh in a concerted way. The fruits of that incubation, additional reading, and my work with children in 1978 (reported in Chapter 8) came to frame the studies that I report in this chapter.[2]

When I returned to Oksapmin in 1980, I wondered whether Yangeh's different use of the body that I observed two years previously was simply an idiosyncratic variation of how adults used the body-counting system. I doubted that. My inkling was that through his recurrent efforts to solve arithmetical problems linked to the burgeoning money economy, Yangeh might be creating alterations in the use of the body form to serve new arithmetical functions. If I was correct, what might these new adaptations be? Might others be innovating as well? Might these adaptations be taken up by others and spread through social networks? How might this process of emergence of new form-function relations be occurring? Collecting the kinds of data needed to engage in this kind of analysis was a serious challenge. My intuition, only in nascent form in 1980 and more developed in 2001, was that the data that I would need to support a serious analysis would require two coordinated analytic approaches, one synchronic and the other diachronic. Each kind of analysis would focus on the properties of the body-counting system as a representational form and the functions the system served in people's activities.

[2] One of my readings was Luria's early study of people living in postrevolutionary Russia (Luria, 1976). Luria used a cross-cohort method to study cognitive changes associated with the effects of collectivization on people's thought. In his study, Luria sampled five cohorts of people at various levels of collectivization, administering the same cognitive tasks to individuals from each cohort. I thought Luria's cross-cohort design was important and clever, but I also had criticisms. For instance, analyses of the collective practices with which people engaged were lacking, and hence the cognitive tasks that he used were not well coordinated with the collective practices. The result was that the study provided little insight into the interplay between genetic processes and processes of collectivization. But I saw much value in a modified version of his methods. Indeed, if one carefully chose segments of a community with varying levels of participation in a collective practice of interest, and if one interviewed individuals from those segments on well-chosen problems linked to targeted collective practices, one could use the results to inform both synchronic and diachronic analyses of shifting form-function relations. From a synchronic perspective, one could glean information on the demographic distribution of form-function relations as people solved the same problem during a time period. From a diachronic perspective, one could make inferences about shifting relations between forms and functions through historical time, as the collective practice became increasingly prominent in the community.

Synchronic analyses would capture the distribution of form-function relations in use of the body system through stratified segments of the Oksapmin community. In microgenetic terms, the analysis would focus on the distribution in a community of the forms used and the functions they served as people accomplished numerical and communicative goals at the many sites of interaction in economic transactions.

Diachronic analyses would require data on the shifting relations between forms and functions through time, as the participation of Oksapmin people in the money economy grew. At core, the focus would be historical continuities and discontinuities of form-function relations. To what extent are these relations reproduced over historical time? If they are reproduced, what are the processes that support their reproduction as the distribution of variant forms waxes and wanes? If they are altered, how are they altered? Are there regulative mechanisms that enable and constrain microgenetic, sociogenetic, and ontogenetic processes that lead to the reproduction and alteration of form-function relations over historical time? What are they?

These were challenging questions, both conceptually and methodologically. In the four studies I report on in this chapter, I illustrate some of the efforts that I made and the findings I produced, both in 1980 (Study 4-1) and in 2001 (Study 4-2, Study 4-3, and Study 4-4). The questions are broad, and I do not pretend that the evidence is complete. But the studies I report provide some of the contours of synchronic and diachronic treatments that I will take up in subsequent chapters and that will become the focus of Part IV, the conclusion of the volume.

The first of the four studies (Study 4-1)[3] was conducted during my 1980 visit; its design draws upon observations that I had made in 1978 with Yangeh. Additionally, I drew upon advice from Tom Moylan, Marshall Lawrence, and Maryl Gearhart as I planned the study.[4] My effort was to understand how people who had varying levels of participation in the money economy were adapting the body system to solve arithmetical problems linked to economic exchanges of the era. In addition to the synchronic concerns of establishing the character of variability in people's microgenetic constructions using the body system, the study would also contribute to a diachronic analysis in two senses. First, it would provide

[3] Note that studies are numbered first by chapter number and then by the order in which they occur in their respective chapters. Thus, Study 4-1 is the first study to be presented in Chapter 4. It also happens to be the first empirical study presented in the volume.

[4] Tom, Maryl, and I returned together in 1980, and Marshall Lawrence was a Summer Institute of Linguistics (SIL) linguist working in the area for many years.

information relevant to an analysis of reproduction and change. By sampling individuals from different cohorts with increasing participation in the money economy, I would investigate how the body-part counting form of representation was being altered and new functions created for the system as people participated in new ways in economic exchange. Second, from my 2001 perspective, the study provided a diachronic anchor in 1980 for the 2001 Studies 4-2, 4-3, and 4-4.

In 2001, changes in the economy were marked as I noted in Chapter 3. In addition the prevalence of Tok Pisin was on the rise. In Study 4-2, my focus broadened to Tok Pisin forms of number representation in addition to the body system. I assessed people's knowledge of both forms as well as variations in the character of people's knowledge of each form, including in my sample cohorts with different levels of participation in the money economy (synchronic analysis). I also examined continuities and discontinuities in these 2001 findings when compared with the 1978/1980 period (diachronic analysis).

In the last two studies I discuss in the chapter, I shift focus from people's *knowledge* (in 2001) of Tok Pisin and body forms to *use* of these forms; my intent again was to use these data to inform synchronic and diachronic analyses. The arena for analysis was trade stores. In Study 4-3, I take up a cross-cohort analysis of the use of representational forms of quantification in trade store exchanges, documenting differential use of quantification in Oksapmin and Tok Pisin by different segments of the Oksapmin community. In Study 4-4, the focus turns to the emergence of arithmetical problems in trade store transactions. The analysis reveals the way the distribution of cognitive work emerges and is accomplished as problems and solutions take form.

As a collection, the four studies share some similarities. First, each was preceded by informal observations – "hanging out" in different venues in which number was used; observations led to discussions with informants, leading to the refinement of some questions and allowing new questions to emerge. Second, all of the studies used a comparative research design, contrasting cohorts with one another.

Study 4-1. Body Arithmetic in 1980

From the mid-1960s through the 1970s, the emerging money economy in Oksapmin had expanded. Currency entered the community from varied sources. Missionaries had provided people with seeds to cultivate new crops and now began to buy back part of people's harvests for export to other markets by small plane. As new sites for mines were identified in central New Guinea, the possibility of working outside of the community and sending back remittances was on the horizon. Finally, Western-style schooling had entered the Oksapmin world in the late 1960s, and with schools came teachers with salaries and perhaps a modest amount of currency and knowledge of currency exchange practices. The teachers were mostly people from other parts of Papua New Guinea but included an occasional westerner (Boram, 1980; see also Lawrence, 2010).

During the 1960s, people used the 27-body-part system as a means to represent cardinal and ordinal relations, putting body parts in one-to-one correspondence with objects. The subsequent introduction of trade stores led to new kinds of numerical problems. Customers and store owners engaged in transactions in which they produced payments and change, occasions that would lead to new problems of arithmetic.

For me, inklings that people might be constructing new functions for the body system linked to emergent problems in economic exchanges came from discussions with Yangeh and Tom. Recall that in 1978, Yangeh was a trade store owner. Tom and I noted that when Yangeh counted shilling-equivalent values, he would stop at the 20th body part, calling it one pound (the equivalent of a K2 note); to count further, he would then begin over again with the thumb (1), in effect adapting the count system to a 20-cycle structure of the currency system (see Figure 29).[5] Study 4-1 was born through informal observations of Yangeh, trade store transactions as well as reflections on change processes.

I designed the study with both synchronic and diachronic analytic concerns in mind. On the one hand I sought to document variation in people's use of the body form to serve arithmetical functions in the community, sampling different population groups (synchronic analysis). Using the same data, I also sought to pursue a diachronic question: Were people in cohorts with greater participation in the money economy using the body form in new ways that were better adapted to the problems that had emerged in collective practices of economic exchange (diachronic analysis)? In other words, was the historical shift towards a cash economy leading to the emergence of new cognitive functions and specialized uses of the body system? I targeted four segments of the population, depicted in Figure 30, each

[5] In a particularly telling account told by an Oksapmin associate in 2001, Indigo and I learned of a store owner (deceased at the time of the 2001 visit) who would only sell to customers who came in with 10t coins. The store owner thereby avoided the arithmetical computations that emerged when making change. While this anecdote must be taken as such, it is consistent with observations of the gradual transformation of the structure of transactions that has occurred in doing business in stores over the targeted stretch of Oksapmin social history.

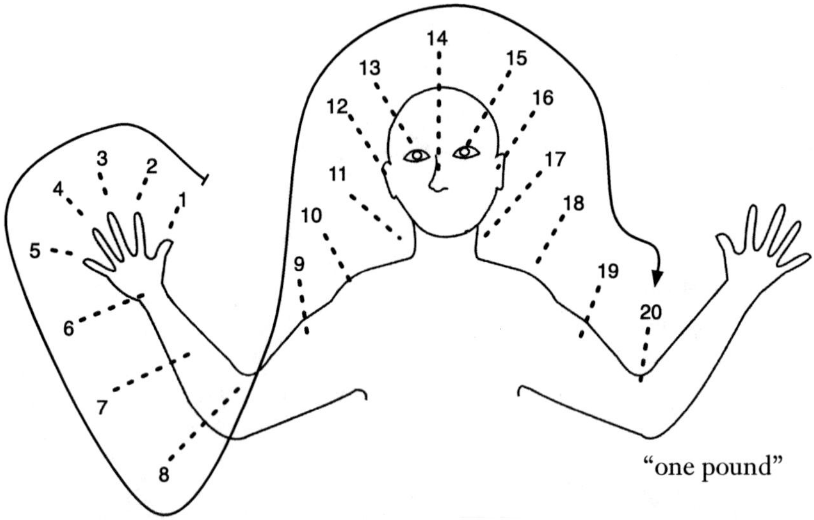

Figure 29. Truncated body-count system in which the 20th position (*tan-kir* (*20*)) was treated as a "*faun*" (pound), since 20 shillings were equivalent in trade value to one pound.

cohort having different levels of participation in the cash economy; groups consisted of elders, unschooled adults (nontravelers), plantation returnees (non-trade store owners), and trade store owners.

To reveal form-function shifts that might inform synchronic and diachronic analyses, I created two problem types, one closer to traditional life and the other linked to collective practices of economic exchange with currency. The principal distinguishing feature between these problems is that the former contained countable physical objects (coins) whereas the latter did not.

The problems closer to traditional practices afforded the use of one-to-one correspondence between body parts and coins. Because the coins were present, addition problems could be solved by the person's counting one group of coins and then including the second group in the count. Similarly, for subtraction, the coins could play a similar role in the problem solution. The coins to be taken away could be counted and removed by an individual, and the remaining coins could be counted to produce the solution.

The problems that more closely resembled problem types that might emerge in cash-based collective practices of exchange were identical to the more traditional problems except that no coins were presented. The fact that coins were absent left the problem solver in a predicament: how to keep track of the addition or subtraction of one term in relation to the other if the

Cohort	Estimates of Relative Experience with Economic Exchange	
	Traditional exchanges	Cash for commodity exchanges
Elders	Greater past participation	Rarely
Nontraveling (Younger adults)	Little past participation	Infrequently
Plantation returnees	Some past participation	~Weekly
Trade store owners	Some past participation	~Daily

Figure 30. Relative extent of engagement with economic exchange practices of cohorts sampled in Study 4-1.

coins were not present to be counted altogether as a sum or as a remainder. Based upon my observations prior to the conduct of the study, I expected that at least some people with a history of participation in the money economy were creating an alteration in traditional uses of the body system. The alteration involved turning correspondences between the body and external objects inward to serve keeping-track functions – to keep track of the addition of numbers or the subtraction of one number from another.

In sum, I expected that traditional people with little exposure to the money economy would rely on external correspondences across problem types. This would enable them to successfully solve the problems when coins were present, but not when coins were absent. In contrast, I expected that the people who had participated in the money economy would show new developments in arithmetical problem solving with the body system, creating internal correspondences to enable them to solve problems with coins absent.

Method

Participants

The participants included 79 people, grouped into four cohorts: (a) 22 trade store owners[6] (age range about 30 to 35 years), (b) 18 plantation returnees

[6] In 1978 and 1980, virtually all clerks were trade store owners. By 2001, some clerks worked for owners while other clerks were owners themselves.

who were not trade store owners (age range about 30 to 35 years), (c) 19 younger adults who did not own trade stores and had not worked on a plantation (age range about 20 to 35 years), and (d) 20 older adults ("elders") who did not own trade stores and had not worked on a plantation (age range about 40 to 50 years). Participants were recruited with the aid of an Oksapmin informant.

Procedure

The interview procedures and accompanying arithmetical problems were developed in pilot interviews. Once developed, the procedures were back-translated. In addition, Tom Moylan and Marshall Lawrence (of the Summer Institute of Linguistics), who were native English speakers with proficiency in the Oksapmin language, provided useful feedback on the interview procedures and the translation process.

In the procedure, each Oksapmin participant was presented (orally) with eight arithmetic problems involving the addition and subtraction of coins. Four of the problems (two addition and two subtraction) were administered with 10t coins present, and four problems (two addition and two subtraction) were administered in the absence of coins. Within each population group, the order of administration of the coin and no-coin problems was counterbalanced, and within this counterbalancing the order of administration of the addition and subtraction problems was counterbalanced.

To avoid repetition of the same problem to the same individual in the coin and no-coin forms of the problems and yet to retain comparability of the findings, two similar problem types were used. These problem types are indicated below by the letters *a* and *b*. For half of the participants within each population group, the *a* problems were administered as the coin problems and the *b* problems were administered without coins; for the other half, the reverse was the case. The problems included the following: (la) 6 + 7, (lb) 6 + 8, (2a) 9 + 13, (2b) 9 + 12, (3a) 16 – 7, (3b) 16 – 8, (4a) 18 – 6, (4b) 18 – 7.

Problems with Coins (More Closely Linked to Counting Problems of Traditional Life). Each participant was presented with a set of 10-toea (10t) coins, and the number of coins varied with problem condition. For subtraction problems the coins were presented in a single linear alignment, with the greater term in the problem defining the number of coins presented. The participant was told, "You have ____ shillings (1 shilling = 1 coin)," as the informant gestured around the body to the appropriate body part and to the coins.[7] "You go to the trade store and spend ____ shillings." Again, the informant gestured around to the appropriate body part. "How many do you have left?" For the addition problems, two sets of coins, each corresponding to the terms in the addition problem, were presented in parallel alignment. The informant told the participant, "You have ____ shillings," the number indicated by gesturing around the body to the appropriate body part as well as by pointing to the first set. "A friend gives you ____ shillings (indication is the same as above); how many do you have altogether?"

Problems without Coins (More Closely Linked to Arithmetical Problems Emerging in Stores). Each of these problems was presented in identical fashion to the problems with coins, except that no coins were present, and indications concerning the number of coins were solely with reference to body parts.

Results

The microgenetic constructions that people produced to solve the problems with and without coins differed markedly from one another; as a consequence, these findings are presented in separate sections.

Problems with Coins

Participants' solutions to the coin problems were coded as one of three types. These included solutions in which a participant used a systematic counting strategy and was correct, solutions in which a participant used a systematic counting strategy but was incorrect, and solutions in which a participant produced a partial solution or in which the informant coached the individual to achieve an accurate answer. Coaching generally happened when the individual did not appear to understand what was being asked in the interview context.

For the addition with coin problems, a systematic strategy consisted of counting one collection and then counting the other collection in a summation of all of the coins (see Figure 31a). In addition to this type of systematic solution, two trade store owners used advanced body-part counting strategies described in the section on problems without coins – these were coded as systematic solutions. A partial or coached solution consisted of all other

[7] Recall that most people called the 10t coin a shilling (*siling*), a name that had been used for the now obsolete shilling (used during most of the colonial era).

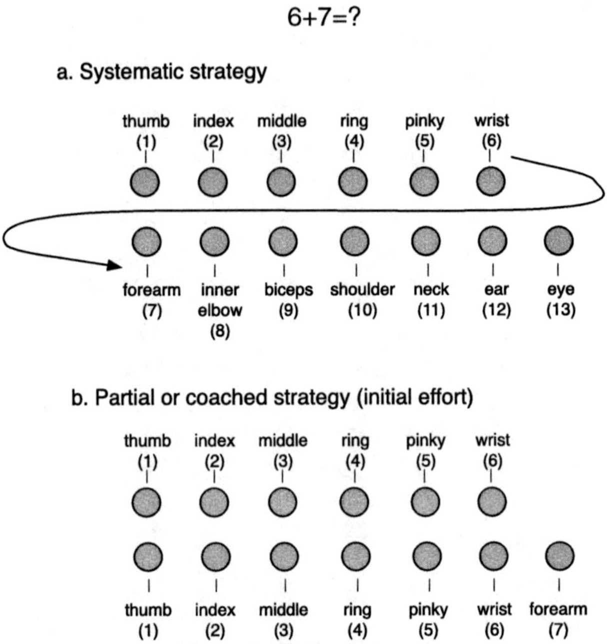

Figure 31. Addition strategies for coins-present problems: (a) systematic; (b) initial effort at partial or coached strategy.

types of solutions in which the interviewer took on a stronger role in structuring the problem and a solution. For example, it was common, especially among some of the older adults, to find that participants would count the two collections separately (see Figure 31b). In this kind of effort, the translator would repeat the problem. After one or more repetitions, if the participant persisted in counting the two sets separately, the translator would intervene and explain to the participant that they had to find out how many there were altogether or not to stop counting when they had completed the first set. Often these interventions led to a correct answer. Regardless, these types of behaviors were coded as partial or coached solutions.[8]

[8] The source of the need for support was unclear for the solutions that were coded as *partial or coached*. These problems (and the interview practice itself) may have been so foreign that what was being asked of participants was unclear to them. But there are other possibilities as well. Perhaps for these individuals a sum of two values could simply be treated as two separate counts, or perhaps the collective efforts of the interviewer and interviewee to produce a sum were simply accepted norms of interaction. Or, perhaps all of these explanations are appropriate.

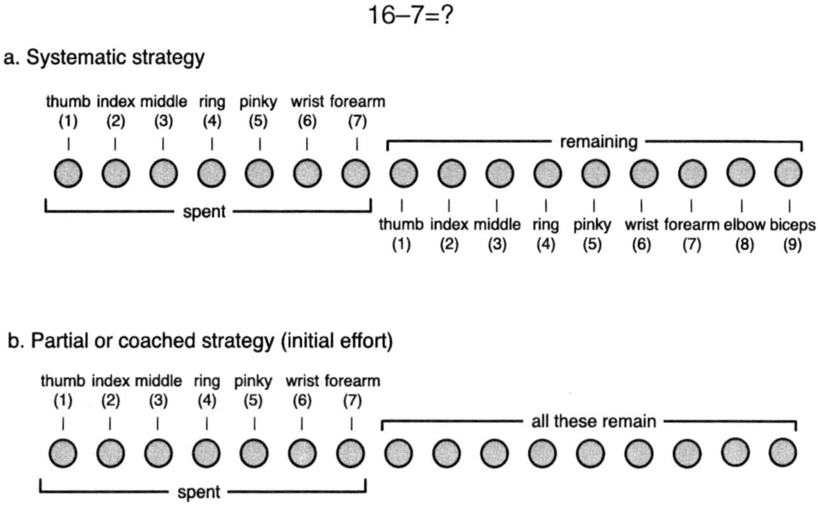

Figure 32. Subtraction strategies for coins-present problems: (a) systematic; (b) initial effort at partial or coached strategy.

For the subtraction problems, a systematic strategy consisted of counting the number of coins that "were spent at the trade store" and then counting the remaining coins (see Figure 32a). Again, a few individuals (three of the trade store owners and two of the plantation returnees) used advanced body strategies (described in the following section), and their use of these strategies was coded as "systematic." A partial or coached solution consisted of all other types of behaviors. Two common partial or coached solutions to the subtraction problems consisted of counting all of the coins and not stopping when the number of coins spent was reached or counting only the amount one spent and stating that the rest remained (see Figure 32b). On these occasions, as on the addition problems, the translator would often intervene and attempt to clarify the problem, sometimes leading to the correct answer.

Figure 33 shows the percentage distribution of solution strategy types over the four cohorts for the two addition problems. The proportion of people achieving a correct systematic solution was greater than 50 percent in each cohort, with all cohorts attaining at least 60 percent correct. For a correct solution, people in each group tended to count all of the coins present using a systematic procedure of a linear count of one set and then the next set; four of the older adults touched coins to the appropriate body parts.

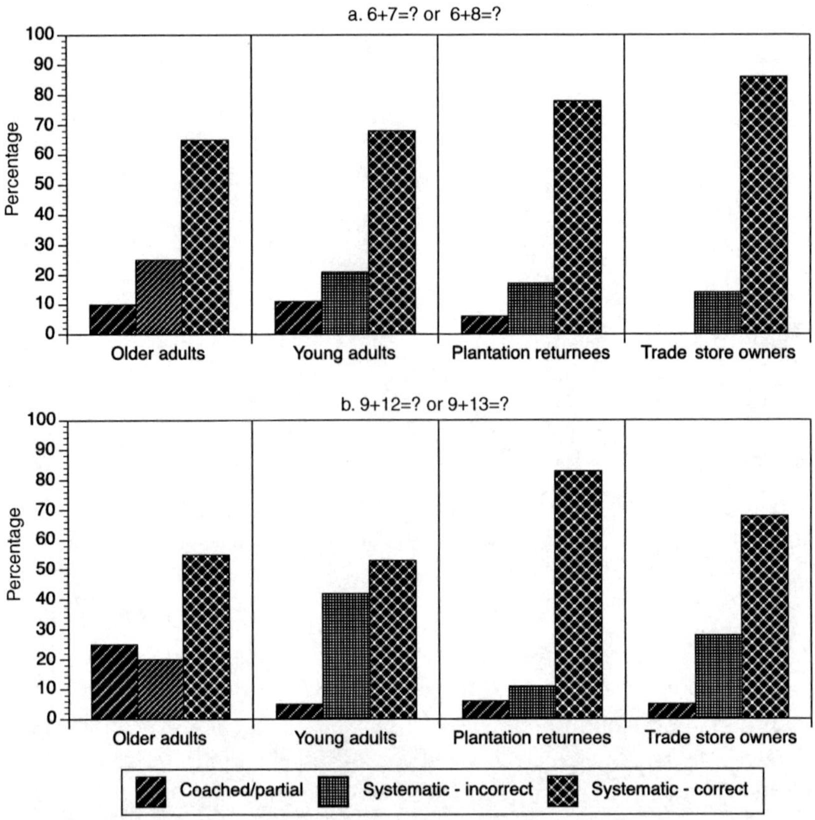

Figure 33. Percentage distribution of approaches to the two addition problem types as a function of cohort (1980).

Figure 34 shows the percentage distribution of solution strategy types over the four cohorts for the two subtraction problems. In general, the performance of the plantation returnees and the trade store owners was similar to their performance on the addition problems, with the large majority achieving systematic/correct solutions. But the young adult and older adult cohorts had difficulty, with many of them achieving solutions that were coached. In particular, about half of the older adult population required coaching to achieve a solution to these problems; a common issue in these coached solutions was that the individuals would count the entire set, then count the number to be subtracted and merely point to the remainder, without producing a specific cardinal number as the answer (see Figure 32). Of course, in some sense these were correct, but the approach differed

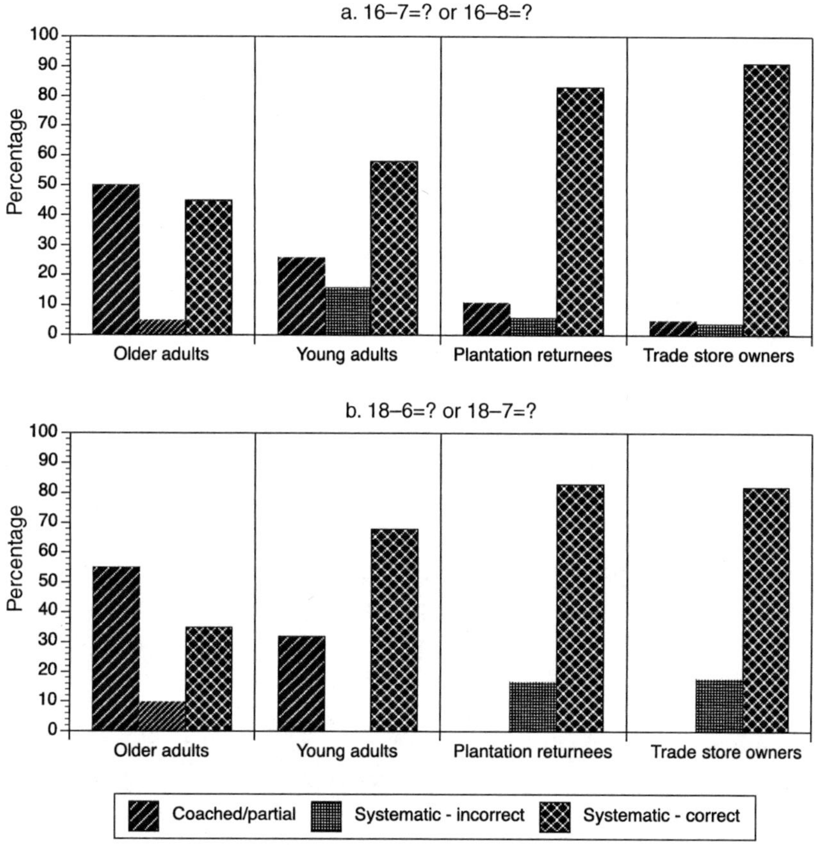

Figure 34. Percentage distribution of approaches to the two subtraction problem types as a function of cohort (1980).

from what had been coded as "systematic correct strategies," and these solutions were coached in the course of the interview, often leading to cardinal representations using the body system.

Problems without Coins

In the course of conducting the interviews, I found people's solution displays for problems without coins particularly fascinating. There was variation in the way the body form was used to solve the problems, which I grouped into five principal types. One was the absence of an overt use of the body, even following probe questions. In this type, a participant would simply state the answer as a body part. The other four types varied in the character of correspondences produced. I first consider each strategy type, and then analyze how strategies were distributed across cohorts.

Global-Enumeration Strategy. The *global-enumeration strategy* is a direct extension of the traditional use of the body system. People did not establish correspondences between body parts themselves, a microgenetic construction characteristic of the other approaches. As a result, in their solutions, people did not keep track of the addition of two terms or the subtraction of one term from another. As depicted in Figure 35, the approach was used on both the addition and the subtraction problems.

For addition, the global-enumeration strategy consists of enumerating body parts beginning with the thumb (1) and ending at a specified point to represent the first term. The count is then continued around the body to represent the addition of the second term, but the individual shows no way of keeping track of where the count stops. For example, to add 6 + 8 (as depicted in Figure 35a), an individual enumerates body parts from the thumb (1) to the wrist (6), and then continues counting on from the forearm (7) upward, stopping at an estimate of the correct answer. Since there is no method of keeping track of when the second term is completed in the enumeration, the person is typically incorrect.

For subtraction, the global-enumeration strategy is manifested as an enumeration or indication of the body parts that would remain after the specified number was spent. For instance, for the problem 16 – 7 (Figure 35b), this strategy type would include someone who (a) enumerates the thumb (1) to the forearm (7) to indicate the number spent, and then (b) enumerates the elbow (8) to the "ear on the other side" (16) to indicate the number of coins that remain. In an important sense, this solution is an accurate one. But it is not a specific representation of the cardinal value of the set.

Double-Enumeration Strategy. The *double-enumeration* strategy turns inward one-to-one correspondences traditionally established between body parts and objects. Correspondences are produced between two series of body parts in order to keep track of the progressive summation (or subtraction) of coins. This strategic approach is depicted in Figure 36.

To add 6 + 8 (Figure 36a), people, in one exemplar of this approach, begin with the first term of the problem, the wrist (6). As the second term, the elbow (8), is added, they keep a record by establishing physical correspondences between the series that constitutes 8 (thumb (1) through inner elbow (8)) and the series of body parts that follow the wrist (6). Thus, people link the following body parts in one-to-one correspondence: The thumb

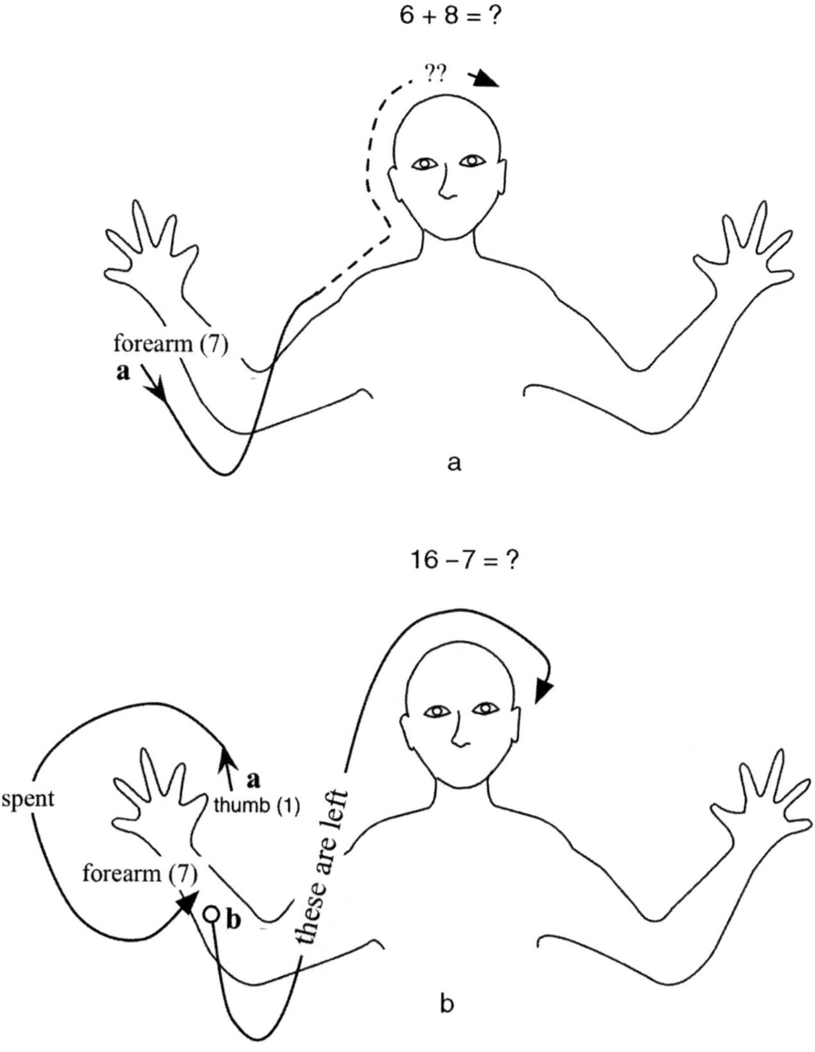

Figure 35. Global-enumeration strategy for (a) addition and (b) subtraction.

(1) goes with the forearm (7), for the addition of 1 to 6; the first finger (2) goes with the inner elbow (8) for the addition of 2 to 6; the middle finger goes with the biceps (9) for the addition of 3 to 6. The process proceeds until the double enumeration ends with the correspondence of the inner elbow (8) with the nose (14) for the addition of 8 to 6. Unlike the global strategy, people typically achieve the accurate answer, nose (14).

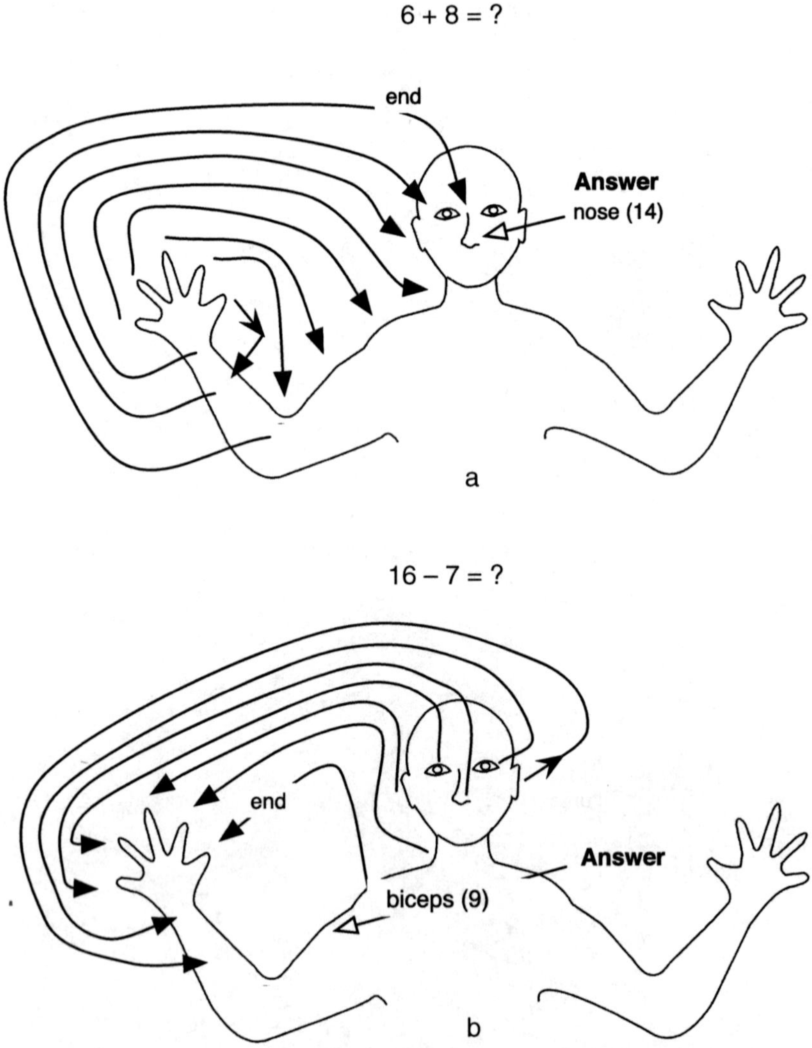

Figure 36. Double-enumeration strategy for (a) addition and (b) subtraction.

To subtract 16 – 7 using the double-enumeration approach, people again create internal correspondences within the body system, using one series of body parts, in this case thumb (1) through forearm (7) (the subtrahend), to keep track of the subtraction from 16 (the minuend). In an exemplar of this process (see Figure 36b), the thumb (1) goes with the ear-on-the-other-side (16), leaving the eye-on-the-other-side (15), solving the subtraction 16 – 1; the index finger (2) goes with the eye-on-the-other-side (15), leaving the nose (14), the subtraction 16 – 2; the middle finger (3) goes with the nose (14) leaving the eye (11), the subtraction 16 – 3. The process proceeds until the double enumeration ends with the correspondence of the forearm (7) with the shoulder (10), leaving the biceps (9), the subtraction of 16 – 7 (and the solution!).

The microgenetic process entailed in the double-enumeration approach shifts the target of the count to the body parts themselves in order to keep track of an addition or subtraction. In this shift in function, correspondences turn inward. Compared to the global-enumeration approach (as well as approaches that rely on the enumeration of physical objects), people's use of the body system in double-enumeration strategies is increasingly autonomous, as it is not dependent upon counting objects themselves but used to produce computations independent of objects, relying on the construction of correspondences internal to the system.

Body-Part Substitution Strategy. The *body-part substitution* strategy (Figure 37) is a further extension of the use of internal correspondences, one in which the words for body-part names become differentiated from body-part positions to support a new keeping-track function. As in the double-enumeration strategy, people establish correspondences between two series of body parts in order to keep track. But rather than establishing physical one-to-one correspondences between two series of body parts, here one series is strictly verbal and the other physical. Consequently, people often produce calculations at a quicker pace.

To add 6 + 8 (Figure 37a) using a body-part substitution strategy, people might put in correspondence the verbal expressions of "thumb (1)" through "inner elbow (8)" as they point to the series forearm (7) through nose (14). Thus, the solution is wrist (6) + elbow (8) = nose (14). Similarly, to subtract 16 – 7 (Figure 37b), people recite the series "thumb (1) to forearm (7)" as they point to the series ear-on-the-other-side (16) to shoulder (10). Thus, the solution is ear (16) – forearm (7) = biceps (9).

In microgenetic processes of the body substitution strategies, people violate norms of communication, calling one body part by the name of another. But the warrant for the violation should be clear to the reader, in that it abbreviates the double-enumeration strategy. Thus, in this microgenetic construction, calling one body part by the name of another appears natural, thus showing a further distancing and increasing autonomy of the body as a number system as distinct from named parts of the body (Werner & Kaplan, 1963).

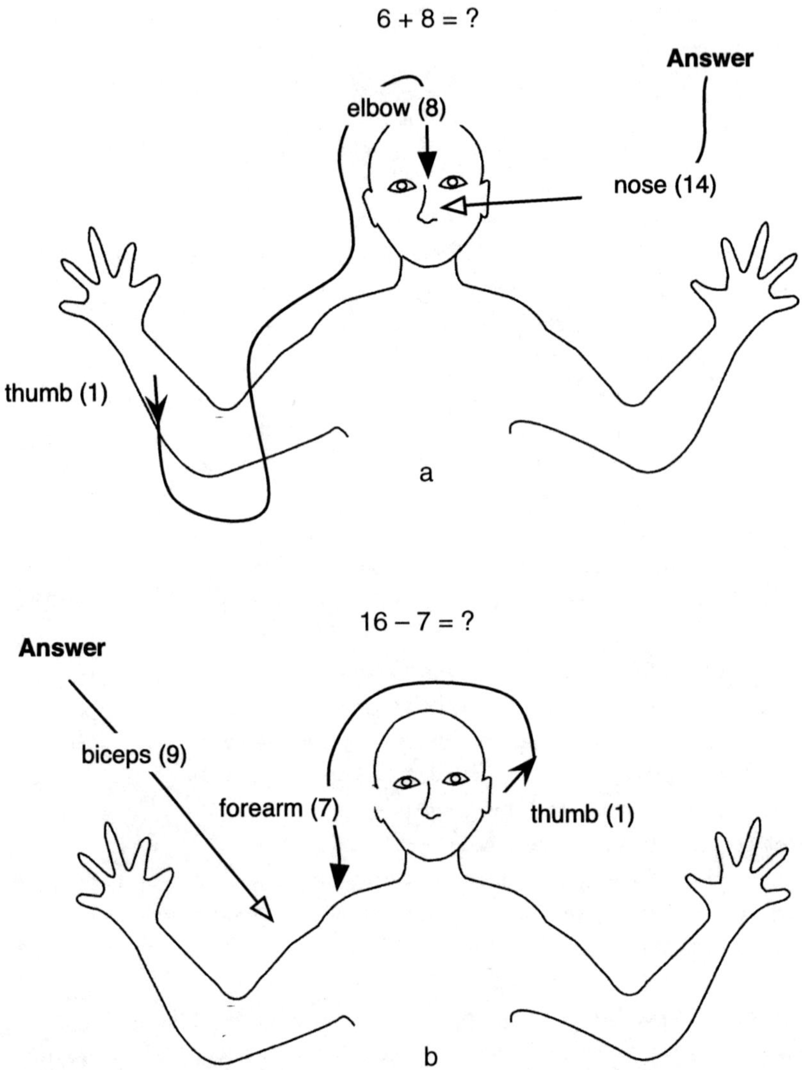

Figure 37. Body-substitution strategy for (a) addition and (b) subtraction.

Halved-Body Strategy. Finally, the *halved-body* strategy (Figure 38) is perhaps the most distant from traditional methods. In this approach, rather than using a subseries of the body-part sequence to keep track, people use each half of the body as a separate register for numerical values, often organized by the 10-cycle structure unit and 10t unit. With this approach, an individual organizes each half of the body around a focal point, often the

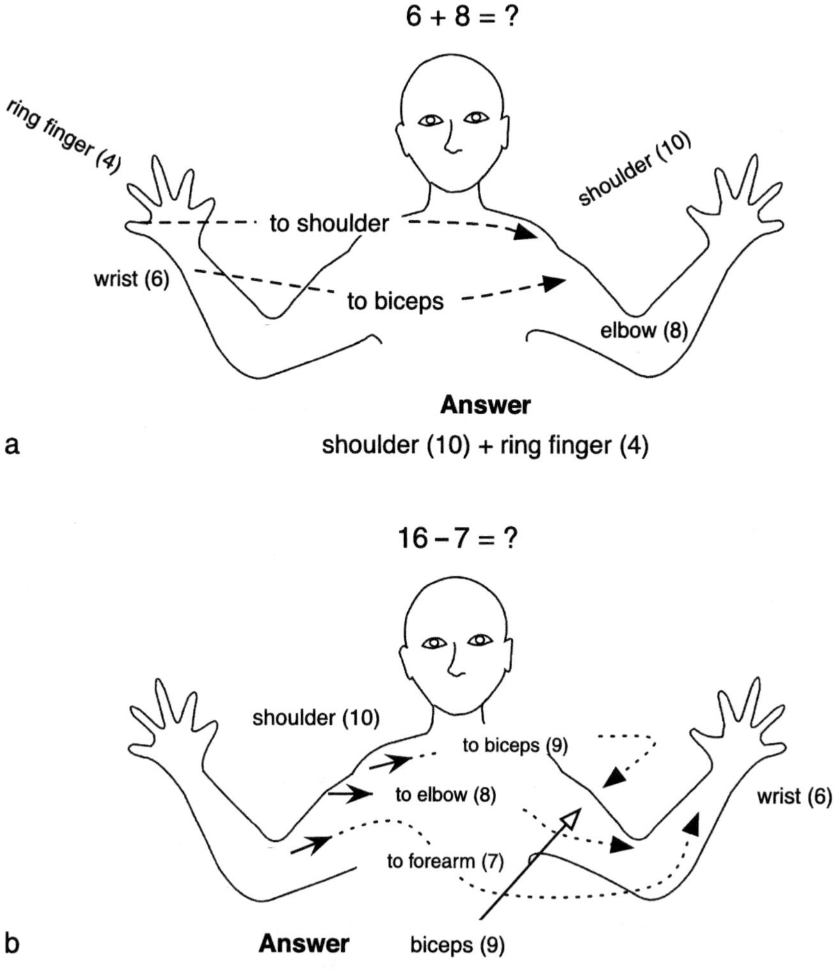

Figure 38. Halved-body strategy for (a) addition and (b) subtraction (a trade store owner's display of subtraction strategy on video found at http://www.culturecognition.com/trade-store-owner/).

shoulder (10). Each arm is then used as an independent register to add or subtract coins. For instance, to add 6 + 8 coins (Figure 38a), an individual might "put" 8 on one arm (inner elbow) and 6 on the other (wrist). The addition would be accomplished by a transfer from the first arm of the 6th (wrist) and 5th (little finger) to the second arm at the positions that complete 10, the 9th (biceps) and 10th (shoulder). Thus, the solution would be 14 coins – shoulder (10), registered on the second arm, and ring finger (4), registered on the first arm (or K1 and 40t).

For subtraction, a halved-body procedure again is constituted by a two-register approach. To subtract 16 – 7 (see Figure 38b), an individual might break down the ear (16) into shoulder (10), on one side, and wrist (6), on the other side. The individual would then indicate that the elements up to the forearm (7) of the first arm (the arm registering shoulder (10)) had been spent, leaving the inner elbow (8), the biceps (9), and the shoulder (10) on the first arm. These body parts would then be "transferred" to the second arm. In this process, the inner elbow (8) of the first arm would be shifted to the forearm (7) of the second arm; the biceps (9) of the first arm would be shifted to the inner elbow (8) of the second arm; and, the shoulder (10) of the first arm would be shifted to the biceps (9) of the second arm. The result of the transfer leaves biceps (9), the solution.

Strategic Approach by Cohort Analyses. People's approaches to solving the addition and subtraction problems were coded as one of five types. These included a covert category and the four strategic approaches described earlier: global enumeration, double enumeration, body substitution, and halved body.

Figure 39 shows the percentage distribution of solution strategy types over the four cohorts for the two addition problems without coins. Covert strategies are not included in the figure but were used in calculating percentage distributions of strategies; thus the percentages of strategies used do not add up to 100 percent within a cohort. An inspection of Figure 39 shows marked differences in the strategies used by cohort. The older adults almost exclusively used global-enumeration strategies on the addition problems, and more than half of the young adults used global-enumeration strategies. In contrast, use of global-enumeration strategies declined among plantation returnees, and use of the internal-correspondence strategies increased. Finally, internal-correspondence strategies were most frequently used by the trade store owners, with the mode being the body-substitution strategy; however, many used the halved-body strategy, especially on the problems with greater numerical values.

Patterns in solutions to the subtraction problems contained in Figure 40 revealed for cohorts a profile similar to that in the addition problems. The older adults used a global-enumeration strategy almost exclusively. For the younger adults, use of the global approach declined, but remained moderate, especially for the 16 – 7 = ? or 16 – 8 = ? problems. The use of

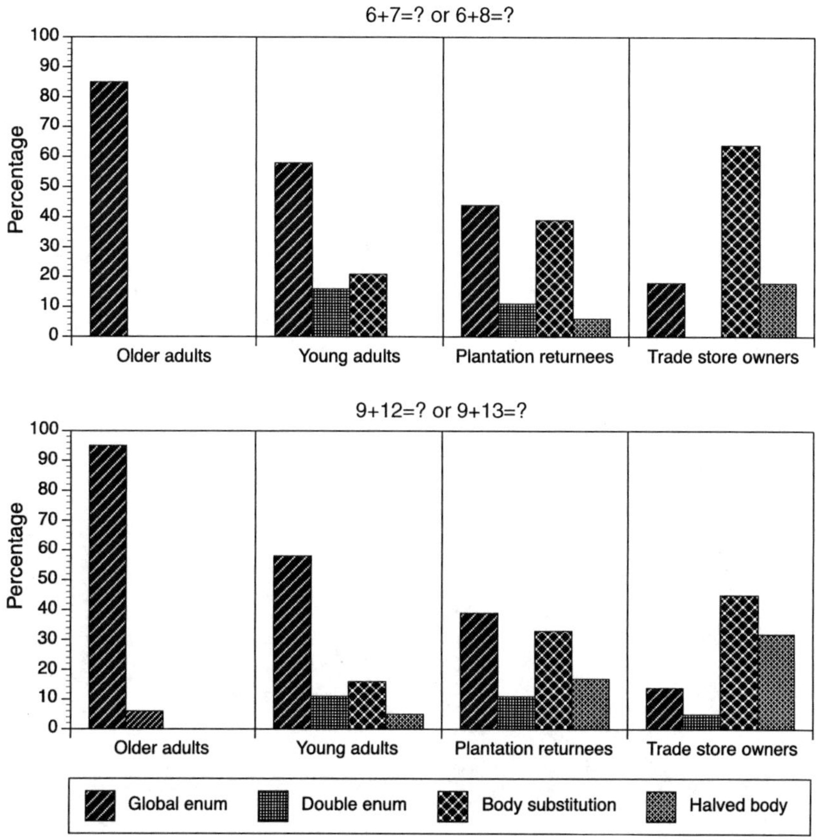

Figure 39. Percentage distribution of strategies to accomplish addition problems without coins present as a function of cohort.

global enumeration further declined among plantation returnees, and it was almost nonexistent in trade store owners. Further, plantation returnees and trade store owners predominantly used strategies in which they constructed internal correspondences between two series of body parts.

The cross-cohort analysis of the problems without coins present revealed a remarkable shift in how the body was used to organize and solve arithmetical problems in the absence of tokens to support a computation. The elder population almost exclusively used the global-enumeration strategy, on both the addition and subtraction problems. On addition problems, this procedure typically led to inaccurate solutions, since the individual did not know when to end an enumeration. On the subtraction problems, the

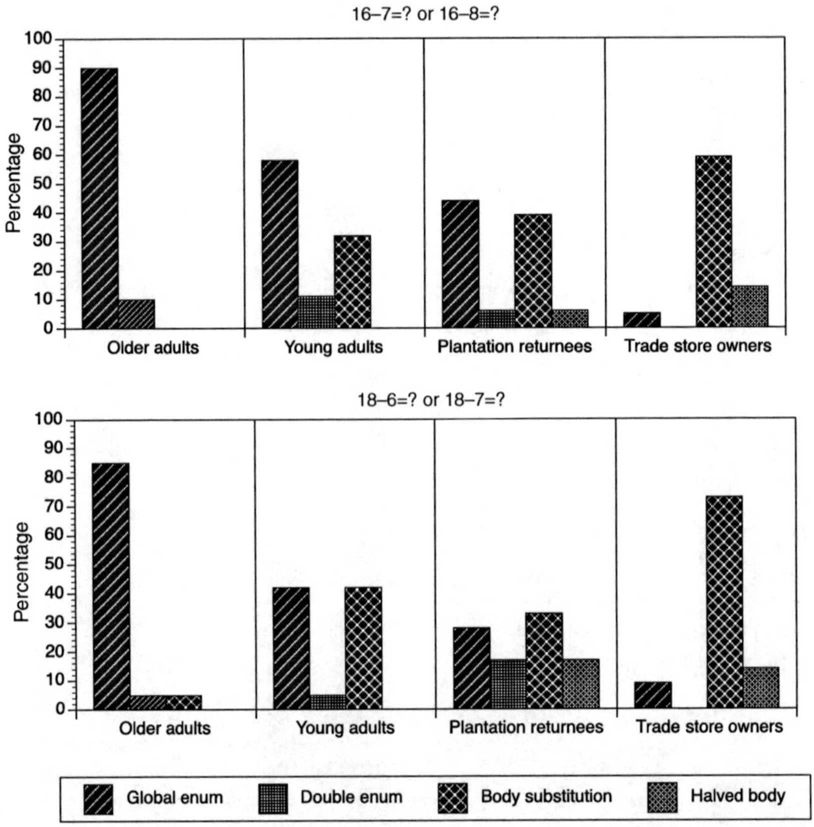

Figure 40. Percentage distribution of strategies to accomplish subtraction problems without coins present as a function of cohort.

global-enumeration strategy typically led to a nonspecific representation of the answer.

Young adults who had not traveled out of the Oksapmin area but engaged in minor commerce with missionaries showed a somewhat different profile. Like the elders, they used the global strategy with the most regularity. But there was some heterogeneity in strategy use in this population, with some use of the other three internal-correspondence strategies on the addition problems, and some use of the double-enumeration and body-substitution strategies on the subtraction problems.

Young plantation returnees showed a further decline in use of the global strategy, and a concomitant rise in the other three strategy types. For both addition problems and subtraction problems, they used the body-substitution strategy with almost as much regularity as they used the global strategy, and a number of these individuals used the halved-body approach.

Finally, the trade store owners used the global strategies only minimally. On both the addition and the subtraction problems, trade store owners favored the body-substitution strategy, but they also used the halved-body approach.

Relations between Solutions to Coin and No-Coin Problems

A final set of analyses provides insight into the way participation in the money economy has led some to extend the body-part forms to serve new arithmetical functions. The analysis directly compares performance on problems with and without coins. For this analysis, all participants were assigned a score on each of the eight problems. Scores were based on whether an "advanced" strategy type was used. For the coin problems, advanced strategies were defined as the use of a systematic strategy with either an accurate or inaccurate outcome. For the problems without coins, an advanced strategy was defined as the use of internal correspondences – the double-enumeration, body-substitution, or halved-body approaches. If an advanced strategy was used for any given problem, a participant was assigned a score of 1. Using these scores, I computed four subscales, with each ranging from 0 to 2: addition with coins, subtraction with coins, addition without coins, and subtraction without coins.

The mean usage of advanced strategies for each problem type as a function of population group is presented in Figure 41. A 4 × 4 repeated measures analysis of variance was used to determine the effects of population group and problem type on people's use of advanced strategies. As expected, this analysis revealed a significant interaction between population group and problem type ($F(9, 225) = 5.29$, $p < .001$), and significant main effects for both population group ($F(3, 75) = 23.04$, $p < .001$), and problem type ($F(3, 225) = 52.27$, $p < .001$). These results conform to the view that with little participation in the money economy, Oksapmin frequently achieve systematic solutions to addition problems when coins are present. As problem types diverge from addition with coins present (the problem type most similar to problems encountered in traditional life), people with little participation in the money economy use less-advanced strategies. As participation with the money economy increases, people create new ways of coordinating relations of correspondence that serve new arithmetical functions.

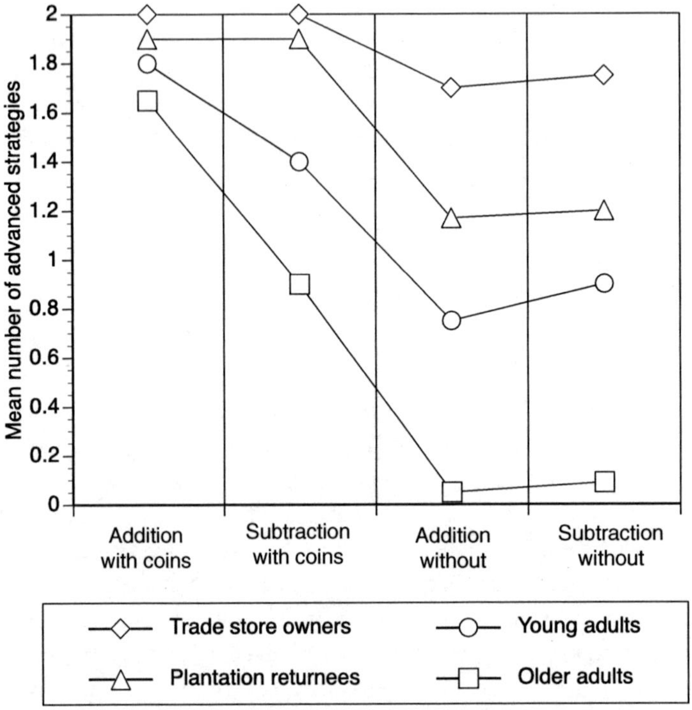

Figure 41. Mean number of advanced counting strategies used by participants as a function of cohort and problem type.

New Developments in Oksapmin Arithmetic: Synchronic and Diachronic Processes

As they are produced, the strategies documented with or without coins are microgenetic constructions. In this process, people transform the body-part count form into a vehicle of numerical representation, and the coins into objects of numerical representation. Nothing inherent in the coins, real or imagined, or in the body parts themselves contains a number, and there is no addition or subtraction operation in the coins or in the body form. Rather, the numerical representations and arithmetical manipulations take form in the moment-to-moment process as individuals turn body parts into the means to accomplish emergent numerical goals.

The strategies used on problems without coins revealed a remarkable development in the function of body parts in people's microgenetic constructions. In the global strategies, people enumerated body parts in counting but did so without keeping track of the addition or subtraction of terms. With shifts in strategy types, body part names increasingly became differentiated from signifiers of particular places on the body. The differentiation is nascent in the double-enumeration procedure, where the important adaptation is the construction of internal correspondences between two sequences of body parts, such that one sequence can be used to keep track of the addition or subtraction of a value, a novel numerical function. The body-substitution strategy is an abbreviation of double enumeration. Here we find a development in which body-part names become dissociated from body-part locations; rather than physically coordinating two sequences of body parts, people call a sequence of body parts by the body-part names of the added or subtracted term. Another development is the halved-body strategy, which uses each arm as a register and then translates body parts across registers to accomplish a solution that is interpretable in terms of kina and toea.

The 1980 study described here can be used both to inform a synchronic analysis of body-part arithmetic during that period of data collection and to support a diachronic analysis of the emergence of arithmetic in this stretch of Oksapmin social history. From the synchronic perspective, the interviews present a fascinating heterogeneity of arithmetical approaches that were being used in the community. Within and across cohorts, people were recruiting the body form to serve varied numerical and arithmetical functions in the presence and absence of coins. Noteworthy is that the heterogeneity of strategies in the Oksapmin population was reduced when we consider variation within particular cohorts. Indeed, almost all of the elders used global-enumeration strategies, and almost all of the trade store owners used strategies that required the construction of internal correspondences.

From the diachronic perspective, the cross-cohort findings suggest that with the rise of a cash economy there was a shift to more complex correspondence strategies over a stretch of historical time. Indeed, with increasing participation in the money economy associated with Oksapmin cohorts, we find a shift from external correspondences that serve numerical functions to internal correspondences that serve arithmetical functions. But these findings in themselves do not provide much insight on the process by which the shift was occurring. Consider two possibilities.

One possibility favors individuals as the principal agents in their own development; thus, the shift in strategies may reflect the independent invention of multiple individuals who had encountered problems with varying levels of intensity over time. In this case, as individuals engaged with the recurrent problems of economic exchange, greater engagement might have led individuals to use their prior achievements to construct strategies of greater and greater sophistication. The distributions by cohort seem to reflect this process.

The other possibility favors a network interpretation. People in the same cohorts may engage in exchanges in which they have an increase in opportunity to observe the strategies of others. Their communication with others may seed like strategic developments.

The present data do not allow us to disambiguate the alternatives of individual construction (independent invention) versus social network propagation. I take the most plausible model as having elements of both. Indeed, the efforts to coach older adults seldom led them to adopt more systematic approaches, suggesting constraints on the acquisition of new coordinations. At the same time, as people participate with others in exchanges, their negotiations and scrutiny with their interlocutors may lead them to make efforts toward novel coordinations of correspondence relations, a process that may result in new strategic approaches in their own development.

I will return to diachronic processes in Part IV of the book. For now, three developments are noteworthy. First, the body-part form has shifted in function as people have increasingly participated in the cash economy; people have turned external correspondences inward to serve keeping-track functions. Second, the shift to internal correspondences is associated with the increasing independence of a number system (body-part names) from physical body parts and physical objects. The greater autonomy of body-part names for number is clear in the differentiation people made between the names of body parts and the body parts themselves – one can call one body part by the name of another to serve the function of keeping track as in the body-substitution or halved-body strategies. Third, the construction of internal-correspondence relations leads to an arithmetic that is differentiated from the manipulation and counting of groups of objects. With internal correspondences, terms of arithmetical problems, like pinky (5) and forearm (7), can be added without five and seven objects present. These three remarkable developments in the social history of the Oksapmin group set the stage for the follow-up studies that were conducted 21 years later.

Study 4-2. The Body and Tok Pisin: Forms Used for Number Representation in 2001

The shift in uses of the body to serve arithmetical functions that I documented in the 1980 study on arithmetic (Study 4-1) led to a host of questions on my return visit in 2001. A central one was whether the innovations in using body parts to serve arithmetical functions had become the normative approach in arithmetical problem solving. Certainly one might expect this to have happened. After all, not only had the money economy grown, but more and more Oksapmin people were active economic participants.

As it turned out, this expectation did not hold up. My first hint of this came just before Indigo, Cliff, Josh, and I arrived in Oksapmin in 2001. As a way station on the trip, we had stopped in Tabubil, a mining town about 80 kilometers (through mountainous terrain) from the Oksapmin valleys. Tabubil had become a melting pot for job-seekers from a wide range of Mountain-Ok communities; they sought employment either with the mine or with the associated institutions and businesses (e.g., as night guards at an international school for administrators' children, store clerks, janitors). Some of the Oksapmin people whom I had been the closest to in 1978 and 1980 had relocated to Tabubil, and by serendipity, I met them soon after disembarking from the flight. This was quite a surprise. Among the old friends was Yangeh, who in 1978 had owned a trade store in Kweptanap; in the 1980 study, he was a fluid user of body-substitution and halved-body strategies – in fact, he was one of my teachers (as I noted earlier).[9]

After a wonderful reunion with a host of Oksapmin people, Yangeh and I returned to his mine-owned Tabubil dormitory room. Here I asked if he would mind my asking him some of the same kinds of number questions that I had been interested in during my earlier visits, and as ever, he was pleased to help. To my surprise, Yangeh did not use the body system in the way he had years before. Instead, he made use of mental procedures for adding and subtracting values, often telling me the result in Tok Pisin or translating the result in body-part terms. I tried to elicit some of the more sophisticated body strategies, but he indicated that he did not use them.

Several days after the Tabubil reunion, Cliff, Indigo, Josh and I boarded a small plane to fly 80 kilometers over mountainous terrain to the new landing strip in Bak, one of the two Oksapmin-inhabited valleys. Bak Community School, where we would stay, was about a 45-minute hike from the landing strip. Our ethnographic work started immediately, as we engaged in extended conversations and interviews with the local people. Though we found an occasional use of the more advanced body strategies, the lack of use of them that I had seen in Yangeh's case in Tabubil seemed to

[9] Yangeh is shown in 1978 in a photograph in Figure 23. He is the second from the left, the man whose leg is visible.

be representative of the population as a whole. Like Yangeh, many people in Oksapmin were using mental procedures for adding and subtracting values or for computing using Tok Pisin or English number words.[10] Perhaps with the greater penetration of schooling and greater mercantile contact with the world outside of Oksapmin, the usefulness of body computational approaches had declined. Indeed, perhaps knowledge of the body system in general had declined.

The purposes of the 2001 study reported here were twofold, each engaging issues related to sociogenesis. The first was related to body counting. Was the reproduction of the indigenous form waning in collective practices of exchange, with other forms, like Tok Pisin numeration, taking on quantifying functions? To the extent the body system was used in day-to-day activity, had its structure become altered in relation to the emergent problems of exchanges, continuing the truncation to 20 that had been documented in 1978 and 1980?

My second purpose was to document the rise of Tok Pisin as new representational forms were becoming used in acts of quantification. In 1978 and 1980, quantification in Tok Pisin was known only by few people. To what extent were people across cohorts knowledgeable about this system in 2001?

To address these questions, I again sampled four cohorts with a concern to represent the changing demographics of the community relevant to patterns of change in quantification using the body system and Tok Pisin. The groups this time around consisted of (a) unschooled elders, (b) unschooled middle-aged adults, (c) schooled middle-aged adults, and (d) schooled adolescents.

Methods

Participants and Setting

The levels of schooling and relative ages of the four cohorts who participated in the study are shown in Figure 42. As in Study 4-1, my concern here was to select groups with varying levels of participation in schooling and economic exchange, collective practices in which new forms of problems involving quantification emerged. The cohorts selected included

[10] Tok Pisin number words and English number words are very similar, and many are identical, though they are spelled differently. For example, "five" in English is spelled "faiv" in Tok Pisin.

Cohort	Schooling
Elders	None
Middle-aged adults	None
Middle-aged adults (schooled)	Grade 6 or more
Adolescents (schooled)	Grade 7 or 8

Figure 42. Selection criteria by cohort for participation in Study 4-2.

15 students in Grades 7 and 8 at Divanap Community School, 19 adults who had completed at least Grade 6 at school, 20 unschooled middle-aged adults, and 20 unschooled elders.[11]

Participants were offered 1 kina as a token of appreciation for taking the time to participate in the interview. Prior to the interview, they were told that we (Indigo and I) were interested in understanding about how people counted and how they talked about currency.[12] Almost all of our participants lived in hamlets in the Tekin and Divanap areas, and the principal site for the interviews was the grounds of Divanap Community School. Secondary sites included the Aid Post in Tekin and smaller hamlets.

Procedures

Designing Interview Procedures. The interview procedures we used emerged from informal conversations with schoolteachers in the Divanap area. The formal interview procedures were written in English, translated into Oksapmin and Tok Pisin, then back-translated by native Oksapmin

[11] To assign people to cohort groups, we collected schooling information from all potential participants, with the exception of the adolescent group, who were all enrolled in a Grade 7 or Grade 8 class at a community school. Assignment to age category of the adult groups was by participants' reports of their ages relative to two target individuals well known in the community, one whom we considered an elder and the other whom we considered middle aged (see Appendix C). We asked all candidates whether they were about the age of either person. If we could not place people in relation to these reference individuals, we asked them if they were older or younger than the two reference individuals. If they were older, they were assigned to the older cohort, and if younger, they were assigned to the younger group. (We dropped at least two people because of ambiguity about whether they should be in the middle-aged group or the elder group.)

[12] Almost all of our participants lived in hamlets in the Tekin and Divanap areas, and a principal site for the interviews was the grounds of Divanap Community School. Secondary sites included the Aid Post in Tekin and smaller hamlets. This is in contrast with the participants for the currency study (Studies 5-1 and 5-2) reported in Chapter 5, who lived mostly in Daburap, Kweptanap, and a few in Tekap. There were some Oksapmin dialect variations across these sites.

speakers. Translators took some minor liberties with the wording of the back-translated procedures to accommodate the needs and comfort of the interviewees. The questions used in the interviews are provided below.

Counting in Oksapmin. Do you know how to count in the Oksapmin language? (If necessary, the interviewer indicates in sequence: thumb (1), index finger (2), etc.) If yes: Show me. If the interviewee counts up to the elbow on the other side (20), the interviewer prompts: Show me the number that's one beyond elbow on the other side (20). If the interviewee counts to pinky (27) the interviewer asks, Do people count beyond the 27th? Show me.

Counting in Tok Pisin/English. Do you know how to count in Tok Pisin/ English? (If necessary, the interviewer starts counting, one, two, three, etc.). (If yes) *Show me.*

Representing Greater Values. In order to explore how individuals produced values considerably greater than 27, participants were asked how to represent values of 50 and 60. "Do you know a way of showing 50 in Oksapmin? Do you know a way of showing/expressing 60 in Oksapmin?" Though the interviewer asked the questions in the Oksapmin language, the numbers themselves were given in Tok Pisin. Some of the unschooled participants could not recognize the Tok Pisin numbers, but others, even those with no fluency in Tok Pisin, could recognize these numbers.

Results

Knowledge of and Variations in the Body-Part Count Sequence

Individuals' recitation of the body-part count sequence was recorded. Figure 43 shows how far people were able to count in the sequence between thumb (1) and elbow (20) before producing an error.[13] The figure shows that the only cohort producing a substantial number of errors was the schooled adolescent population: More than half of the schooled adolescents made errors, and most of these errors occurred between the shoulder (10) and biceps on the other side (19). Only one elder made a count error. These findings point to the decline of body-count system knowledge among the generation of adolescents who were attending school in 2001.

In order to produce data on the extent to which people had taken up elbow (20) as a privileged point, we recorded the body part at which

[13] The reason that 20 was used as the ceiling for errors was that some treated 20 as the greatest number in the body-counting system. To equate the possibility for error across individuals (some of whom counted past 20), we recorded errors only up to 20.

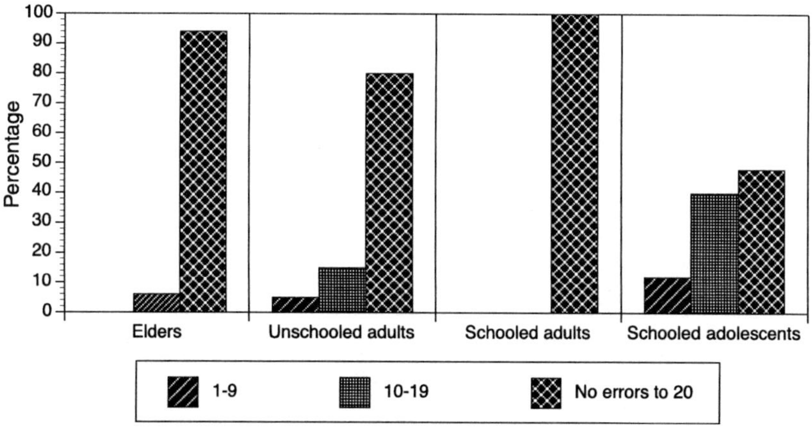

Figure 43. First error in count of first 20 body parts as a function of cohort.

individuals stopped their display of the count system. Figure 44 contains the body-part name at which individuals stopped their counts. The figure shows that most people stopped their count at either the 20th or the 27th position (a few spontaneously counted beyond the 27th position). The schooled adolescents, again, were the exception – 28 percent did not complete a count to the 20th position, the elbow (20). Figure 44 also reveals that among schooled and unschooled adult cohorts, stopping on the 20th body part was common – more than 40 percent of the unschooled adults and more than 50 percent of the schooled adults stopped at this point. In contrast, for most of the elders and the schooled adolescents, the stopping point was the 27th body part in the series, though as noted in prior remarks about Figure 43, the schooled adolescents' knowledge of the system was weak, with many making mistakes in their count.[14]

For those who did not count beyond the 20th position, I reasoned that they may well have stopped not because they thought that counting

[14] There are two plausible reasons for the adolescents' display of the pinky (27) as the last body part of the count system. First, Indigo and I observed some peer coaching on the school grounds in the count system among the adolescents due to our presence (these interviews were conducted at a bush school where peers were in close proximity during the day). The effects of peer coaching were probably limited and the knowledge gleaned was shallow, so the findings for the adolescent group may well be an overestimate of their knowledge of the body counting and their identification of the pinky as the 27th body part. A second possible factor is that the body sequence was being reintroduced to young schoolchildren in the new elementary schools in the community (see Chapter 11), and young children were learning to count to the 27th body part. This procedure may well have spread to the adolescent population.

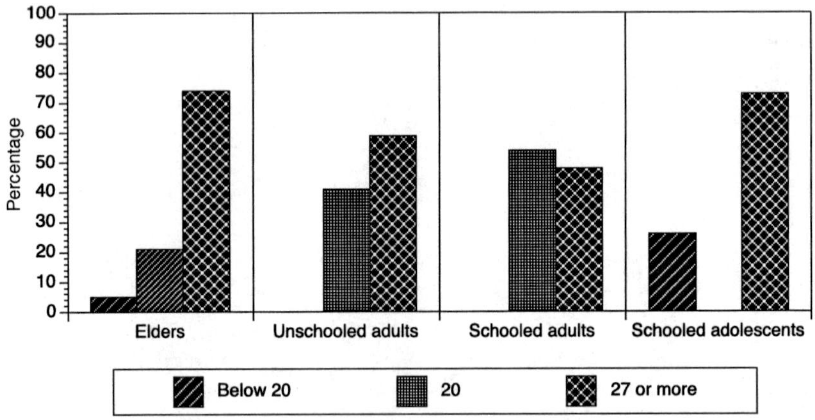

Figure 44. Stopping values as a function of cohort.

beyond the 20th position was not possible, but that this was the system that we were trying to elicit from them. Therefore, we asked these participants whether they knew how to count beyond the elbow (20), and if they did, to show us. We pooled these elicited counts beyond the elbow (20) with the counts of people who spontaneously counted beyond the elbow (20) and analyzed the variations in the counting trajectories.

We noted two types of variants among counts beyond the elbow (20). One type was the use of body parts from the 21st through the 27th positions. The other was the use of body parts following the 27th position.

1. Body Parts to Count from 21 through 27. We coded four different approaches to body-part counts for positions from the 21st through 27th.

a. Standard: Thumb (23) → Pinky (27). One variation was, of course, the "standard" approach, as depicted in Figure 45. Standard: thumb (23) → pinky (27). This was the system that Tom Moylan and I had learned in 1978. In this count, the ending of the count is not symmetric with the beginning of the count from the thumb (23) to the pinky (27), as depicted in Figure 45. (Recall that the body-count sequence begins with the thumb (1).)

b. Word-Body Relation Shift: Begin Again after 20, Using "Thumb" (1) to Refer to Forearm (21). A second variation for finishing the count to 27 is depicted in Figure 46. The individual calls the forearm (the 21st body part in the series), thumb (1), and proceeds on to the 27th body part using the names for successive body parts as index finger (2) through forearm (7). In this case, vestiges of the 20th position in the stopping point of body parts are evident, as noted in the 1978 and 1980 studies. But unlike what I

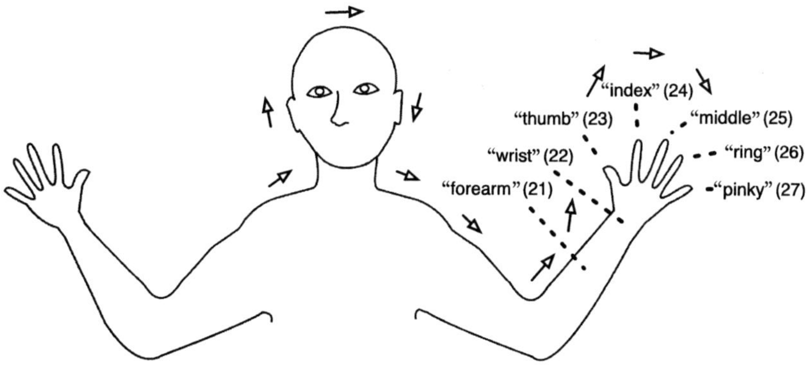

Figure 45. Standard: thumb (23) → pinky (27).

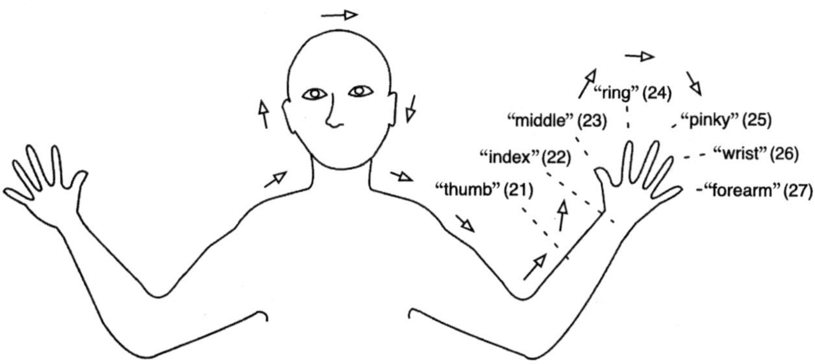

Figure 46. Begin again after 20: forearm (21) called thumb (1).

observed in 1978 and 1980, in 2001 the innovation of the 20th position as a marked place in the system was invoked without reference to currency. Further, the innovation was manifest as a particular trajectory on the body, in which one calls one body part by the name of another; perhaps it had its origins in the double-correspondence procedures documented in 1980 (see Figure 36).

c. Symmetric: Pinky (23) → Thumb (27). A final variation that does not reflect the termination at the 20th body part is depicted in Figure 47. In this variation, the order of the final fingers is rendered symmetrical to the start of the count. Rather than proceeding from thumb (23) to pinky (27), the traditional asymmetric ending, the person completes the sequence of 27 body parts by proceeding from pinky (23) to thumb (27), creating a symmetric ending with the nose (14) at the midpoint.

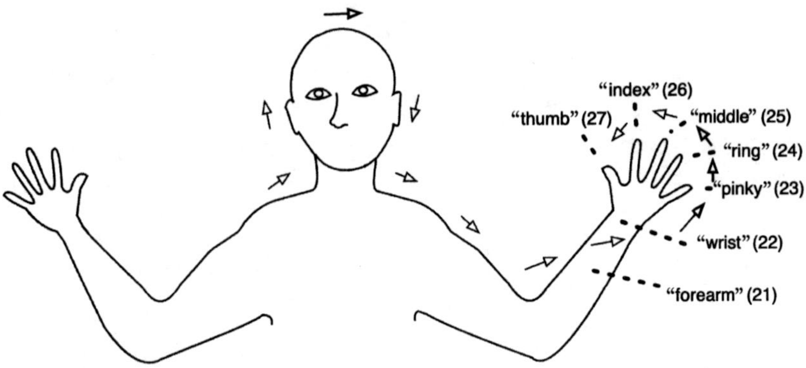

Figure 47. Symmetric: pinky (23) → thumb (27).

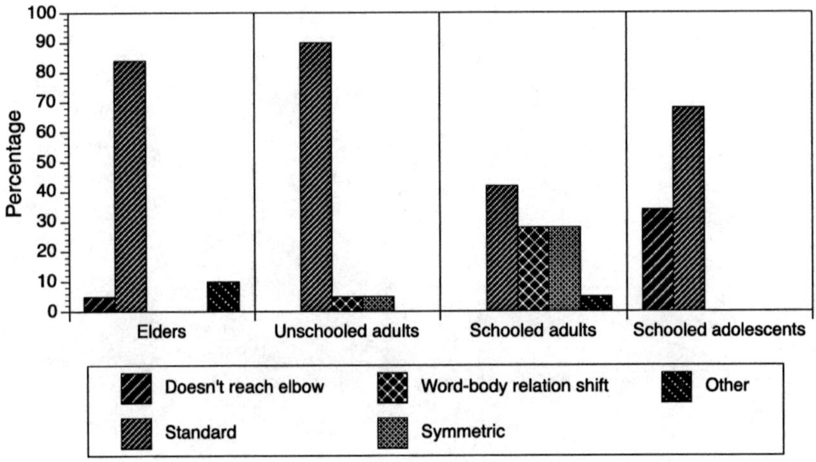

Figure 48. Distribution of body-part counting trajectories after elbow (20) as a function of cohort.

Figure 48 shows the distribution of count trajectories after elbow (20) as a function of cohort. Though a large majority of elders and unschooled adults used the standard procedure, Indigo and I observed 10 percent to 15 percent of these individuals using two of the nonstandard variations. The population that shows the most variability is the schooled adults, with about 25 percent using symmetric variation (pinky (23) → thumb (27)) and about 25 percent using the standard trajectory but calling these body parts by the names of the first body parts in the traditional count sequence. The

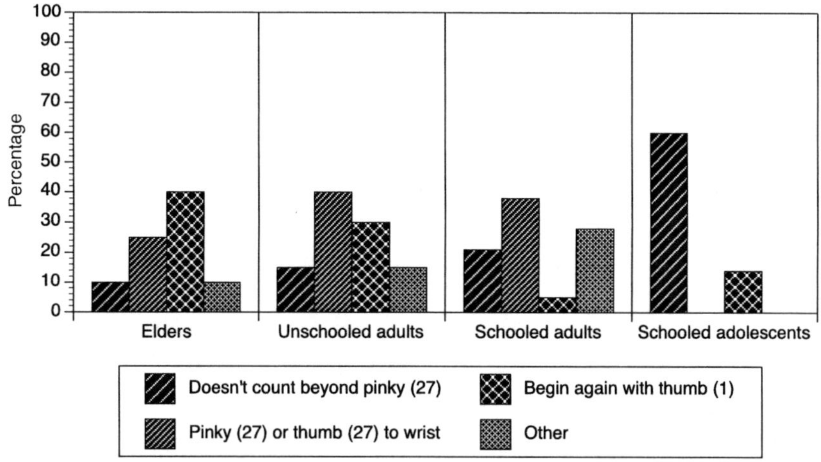

Figure 49. Percentage distribution of body-part counting trajectories after pinky (27) (or thumb (27)).

later variant points to the sustained dissociation between body-part names and locations in Oksapmin semiotics for number that was observed in 1980 in the body-substitution strategies (Study 4-1).

2. Body-Part Counts beyond 27. To determine whether there were alterations in what I understood to be the traditional counting practices of looping back around to the wrist (28) to identify the 28th position (see Figure 12), Indigo and I recorded the counts of people who spontaneously counted beyond the 27th position (typically the pinky (27)). Further, for those who did not count beyond the 27th position, we asked them whether it was possible for them to count further using the body system and included these elicited performances in the analysis. Again, responses varied and we show the principal variations in Figure 49.

Some people indicated that the 27th body part was the final body part in the count or that they did not know how to count further. This response was dominant among the schooled adolescents, but much less common in the other groups. A substantial minority of elders, unschooled adults, and schooled adults displayed what was common in the 1978/1980 study, looping to the wrist (28) after the pinky (27). Finally, a substantial number of elders and unschooled adults indicated that when the 27th body part is reached, one begins a count anew at the thumb (1).

Indigo and I found that a few individuals spontaneously used the expression *fu* when demonstrating a count to us. For example, upon completing a count to the 27th body part, one elder exclaimed *"tit fu gona!,"* holding up both fists in an exclamatory gesture. Another did something similar for the 20th body part, elbow (20). I note additional observations of spontaneous use of *fu!* in Chapter 5, and *fu* becomes the topic of focused inquiry in Chapter 6.

Knowledge of the Count Word Sequence in Tok Pisin/English

As indicated in the methods section, Indigo and I planned to ask all individuals to count in English/Tok Pisin, stopping them if they counted to a ceiling value of 41. But there were problems with this procedure. Many participants were fluent Tok Pisin speakers, and we were therefore concerned about asking foolish and "off-putting" questions. As a result, we only asked individuals who did not have fluency in Tok Pisin to count in Tok Pisin. This group included elders and some of the unschooled adults. Because we did not ask the Tok Pisin count word questions of all of the unschooled adults, I report findings only on elders.

We found that elders' knowledge of count words in Tok Pisin was limited. Only 10 percent of our sample of elders could count to 41 in Tok Pisin. Thirty-five percent knew the number words to 20 or less. Fifty percent could not count in Tok Pisin/English at all. Several who did count in Tok Pisin also made reference to the conventional body parts as they counted as if to communicate the meaning of Tok Pisin/English number words. In other words, the interviewee's count of "one, two, three..." co-occurred with pointing to the thumb (1), first finger (2), middle finger (3), et cetera.

Summary

The findings of this study revealed variation in knowledge of body counting and Tok Pisin number words across cohorts. With regard to body counting, the adults (elders, unschooled adults, and schooled adults) but not the adolescents made few errors when counting with the body to the 20th position, elbow (20). Further, when using the body-counting system, the elders showed clear evidence of using all 27 body parts, while other cohorts often used the 20th body part, elbow (20), as the terminating point. The unschooled cohorts (elders and unschooled adults) but not the schooled cohorts (both adults and adolescents) maintained the traditional trajectory of the last five body parts in the counting sequence: thumb (23) → pinky (27). All cohorts showed variability when counting beyond the 27th position. With regard to knowledge of the Tok Pisin register, the findings were more uniform. Most of the cohorts appeared fluent in Tok

Pisin number words through 41 (though again, we thought it presumptuous to ask people with some competence in Tok Pisin to count to 41). The exception was the elders, who showed limited knowledge of Tok Pisin number words.

From a diachronic perspective, the findings indicate different trajectories in knowledge of body counting and Tok Pisin numeration. On the one hand, unlike in my 1978 and 1980 visits, the Tok Pisin count register in 2001 was used with fluency by the younger populations. Indeed, knowledge of Tok Pisin is no longer restricted to men who have worked for stints outside of the Oksapmin community. On the other hand, in 2001, knowledge of the traditional body-count system was still known with some level of fluency in the adult population, but less so in the schooled population. Further, in the unschooled and schooled middle-aged populations the system appeared to be not only reproduced but also altered, with an increased variety of forms of counting. The alterations included: (a) the use of the elbow (20) as a terminating point in the body-counting system, (b) the use of variant trajectories of the positions from the 21st through 27th body parts, and (c) the use of variant approaches in how to count numbers beyond pinky (27) (or, alternatively, thumb (27)).

Why the increased variation in the body system? One source may well be a process of collective forgetting of the canonical forms, due to decline in use. If this were the case, the variants people displayed would be generated in their on-the-spot efforts to "remember" how to use the body system. But this could not be the complete answer. The variants were not random; they took particular forms and were used by many. Further, many individuals who reported nonstandard endings were adamant that their way was the only correct way. Thus, there may well be semidurable variants emerging through sociogenetic processes in the larger population.

Study 4-3. Languages of Quantification Used in Stores in 2001 and the Accomplishment of Emergent Arithmetical Problems

It is one thing to describe knowledge of cultural forms of quantification in a community, like body counting and Tok Pisin number words as was accomplished in the prior study. It is quite another to describe the frequency of use of these forms in everyday activity in a population. It was towards the end of documenting frequency of use that Indigo and I observed interactions in trade stores in 2001. But our efforts met with problems. First, we believe our presence changed what occurred, perhaps affecting people's choice of language as well as social interactions. A second and related problem in an observational study was that I wanted to draw a sample of observations from many sites that might contribute to sociogenetic analyses involving languages used for quantification, so the time investment to sample many sites would make such an observational approach impractical. Third, even if we could manage to be at multiple places at a single time, our lack of fluency with the Oksapmin language limited our ability to make differentiated observations about naturally occurring talk in interaction. Fourth, the exchanges were infrequent and spread across a day, making collection of sufficient numbers of observations impractical.

To circumvent the difficulties, we used a different approach to data gathering. Indigo and I recruited the help of store owners and clerks to keep records of every transaction within a specified period of time. Records would include language of quantification used by each customer, cohort of each customer, and the grade that each customer had completed in school. We collected the records immediately after the end of the recording period, and Indigo and I debriefed the owner about selected records.

Methods

Participants

We chose trade stores in Bak, one of the two Oksapmin valleys, as principal sites for the study. Participants included both store clerks and customers.

Store Clerks. We recruited nine individuals who worked in one of seven trade stores located in Bak. Store clerks varied in age from adolescents to late thirties. Most of our sample consisted of store owners who worked as clerks in their own stores.

In our recruitment procedures, Indigo and I approached clerks at their stores. We explained that our purpose was to understand economic transactions that occurred in stores where people made or considered making purchases. We provided printed record sheets, and we asked the clerks whether they would help us by recording information about the transactions occurring in their stores. We indicated that we would pay clerks for their efforts. Bak trade stores, the locations of which are depicted in Figure 50, became the sites for record-keeping.

Customers. A tally of clerks' records revealed more than three hundred clerk-customer transactions in the time allotted for data collection. Clerks

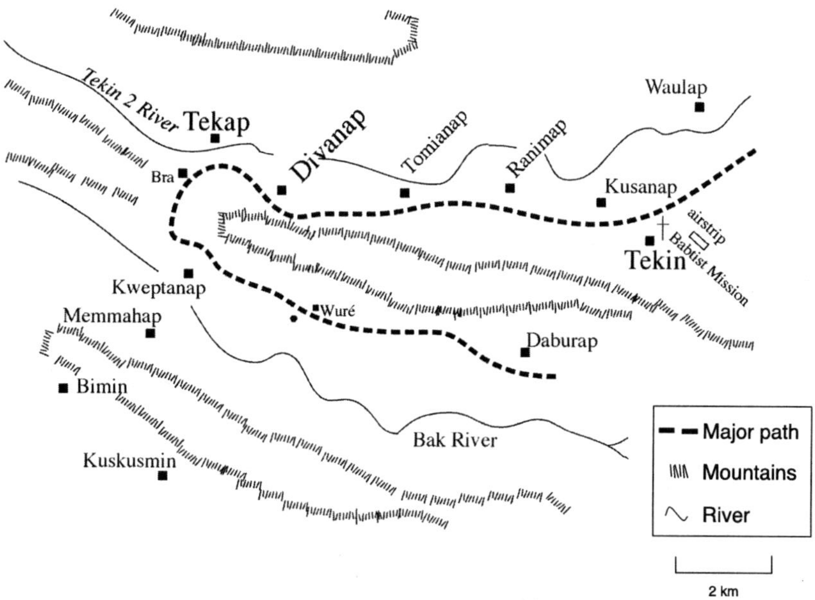

Figure 50. Map showing trail from Tekin to Daburap and store locations in Bak Valley (2001).

recorded information about each customer at the time of the transaction using the printed form described below. Information about customers was used to assign each customer transaction to one of four mutually exclusive customer groups: elders (15 transactions), older adults (65 transactions), younger adults (102 transactions), and youths (123 transactions). Schooling varied within these groups, with the exception of the elder group, in which no one was schooled.

Procedures

Transaction Recording Form and Its Use. Clerks were asked to keep records on a printed form contained in Appendix D. Each row of the form's eight printed rows was used to record information about a single purchase. Each column of the form's 10 preprinted columns was used to document information related to that purchase. In presenting the chart to clerks, we reviewed each column, explaining the values that were to be entered for each transaction. If a customer requested an item, paid for the item, and then requested and paid for a second item, the exchanges were to be recorded as two transactions.

Of import for the present analysis were the following seven columns on the transaction recording form (see Appendix D):

Name of customer. The clerk wrote the customer's name.
Age of customer. The clerk was required to approximate the customer's age by circling which of three listed individuals the customer was closest in age to. The three individuals were well known to the clerks. We considered one as a younger adult, another as an older adult, and the third as an elder. Two additional categories were also included, young child and school-age child (preadolescent and adolescent).
Grade completed in school. The clerk was directed to circle one of three categories: no school, Grades 1–5, Grade 6 or above.
Language used for number. The clerk was directed to circle one of three categories: Oksapmin, Tok Pisin, Mixture of Oksapmin and Tok Pisin.
Number of items purchased. The clerk was directed to circle one of seven categories: 0, 1, 2, 3, 4, 5, 6+.
Amount of money that the customer offered for the goods. The clerk was directed to enter the amount of money the customer presented for payment.
Amount of money that the item(s) cost. The clerk was directed to enter the cost of the item or items.

Prior to data collection, Indigo and I reviewed the form with each clerk to prepare for record-keeping. In this procedure, we engaged in a mock transaction in which the clerk was to use the form to record a transaction with our Oksapmin research assistant. The assistant or one of us made a mock purchase and the clerk filled out a row under our supervision. We answered questions as they emerged and we repeated the process, if necessary, to clarify the recording task. We visited stores during the time that individuals were using the forms, observing their use and checking whether clerks had any problems with the forms. Most clerks had few if any problems, and the problems that they encountered we corrected.

Debriefing. Indigo and I retrieved the record forms shortly after the clerks finished using them (in the late afternoon or the next morning). We checked over the forms with the clerks individually, asking for clarification of questionable entries as a check of how faithfully the participant had followed our procedures. We made some changes in entries in collaboration with the clerk on questionable entries. On one occasion, we discarded a day's entries and asked the clerk to repeat the day's efforts.

Whenever possible, we selected three records from each clerk's diary of records for a follow-up debriefing interview. One record involved a transaction with a community elder, the second with a middle-aged adult with schooling, and the third with a middle-aged adult without schooling. Sometimes a clerk's record sheets contained no individuals from a targeted group. For groups that were represented, we tried to select a recent transaction to discuss in the interview. We avoided interviewing clerks about transactions that had occurred many hours earlier (except in small trade stores with very few customers).

We asked the clerks to describe each of the three targeted transactions. The focus of the interview was on eliciting a "blow-by-blow" description that could be used later to describe prototypical transaction structures: "After entering your store, what did she/he do first? (Did she/he say anything to you? What?) What did you do? What did you say? What would he/she do next?" This style of elicitation continued until a full description of the exchange was completed. We focused on the following issues that occur in transactions: the customer considers items, the customer considers prices, the customer selects items, the customer pays for items, the clerk and/or customer break denominations, the clerk and/or customer pay with exact change, the customer receives change, the customer receives credit.

Results

Transactions and Amounts Tendered

The total number of recorded transactions varied across stores. The largest number of transactions at a single store was 104 and the smallest was 2. For purposes of analysis, we pooled all transactions, regarding them all as occurring in Bak Valley and displaying common properties. To determine an overall sense of the amount of cash and goods exchanged across transactions, we computed amounts paid for items, number of items in a purchase, and whether change was tendered across all of the recorded transactions. Costs of items for purchases varied from zero kina to K36,[15] and the median purchase was K1.50. The number of items purchased also varied from no items to more than six items; the median number of items

[15] Store clerks included occasions in which customers did not make a purchase. Generally, such exchanges were either discussions of item prices or discussions related to payment of credit due.

purchased was one. Finally, for 45 percent of the transactions change was tendered; in the transactions in which change was not tendered, either the customer produced exact change or, on a few occasions, the customer bought goods on credit.

Language of Quantification in Purchases

To analyze quantitative language use in transactions by cohort and schooling, Indigo and I coded the language used for quantitative representations, drawing on the codes from the diary record sheets – "Oksapmin only," "Tok Pisin only," and "mixed." It was unusual for Oksapmin speakers to use Tok Pisin when speaking to other native Oksapmin speakers, except for quantification.

"Oksapmin only" indicated that the transaction was accomplished using only indigenous expressions for quantity. This category included references to body parts used in the conventional body system, references to small numbers with the five conversational number words, and references to currency units using traditional language. (In the latter case, for example, the 50t coin is often called *gangasi*, meaning the pointed one, referring to its cornered points. A K2 note may be referred to as *haben*, meaning "leaf of tree.") The following are transactions described by trade store clerks in which customers used indigenous quantifications only.

> *Clerk: KUB, Customer: STE (Unschooled Adult).* STE entered the store, asking for a bag of rice and a can of tinned mackerel. He already knew the prices of these items, because he was a frequent customer at the store. (Prices were posted beside each item.) The total cost of the items was K7.20, but neither the clerk nor STE said the values aloud. STE handed the clerk a K10 note, and told the clerk that he expected "*haben tansi kir hai*" back in change [K2.80, literal translation: "leaf of tree (K2) and elbow (8 10t coins)']. The clerk gave STE his change, which he did not count but put directly in his pocket.
>
> *Clerk: ROB, Customer: HAN (Elder).* Two days earlier, HAN had come into the store and asked for noodles. At that time the clerk told him the price. On the day of the study, HAN returned and asked for a package of noodles, handing the clerk a K5 note. The clerk gave him back K4, without counting it out. The clerk explained that HAN probably remembered the price of noodles from his previous visit to the store. HAN returned a few minutes later, asking for the price of a cooking dish, and ROB told him in the indigenous language, "*imah kak-si, wan faun tana*" [K22, literal translation: "pig head (twenty-kina note), one pound (the equivalent of K2)." The phrase

"pig head" refers to the picture of a pig's head that appears on the K20 bill.] HAN turned to his wife, who was standing next to him, and asked in Oksapmin language for two kina. She handed him a K2 note. After examining the cooking dish, HAN went back outside. A few minutes later, he re-entered the store, having decided to buy the dish. He put a twenty-kina note and a K2 note on the counter. The clerk took the money without comment and handed HAN the dish.

"Tok Pisin only," as the name implies, refers to the use of quantitative language that only included Tok Pisin in the transaction.

Clerk: RUT, Customer: SPA (Unschooled Adult). SPA entered the store and asked the price of a Coca-Cola (Note: In this store, prices are posted beside each item.) The clerk answered in Tok Pisin, saying "*tu kina, tuwenti toea*" [K2.20]. SPA then asked for the price of a packet of biscuits. The clerk again answered in Tok Pisin, saying, "*wan kina, forti toea*" [K1.40]. SPA gave the clerk two K2 notes, then asked her the total cost of the two items. The clerk again answered in Tok Pisin, saying "*tri kina, sixti toea*" [K3.60]. She took the two bills, told him that his change was "*forti toea*" [40t], and returned his change. SPA put the change in his pocket without counting it. Notice that SPA never used any number words in either language, and spoke only in the Oksapmin language. The clerk spoke primarily in the Oksapmin language, only using Tok Pisin for the number words.

Clerk: NIC, Customer: MER (schooled adult). MER came into the store and asked for the price of a packet of rice. (In this store, prices are not posted). The clerk told MER the price in Tok Pisin: "*fo kina, eiti toea*" [K4.80]. MER asked to buy three packages of rice but did not ask for the total cost. He gave the clerk a K20 note. Neither man said the total cost of the three packets of rice. When the clerk gave the change to the customer, they both counted it out together. (The clerk may have used a calculator (rare) to compute change.)

Finally, "mixed use" was defined as the use of both Tok Pisin and indigenous methods of quantification in the transactions.

Clerk: THO, Customer: ERI (schooled adult). ERI entered the store and said that he wanted to buy "*tupela batteries*" ["*tupela*" is the Tok Pisin word for the cardinal number 2]. ERI did not ask for the price and presumably read it from the sign beside the batteries. ERI said the total cost himself in Tok Pisin, as "*faiv kina, forti toea*" [K5.40]. Then ERI counted out the correct change, listing the coins and bills in the Oksapmin language: *haben tan tita, haben tan tita, temsi tan tita, gamen tan tita, gamen tan tita* [K5.40, literal translation: "flat leaf of tree (K2 note), flat leaf of tree (K2 note), flat one with hole (K1 coin), married flat one (20t coin), married flat one (20t coin)"]. Once ERI had counted out the correct amount, ERI handed all the money to the owner, who put it away.

These narratives illustrate that trade store transactions are collective accomplishments, as the customer and the clerk each accommodate the communicative moves of the other (a process that is my focus in Study 4-4). In this give and take, each makes particular choices in the language that is used for number in the interactions. On some occasions, the customer appears to play a leading role in language choices in the interactions. For example, when STE (customer) speaks Oksapmin to KUB (clerk) to communicate how much change he should receive, STE appears to take the lead. On other occasions, the clerk appears to determine the language used for number. When SPA (customer) uses the Oksapmin language to ask RUT (clerk) for prices of Coca-Cola and a packet of biscuits, the clerk (RUT) answers in Tok Pisin, telling SPA his change. The clerk's use of Tok Pisin anticipates the customer's fluency in Tok Pisin numeration. Clerks and customers know one another well in Oksapmin communities and are generally knowledgable of one another's language and arithmetical fluencies. (In these transactions, the computation is notably absent, accomplished outside the context of the interaction and brought forward from prior transactions involving a similar exchange.)

In order to document generational shifts in quantitative language use and relations with schooling, Indigo and I partitioned records by both age cohort (youth, younger adult, older adult, and elder) and level of schooling within each cohort (no school, Grades 1–5, and Grade 6 and beyond). We then analyzed the transaction records as a function of language use by cohort and level of schooling. The results of this analysis are shown in Figure 51.[16]

[16] Note that if a single customer engaged in multiple transactions in a short period of time, each transaction would be recorded separately, even though such sequential transactions might be regarded as a single purchase. We noted that 7 percent of the purchases that included two or more items were sequential transactions. In our data records, eight individuals made two successive purchases and one individual made three successive purchases. For most of the sequential transactions, customers used *indigenous only* (68 percent). Further, most had no schooling (56 percent) or little schooling (44 percent). Of the customers who purchased more than one item, 4 were elders, 31 were older adults, 51 were adults, and 35 were youths. Of this group, 25 percent of the elders, 10 percent of the older adults, 8 percent of the adults, and 3 percent of the youths made sequential purchases.

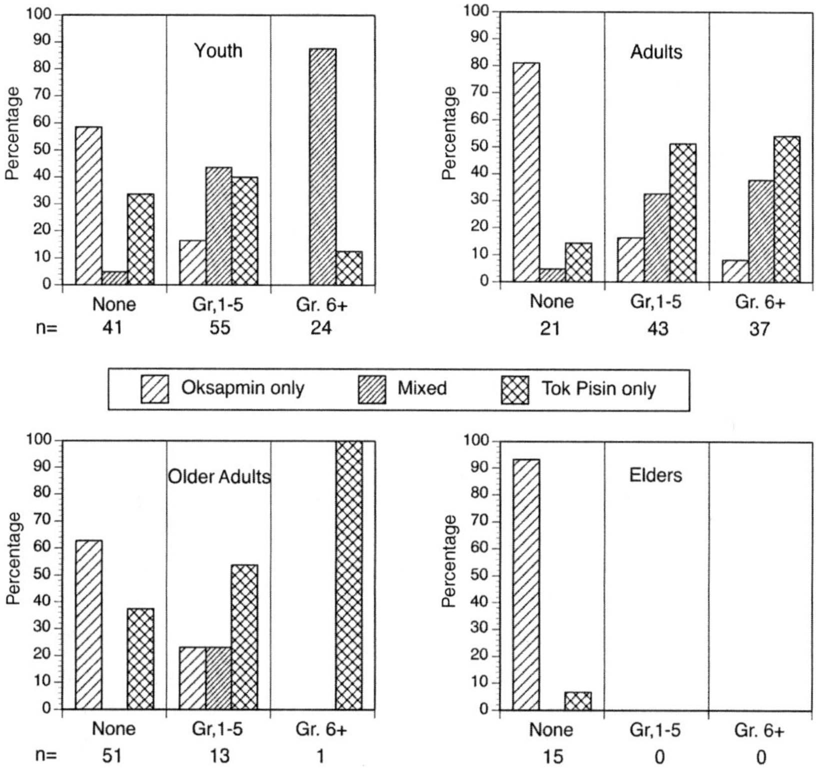

Figure 51. Percentage distribution for language of quantification as a function of age cohort and schooling level.

Inspection of Figure 51 reveals several features of language choice in Oksapmin trade stores. Perhaps most striking is the diversity of quantitative-language use. Both indigenous practices of quantification and practices associated with Tok Pisin are well represented in store transactions, though frequency of these practices differs by cohort and level of schooling. In the populations that have had no access to schooling, there is predominant use of only indigenous methods of quantification. Indeed, for the "no school" categories, all age cohorts use "Oksapmin only" as the principal category, whether youths (58 percent), younger adults (81 percent), older adults (63 percent), or elders (93 percent). With a Grade 1–5 level of education, there is a shift to use of Tok Pisin, though all populations retain some level of "indigenous only" language use. With Grade 6 or greater education, "indigenous only" is no longer the case among youths, and less than 10 percent of the transactions used "indigenous only" in the younger adults with Grade 6 or greater education.

In the Grade 6+ transactions, the difference between youths and younger adults is noteworthy. Many of the adults have visited or lived in the mining town of Tabubil, where many Oksapmin have sought gainful employment. It may well be that being engaged in economic exchange in the town, which is a melting pot for many different indigenous groups that do not speak Oksapmin, leads to using a common system of quantification, Tok Pisin. Note that only one older adult had a Grade 6 education. This level of education in the older cohort was an anomaly, and that adult used Tok Pisin quantifiers exclusively in the trade store.

Summary

These findings show a remarkable heterogeneity of language forms used for quantification. Though indigenous forms and Tok Pisin forms are used side by side, the trajectory of change is toward increased use of Tok Pisin, a language associated with school participation. This heterogeneity of language forms complements the findings of Study 4-2, which showed that elders displayed consistent knowledge of the traditional counting system and limited knowledge of Tok Pisin count words. At the same time, the younger schooled population showed a weak knowledge of the body-counting system and strong knowledge of Tok Pisin count words.

Study 4-4. Emergent Arithmetic in Store Transactions in 2001

I mentioned in the introduction of the previous study (Study 4-3) that Indigo and I met with difficulties in producing systematic observations of clerk-customer transactions in trade stores, and hence we made use of the clerks' record-keeping to document language of quantification. But those data, while revealing, were limited in detailing the cognitive work in which customer and clerk jointly engaged in the back-and-forth of exchanges. To provide additional data that would bear on the way clerk and customer generated and accomplished the collective problems of exchanges, we developed an additional interview procedure.

In the procedure, Indigo and I presented the clerks with scenarios of customers entering clerks' stores. In the scenarios, the customer would request particular items and present the clerk with currency for payment of the price. We then queried the clerks about what would likely ensue in the transaction, based upon their prior experience. To reveal ways in which exchanges and emergent problems might vary as a function of customer cohort, we systematically varied who the customer was in the scenario. In one case, the clerk was told that the customer was an elder; in the other, the clerk was told that the customer was a schooled adult. Further, in each customer condition we included two scenarios, one in which the customer presented exact change, and in the other, change was required. The expectation was that we would find systematic differences when clerks were interviewed about the elder versus the schooled adult as customer. These differences would be manifested in the different ways the clerk described interactions in the exact change presented and change required scenarios.

Methods

Participants and Location

The nine trade store clerks who participated in the previous study (Study 4-3) were interviewed again for the present study. Each interview was conducted in each clerk's respective store.

Procedure

Clerks were each told each of the four stories. As noted, two stories involved an elder as a customer and two involved a schooled adult as a customer. In each story, the narrative began with the customer arriving at the store and purchasing the same two items. In the "exact change not possible" condition the customer (either an elder or a schooled adult) had only a K5 note (shown to the clerk). In the "exact change possible" condition, the customer had 5 kina, but in change (also shown to the clerk). Regardless of condition, the focus of the interview following each story was on how a typical purchase was accomplished by a customer (e.g., language used, how payment was made and how return of change was accomplished). The order of the stories was counterbalanced: For one half of the interviews, we began with the scenario involving the elder, and for the other half, we began with the scenario involving the schooled adult. In both scenario types, the story involving the "exact change not possible" condition was followed by the "exact change possible" condition. The variation in the structure of the four story conditions is depicted in Figure 52.

To make the interviews with clerks further correspond to their actual experience with customers, Indigo and I chose for the stories items that each individual clerk had in stock, located on the clerk's shelves. Though

Exact change possible?	Customer in Scenario: Elder	Customer in Scenario: Schooled Adult
No (K5 note presented)	Story 1	Story 2
Yes (Multiple lesser denominations presented)	Story 3	Story 4

Figure 52. Critical elements of the four stories presented to trade store clerks: exact change possible? and cohort of the customer.

this meant that the particular interview materials would vary slightly across clerks, we used several criteria to ensure comparability of items across interviews: For every clerk two items were selected that together cost more than K4 and cost less than K5; further, the price of the selected items had to be payable with tokens available in the multidenomination (exact change) condition. Throughout the interview, customers (elder, schooled adult) were portrayed as wanting to buy the same two selected items.

Story 1: Exact Change Not Possible (Single Denomination [K5 Bill]), Community ELDER. We used the following line of questioning for this condition:

> Suppose an elderly person came to your store, and s/he had this amount of money.

One K5 note was placed in front of the clerk, and the clerk was asked to suppose that the elderly customer wanted to buy the two items that had been selected.

> What does the elderly person do first when they enter your store?... Do they say anything to you? ... What?... What do you do?... What do you say? ... What would s/he do next? ...

This line of questioning was repeated with appropriate probes until a narrative of the exchange was completed.

Story 2: Exact Change Not Possible (Single Denomination [K5 Bill], Schooled ADULT. The same procedure was used as that for the community elder, single-denomination condition. However, "schooled adult" is substituted for a "community elder."

Story 3: Exact Change Possible (Multidenomination), Community ELDER. The same procedure was used as that for the community elder single denominations condition. However, in this case we laid out one 2-kina note (K2), six 10t coins (K.60), two 20t coins (K.40), two 50t coins (K1), and one 1K coin (K1) to represent the money carried by the community elder. The total of the money laid out was 5 kina.

Story 4: Exact Change Possible (Multidenomination), Schooled ADULT. The same procedure was used as that for the community elder, multi-denomination procedure. However, "schooled adult" was substituted for a "community elder."

Results

Language Used for Number

In all the interviews, we asked participants what language the customer of the targeted cohort would typically use to express numbers: would it be Oksapmin language, Tok Pisin, or both? Answers corroborated the results of Study 4-3. All participants but one stated that they expected elderly customers to use only Oksapmin language, regardless of whether or not they had a single denomination (a K5 note) or multiple denominations adding up to 5 kina. (The one clerk who stated that he expected elders to use Tok Pisin was a young high school student, who said he did not know how to express numbers and prices in Oksapmin language. In order to communicate with this clerk, the elderly customers would have to use Tok Pisin.)

The clerks' judgments about language use for schooled adults also corroborated the findings of Study 4-3. Clerks expected their schooled adult customers to use predominantly Tok Pisin, or a mix of Tok Pisin and Oksapmin language.

Structure of Transactions

In the interview narratives we found that clerks reported that individuals accomplished a purchase in two ways: They would either (a) purchase both items in a single transaction or (b) purchase items sequentially, first purchasing one item and then purchasing another. Both of these transaction

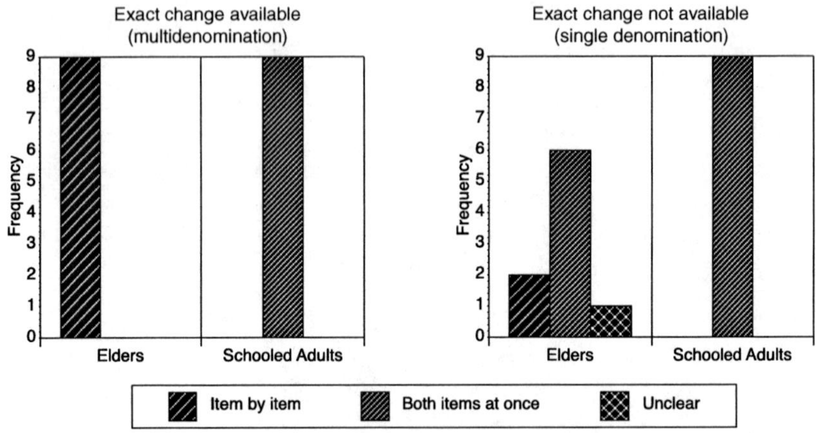

Figure 53. Frequencies of payment structures reported by clerks as a function of currency condition and customer cohort.

types were reflected in the analysis of trade store diary records from Study 4-3.

Frequency of reported payment structures is reported in Figure 53. These results are presented as a function of customer cohort and currency condition. For the schooled adult customer in both currency conditions, all of the clerks (nine of nine) described the transaction structure of "both items at once." This transaction structure required the production of a sum of the prices of the two items. For the elders, in marked contrast, transaction type varied over currency condition; all of the clerks reported for the elders transactions of the "one item at a time" type in the "exact change possible" condition. However, in the "exact change not possible" condition, reports were less consistent. Some clerks reported "item by item" purchases by elders in which change was returned for each item separately; others reported "both items at once" purchases; one narrative description was unclear.

Clerks' accounts of the transactions have implications for the kinds of numerical problems that would emerge for clerk and customer in producing payment and calculating change. When exact change was not available, all clerks reported that customers were more likely to buy both items in a single transaction with a single overpayment and change (though there were a number of exceptions in reports about elders). However, when exact change was available, clerks reported that schooled adults were likely to purchase both items at once, whereas elders were likely to purchase items one at a time.

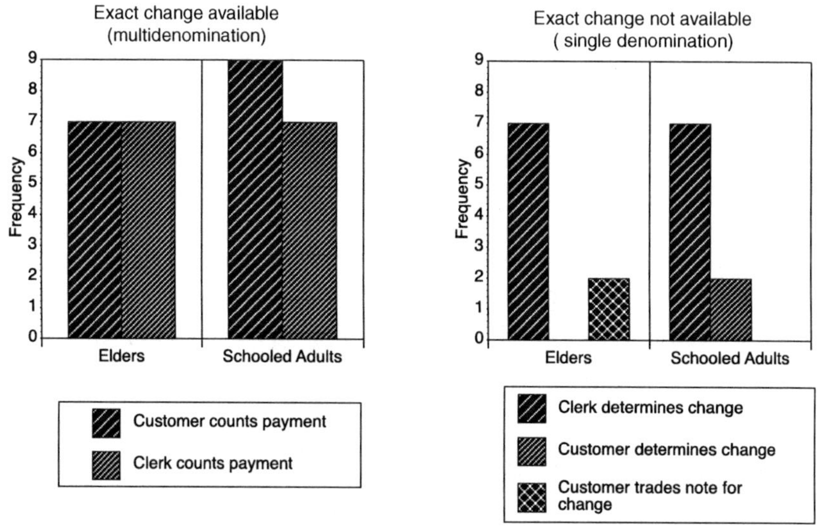

Figure 54. Frequencies of payment and change types reported by clerks as a function of currency condition and customer cohort.

Approaches to Quantifying Payment and Change

To understand the kinds of arithmetical problem solving that might emerge in transactions, we focused on clerks' reports of customers' engagements in making payments and determining change. I was particularly interested in the different problems that might emerge for customers as a function of their cohort. For example, did trade store clerks typically report that they performed all the arithmetical work involved in a transaction, or did their customers solve some of the emergent problems? If customers did participate in the solution of arithmetical problems, did participation differ as a function of customer cohort according to clerks? Data for reports on typical schooled adults and unschooled elders are presented in Figure 54.

Exact Change Available. When exact change was available for the purchase, clerks reported that both customer cohorts did some of the arithmetical work in the purchases. All of the clerks reported that schooled adults typically counted their payment (typically for both items) before presenting it. Seven of nine clerks reported that elders would typically count out their payment (recall that elders' practice was to purchase a single item at a time).

Exact Change NOT Available. When customers did not have exact change, clerks were cognizant that customers had a number of choices. Customers could either ask for change and then pay for the items separately, or could simply pay with the K5 note; these choices led to different kinds of arithmetical problems. If customers exchanged a note for change, they created the equivalent of the exact change possible condition. Two of nine clerk participants reported that community elders often asked for change when they entered a store. In contrast, no clerks reported that schooled adults would ask for change for their note before making payment.

When a customer pays for items with a single K5 note, the clerk must return change. Clerks reported that they were the ones most likely to do the arithmetical work of determining change. For both elderly and schooled adult customers, seven of nine clerks reported that they themselves determined the change. The remaining clerks reported that the schooled adult customers most likely determined the amount of change for themselves; however, as noted previously, two of the clerks reported that elderly customers would exchange their bill first in order to pay with exact change.

Notes and a Summary

Problems with conducting an observational study of clerk-customer interactions were pronounced in the trade store setting. Indigo's and my presence appeared to influence what occurred, and as a result, we all but abandoned the initial observational approach, reserving real-time observations to informal notes when we ourselves happened to be in stores. We pursued an alternative technique to systematic observation, asking clerks for prototypical behavior of customers in transactions. To be sure, there are clear threats to validity of the method we used. Asking clerks to abstract from a multitude of exchanges to report a particular pattern of interaction for a category of customers must have been a challenging task for the clerks, and there was considerable room for error. Whether the interviews captured key elements about emergent problems in exchanges and the way these problems might have varied over customer cohort merit scrutiny.

From the outset, I was concerned about threats to validity using the interview method, and we made various efforts to mitigate validity threats. First, we endeavored to make the interview protocol mirror structural elements of emergent clerk-customer transaction problems. To this end, we selected commodities for our interview tasks that were actually found in each of the clerks' stores, and we used tasks that were typical, linked to the kinds of exchanges observed in everyday transactions. Second, we triangulated these results with those of other studies.

For instance, in Study 4-3, I reported structural features of transactions involving exchange of currency for commodities. We made sure that the problem types we would use in interviews with clerks were represented in the documented exchanges, and the findings produced in the present study were consistent with the findings reported in Study 4-3. Third, the interviews were conducted with nine clerks, and the findings across clerks were quite consistent, corroborating the appropriateness of any individual clerk's description.

Despite the fact that the reports were removed from direct observation, the results produced by the technique offer a useful perspective on the character of emergent problems and goals that individuals structure in trade store exchanges. Even though the nominal exchange problems were identical, customers from different cohorts appeared to be engaged with different numerical and arithmetic goals as they conceptualized and accomplished exchanges. In their roles as customers, elders and schooled adults made choices of various kinds that have implications for the kind of numerical problems that arise in transactions and hence the numerical goals and means they construct to accomplish them. These choices include: (a) purchasing items sequentially as opposed to single items, (b) trading larger denominations for smaller denominations prior to a purchase, and (c) choosing to engage in arithmetical operations themselves or leave this work to the clerk. In the process of give and take in store transactions, clerks became brokers for diverse members of the Oksapmin community, enabling wide variation in the mathematics that customers accomplish in transactions.

Notes across Studies

In the four studies of this chapter, my focus has been on Oksapmin body-part and Tok Pisin representational forms for number along with their quantitative functions as forms are used in collective practices of economic exchange. The findings from the studies are rich with information for synchronic and diachronic analyses of the reproduction and alteration of these representational forms and the functions they serve.

Synchronic Uniformities and Variation in Form-Function Relations

As a group, the studies in this chapter reveal *synchronic* uniformity and variation in number representations. Study 4-1, conducted in 1980, focused on uniformity and variation in the microgenetic constructions that individuals produced as they used body counting to solve arithmetical problems. On the arithmetic problems with coins present, most people extended their use of the body system, using body-counting strategies to establish one-to-one correspondence relations between body parts and coins. But when coins were not present to count, a problem type linked to exchange with currency, there were marked variations across cohorts in how people solved the arithmetical problems. Cohorts with greater participation in the money economy constructed correspondence relations internal to the body system to keep track of the addition or subtraction of numbers in their use of double-enumeration, body-substitution, and halved-body strategies. Cohorts with less participation in the money economy more frequently used global-enumeration strategies, and thus tended not to construct an effective means to keep track of the addition or subtraction of terms.

Conducted 21 years later, the 2001 studies also revealed considerable synchronic uniformity and variability in the population. Study 4-2 revealed that most of the population across cohorts showed knowledge of both body-part and Tok Pisin forms of number representation. But there was also substantial variation. Elders showed greater knowledge of the body-part counting system, while schooled and younger populations showed greater knowledge of Tok Pisin numeration. There was also variation in what was taken as the canonical body system. Variations included differences as to the stopping point of the system (the 20th or 27th position), the body part following a complete cycle, and the trajectory for enumerating the last seven body parts.

Diachronic Continuities and Discontinuities in Form-Function Relations

From a *diachronic* perspective, the studies document continuity and discontinuity over historical time in forms and functions of numerical representations. From 1978 through 2001, there was continuity in the use of the body form to serve counting functions. In 1978 and 1980, people's use of body counting to produce cardinal and ordinal values was common, as I documented in Study 4-1 (1980) in the coins-present arithmetic conditions: All cohorts used the body system to count and produce cardinal representations of coins. In 2001 people's knowledge of the body system was again documented in Study 4-2 (conducted in 2001), in which most of the cohorts sampled displayed fluent knowledge of the system. Another illustration of continuity is related to the stopping point at the elbow on the other side (20) that was just emerging in 1978 – I noted that Yangeh and perhaps other trade store owners identified the 20th position as a "pound." In 2001, for many the elbow (20) had become the point at which the body system was truncated (Study 4-2).

But discontinuities in knowledge and use of the body and Tok Pisin representations were also marked between 1978/1980 and 2001. The 1980 synchronic variability in use of body-part counting to solve arithmetical problems (Study 4-1) may reflect ongoing processes of diachronic change. For example, the coins-absent condition allowed insight into ways that people were extending the body to solve problems that emerged in trade store transactions when coins were not present to count, leading to new functions for the system associated with economic exchange. Other discontinuities were reflected in 2001 through Study 4-2's analysis of the varied endings of the system and the common use of elbow (20) as the stopping point.

Studies 4-3 and 4-4 shed some light on supports in the community for the interplay between continuity and discontinuity through diachronic time. In transactions in trade stores, there were many ways to accomplish successful transactions through use of the body system or Tok Pisin. As a customer, one need not produce an arithmetic computation – indeed the cognitive work to accomplish these problems could be interactively accomplished in joint activity with the store clerk. Yet, by participating in stores, individuals had opportunities to engage in new kinds of collective problems as well as engage in negotiations with others who had generated new uses for cultural forms of representation.

In this chapter, I have documented form-function relations in the microgenesis and sociogenesis of body-part and Tok Pisin representations for number linked to economic exchange. I have written little about the semiotic forms people use to represent the objects of these exchanges – the currency tokens themselves. Another strand of shifting relations between semiotic forms and the functions they serve central to collective practices of exchange is the way in which people name currency tokens and currency values in communication and arithmetical problem solving. I take up this focus in Chapter 5.

5 Reproduction and Alteration in Currency Token Representations

Recall that in Chapter 4, I described my serendipitous reunion in Tabubil in 2001 with Yangeh and others, as Indigo, Josh, Cliff, and I were en route to Oksapmin. In the context of the back-and-forth conversation with Yangeh in his Ok Tedi mine dormitory room, the conversation moved to topics related to the representation of currency, a conversation that would come to seed the two studies described in this chapter. The first concerns the emergence, reproduction, and alteration of representational forms referring to coins and currency notes of varying denominations (Study 5-1); the second concerns the shifting functions these names have served in recent Oksapmin social history (Study 5-2).

Two themes stood out for me in that Tabubil talk with Yangeh related to currency representations. The first concerned the representation of currency tokens and token values. In the conversation, I was reminded of my long ago impressions in 1978 and 1980 about the Oksapmin names for tokens and their motivated character. Unlike words that bear an arbitrary relation to their referents (like "dog" to a dog), many of the Oksapmin names for currency tokens are linked to the objects to which they refer. The symbol was in a certain sense "motivated" by its referent, an idea related to Werner and Kaplan's (1963) treatment of symbol formation. In the Oksapmin language, for example, *kwekor* refers to coins, and its literal meaning is the "sound of rustling stones," similar to the sound of jingling coins. The name for the current 2-kina note (K2) is *iratarin* or *haben*, both of which refer to a particular kind of leaf; a currency note bears some resemblance to a leaf in texture, size, and shape. *Gangasi*, the name for the 50t coin, literally means a thing with points, and this seven-sided coin has pointed edges. Other names have more cryptic links that I explore later in the chapter. For example, *yota*, the conversational number word for two, and *gamantan*, the literal meaning of which is "married-flat," both are used to refer to the 20t coin.[1] How did these names come to be used by Yangeh? A related topic that piqued my interest was Yangeh's use of Tok Pisin names for the same tokens, names that showed no motivated character with their referents. What was particularly fascinating was that when Yangeh used Tok Pisin, he appeared to use different units for the worth of tokens than when he used the Oksapmin language. For example he referred to the 10t coin as "ten" in Tok Pisin but "thumb (1)" in Oksapmin, a tenfold difference in the face value of these representations. Were these ways of referring to tokens and their monetary values unique to Yangeh? I didn't think so. My expectation was that the names he (and others in Tabubil) used indexed a widespread and complex sociogenetic process at work.

The second theme in the conversation had to do with arithmetic using multidenominational currency tokens. When Yangeh engaged in arithmetic to add and subtract multidenominational values, whether using the Oksapmin language or Tok Pisin, his solutions seemed remarkable. Admittedly, I could not follow many of his computations well, given that it had been many years since my prior visit to Oksapmin. But he was fluent with the two language systems and fluently moved across the different units across systems.

Several days after the conversations with Yangeh, Josh, Cliff, Indigo, and I arrived in Oksapmin. The questions about genetic processes in names for and computations with tokens that the conversation with Yangeh had seeded became more refined through discussions with people. Indigo and I piloted and refined interview approaches, modifying candidate tasks to maximize insight into variations in microgenetic constructions linked to problems of daily life. Further, as in the studies of Chapter 4, we made use of cross-cohort designs that would provide data to support synchronic and diachronic analyses. The cohorts targeted included elders, unschooled middle-aged adults, and schooled middle-aged adults. The elder cohort provided a perspective on earlier naming practices whereas the schooled cohort provided a window into more recently emerging practices. The unschooled middle-aged cohort provided a link between the two extremes. In addition, to support a diachronic analysis, I was able to anchor the 2001 data in earlier 1978 and 1980 observations.

[1] As noted later, adults and elders reported to me that the 20t coin was a marriage of two shillings.

Study 5-1. Shifting Ways of Naming Tokens

As I mentioned in Chapter 3, Patrol Officer Hicks reported in 1962 that tokens did not catch on as a medium of exchange until about 25 years after the Hagen-Sepik patrol made first contact in 1938 (Officer Hicks' report). It was about this time that a trade store was opened at the Oksapmin Patrol Station and the conditions emerged for tokens to take on exchange value. But I could find no records of naming practices for tokens in Officer Hicks' or other reports from the era, and no reports on naming practices in prior eras. It may have been that early in their circulation, tokens would travel from one person to another without being named, perhaps passed as curious objects. But with the presence of tokens in the community, the conditions were ripe for the emergence and propagation of two ways of naming that are evident in language use today. One way of naming involved the innovative use of the Oksapmin language itself, as people at different sites engaged with coordination problems, seeking to orient their interlocutors to their communicative intents. The other way of naming involved the alien languages of English and Tok Pisin, as people adopted the naming practices of foreigners, whether patrol officers, missionaries, school teachers, or others.

Some Preliminary Notes on Naming Tokens in the Oksapmin Language

What might be the origins of names for tokens in the Oksapmin language that referred to physical features of their referents? I considered the possible relevance of research in developmental psychology that identifies a lack of differentiation between names for things and things themselves as an early stage in a developmental sequence in symbol formation. For example, in their writings, Piaget (1970), Vygotsky (1978, 1986), and Werner (1963) noted that children often do not appreciate the arbitrary relation between names and their referents. Thus, in the case of child development, a train is "choo-choo," an onomatopoetic mimicking of its sound – what Werner and Kaplan (1963) refer to as a "mimetic facsimile" (p. 47). A dog may not be called "cat," because "a cat is a cat, not a dog." Extending such developmental formulations to traditional groups, some researchers have claimed that similar patterns of naming by adults are evidence of a concrete mentality, perhaps characteristic of "primitive" societies (e.g., Hallpike, 1979; Levy-Bruhl, 1979; Luria, 1976; Tylor, 1920). Thus, the motivated relation between Oksapmin names and the objects to which they refer in the Oksapmin language could be taken as a "symptom" of such a lack of symbol-referent distance, a limitation associated with a "concrete mind." But from the standpoint I develop here, the assumption of a concrete mentality can explain little about the dynamic processes whereby names emerge and propagate in a community. Following the remarks on sociogenesis as well as Werner and Kaplan's (1963) treatment of addressor-addressee relations in symbol formation (presented in Chapter 1), I pursue a different line of argumentation.

I take names for tokens as likely to have emerged from people's efforts to solve coordination problems – that is, of coordinating actions and semiotic references with one another (Clark, 1996; Lewis, 1969) in face-to-face exchanges. In the case of economic exchange, when two interlocutors use the same word forms to refer to tokens in the community, they signal their intended meanings through the use of these common words, and communication may appear seamless; they simply reproduce prior expressions in ways that they have used before. But when the tokens have no names, coordination problems lack solutions based on precedent, and speakers must improvise in the flow of interactions.

Sometimes the solution to a coordination problem may be a simple gesture, like the deictic pointing to an object for which there is no common name. Other times, a speaker may draw upon words conventionally used to serve other functions, recruiting them as tropes to communicate their intended meanings for objects with no common names. Such tropes may involve reference to physical qualities that orient an interlocutor's attentional focus – examples include "hole," to refer to a K1 coin (which has a hole at its center), or "pointed" to refer to a 50t coin (which is heptagonal, with seven points). Tropes of these sorts would certainly be reasonable strategic communicative moves in the early penetration of currency tokens into the community. Other times, tropes may communicate the numerical exchange value of a coin with reference to prior and known tokens: For instance, there is the use of *tit* (conversational word for "one") for a 10t coin or *kat-hai* (shoulder (10)) for a K1 coin, with each referring to units of the now obsolete shilling, to communicate a reference of numerical currency value of the respective tokens. To the extent to which tropes are taken up by others and propagate (and perhaps undergo processes of specialization) in communicative acts, they may account for the motivated character of the collective forms for representing tokens.

These conjectures imply a twofold process in communicative and problem-solving activity, one intended by interlocutors and the other not. First, the intended: in the process of reproducing and altering forms to accomplish communicative and problem-solving functions, people are working to communicate and accomplish local goals, such as to refer to a currency token or to produce a summation of token values in public displays. At the same time, however, as tropes are reproduced by others, people are unwittingly participants in building collective representational forms adjusted to the extant problems of their times, whether these systems involve body-part counting forms, the subject of the last chapter, or forms for referring to tokens, the focus of the present chapter.

Some Preliminary Notes on Naming Tokens in Tok Pisin

In contrast to the Oksapmin language, in which there were no ready-made names for currency tokens, Tok Pisin has a system of token names entrenched in Western institutions and in a written language supporting a certain durability to these names. Tok Pisin token names are generated using a Tok Pisin number word that modifies a unit of kina or toea. Thus, in Tok Pisin, names for coins include *one toea*, *two toea*, *five toea*, and so on, and names for notes include *two kina*, *five kina*, *ten kina*, et cetera.[2] The missionaries, patrol officers, and later trade storeowners, who were the principal possessors of the tokens and commodities, supported a hegemony in use of these names. Indeed, they were the holders of commodities and the tokens themselves.

The penetration of the Tok Pisin language and currency-token names into the Oksapmin community has been slow, due perhaps to the remoteness of the Oksapmin valleys. On my first visit there in 1978, Tok Pisin and English were used by only a small segment of the population. Principal users were those who had traveled out of the area for work or those who attended school. By 2001, with the increase in schooling and travel by some to Tabubil and other locations outside of Oksapmin, people's knowledge of Tok Pisin and English had increased substantially. As in the case of the analysis of token naming in the Oksapmin language, I viewed an analysis of variation and uniformity of Tok Pisin token names as an additional opportunity to explore sociogenetic processes.

My purpose in this study is to explore uniformity and variation in names for tokens used across different segments of the Oksapmin community. The contrast between the origins of names in the Oksapmin language and in Tok Pisin makes for an extraordinary opportunity for providing data that bear on the sociogenesis of naming from both synchronic and diachronic perspectives.

[2] Though the Tok Pisin spelling for these names differs from English, the Tok Pisin pronunciation of these names is similar to their pronunciation in English.

Methods

To analyze naming practices in the Oksapmin and Tok Pisin languages with an eye towards understanding sociogenetic processes, Indigo and I queried people about names for each of the eight principal currency tokens in circulation in 2001, first in the Oksapmin language and then in Tok Pisin.

Participants. Fifty-seven people participated in this study, including 18 elders, 22 unschooled adults, and 17 schooled adults (schooling ranged from Grade 5 through Grade 10). The participants lived in hamlets close to the Bak Community School. We contacted potential participants via local informants and through the school staff.

A procedure similar to that reported in Chapter 4 was used to assign participants to the cohort categories. In this procedure, Indigo and I asked participants whether they were closer in age to one or the other of two well-known individuals in the community, one whom we regarded as elderly and the other, as middle aged. If participants indicated that they were closer in age to the elder, they were assigned to the elder category. If the participants indicated that they were closer in age to the middle-aged adult, they were assigned to either the unschooled-adult or schooled-adult category as a function of their schooling history. Both unschooled and schooled adults had similar levels of participation in economic exchange with currency.

Interview Procedures. Prior to the study the interview procedures were back-translated into the Oksapmin language and Tok Pisin by people fluent in each language. Indigo and I interviewed all participants through a translator; we spoke in Tok Pisin and the translator spoke in the Oksapmin language.

CURRENCY IDENTIFICATION TASKS IN THE OKSAPMIN LANGUAGE AND IN TOK PISIN. Each participant was presented with eight different currency tokens one by one and asked to name them in the Oksapmin language. We then repeated the procedure, asking for the names in Tok Pisin. The currency tokens included coins with values of 10 toea, 20 toea, 50 toea, and 1 kina, and notes with values of 2 kina, 5 kina, 10 kina, and 20 kina.[3] We recorded all responses in writing; we also videotaped a subset of these interviews.

[3] Appendix B (Figures 141 and 142) contains photographs of currency tokens used in the studies referred to throughout the volume. The figures include photographs of the reverse and obverse sides of tokens, as well as notes on color, dimensions, and date of issue.

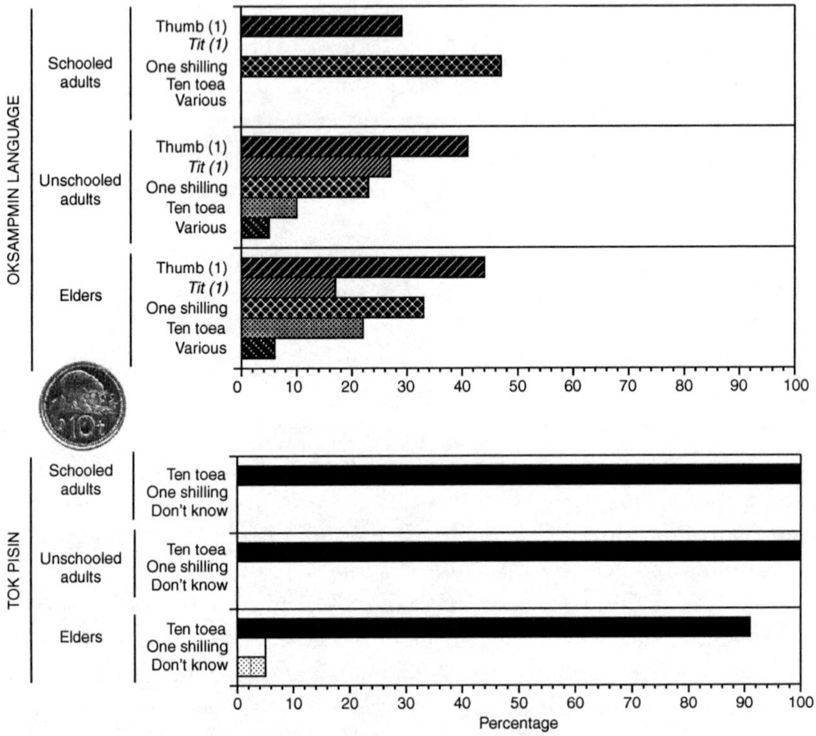

Figure 55. Percentage distributions of names used for the 10t coin in the Oksapmin language and Tok Pisin by cohort (2001).

Results

Naming Coins. Figures 55 through 62 contain the percentage distributions of names that participants provided for each of the currency tokens. Figures 55 through 58 contain distributions for coins (10t, 20t, 50t, K1), and Figures 59 through 62 contain distributions for (K2, K5, K10, K20). To better interpret the figures, it will be useful to note several points.

First, the upper distribution in each of these figures shows the Oksapmin names people used for tokens as well as the percentage of speakers in each cohort who used the listed name. The lower distribution does the same for Tok Pisin. For example, the upper distribution in Figure 55 shows that for the 10t coin, the elders most frequently used the Oksapmin expression for thumb (1) (translated from *tipana (1)*), whereas schooled adults most frequently used the expression "one shilling" (translated from Tok Pisin *wan*

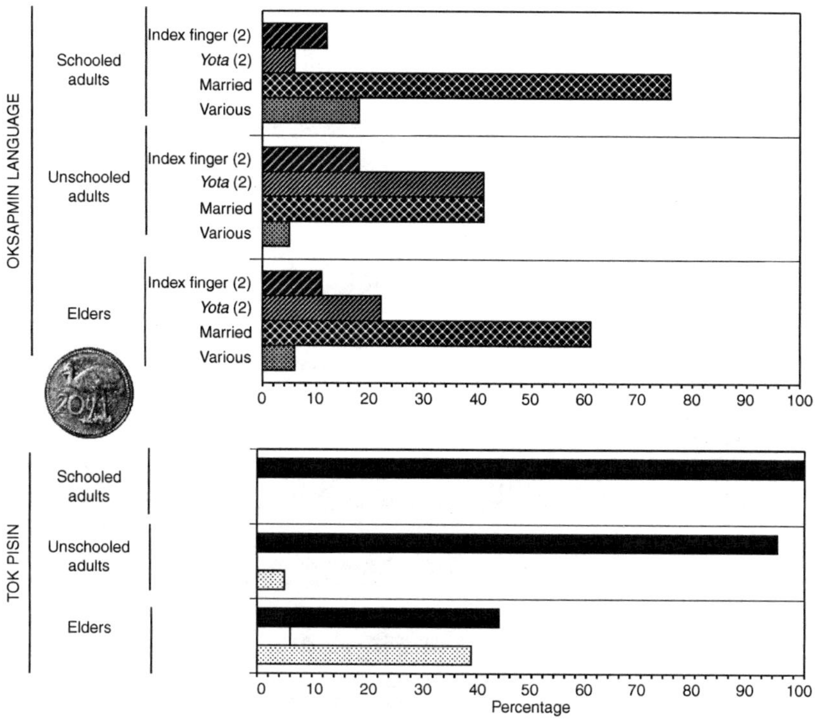

Figure 56. Percentage distributions of names used for the 20t coin in the Oksapmin language and Tok Pisin by cohort (2001).

siling). Note, as one might expect, there is considerable overlap across these two cohorts in the names used for the 10t coin. Some schooled adults used "thumb (1)" and some elders used "one shilling."[4]

Second, two entries recur throughout the distributions presented in Figures 55 through 62 and these are usefully noted here. The term "various" was used to denote names that were idiosyncratic to one individual. Thus, if there is a high percentage of "various" in a cohort, then there was considerable idiosyncratic reference to the token in language use in the population. "Don't know" was recorded when an individual indicated that they did not know the name of the indicated token in the language of interview

[4] Note that the expression "shilling" (*siling*) has its origins in Tok Pisin, though people responded with this expression when they were queried to supply an Oksapmin name for tokens, evidence for a borrowing of the Tok Pisin expression into the Oksapmin lexicon. Note also that sometimes the same expression was used when queried for a Tok Pisin name for tokens (see lower distribution in Figure 55).

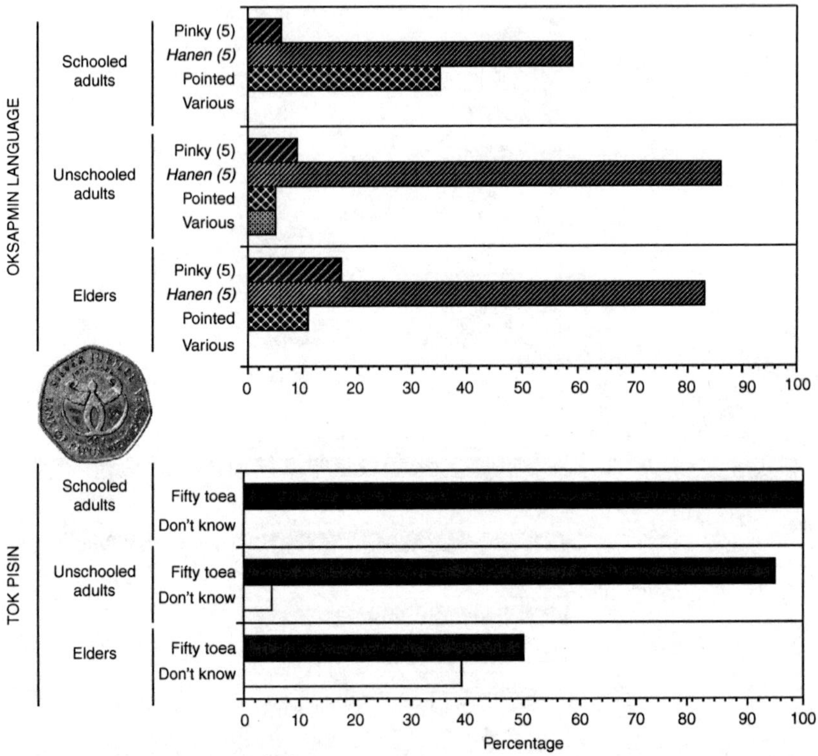

Figure 57. Percentage distributions of names used for the 50t coin in the Oksapmin language and Tok Pisin by cohort (2001).

focus. Thus, in the Oksapmin language, about 5 percent of the elders and unschooled adults used idiosyncratic names for the 10t coin (upper distribution, Figure 55), and about 5 percent of the elders indicated that they did not know the name for the 10t coin in Tok Pisin (lower distribution, Figure 55).

Across the figures, the reader may note marked similarities and differences. Across figures, for example, there are multiple names for the same token in the Oksapmin language. In contrast, in Tok Pisin, there is generally only a single name for any given token. Noteworthy in the Oksapmin names we find varied tropes, references to body parts, and conversational number words (such as *tit* (1), *yota* (2), *yetir* (3), *yota yota* (4), or *hanen* (5)).

Across Figures 55 through 62, a number of trends stand out. I point to these trends below, first with reference to names in the Oksapmin language and then to names in Tok Pisin.

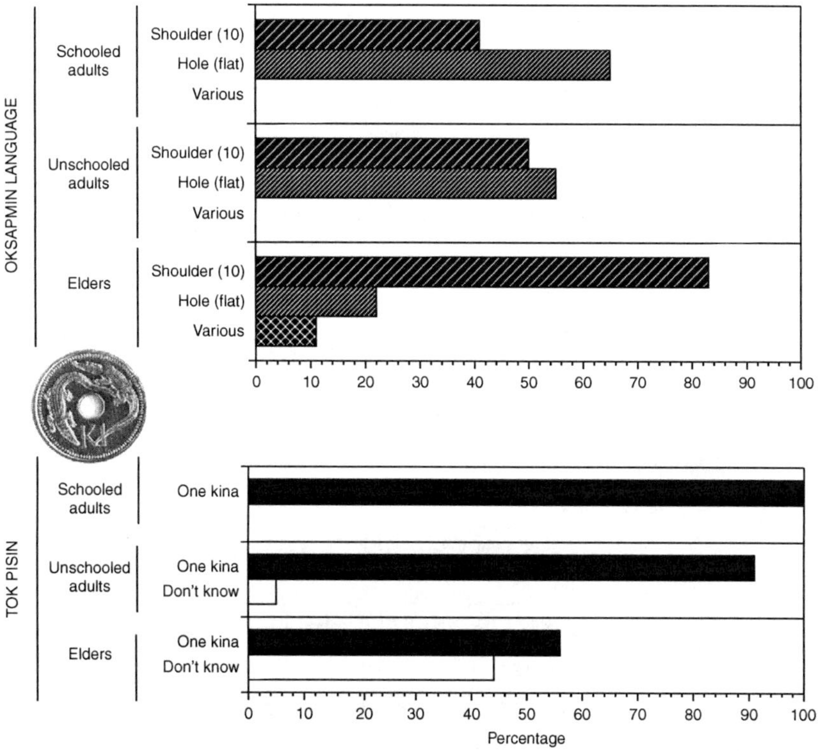

Figure 58. Percentage distributions of names used for the K1 coin in the Oksapmin language and Tok Pisin by cohort (2001).

The Oksapmin Language: Properties of Token Names

1. Numbers and Tropes. People used numerical expressions in the form of both references to body parts and the five conversational number words. They also made use of varied tropes, metaphors and other kinds of figural expressions to name tokens. For six of the eight tokens, at least some people from each cohort used a trope. These included *gamantan* (married) for 20t coin (a marriage between shillings, or 10t coins), *gangasi* (pointed) for the 50t coin, *temsi tana* (flat-with-hole) for the K1 coin, *iratarin* or *haben* (leaf) for the K2 note, and *imakaksi* (pig's head) for the K20 note.

2. Units. Many people named tokens with the Oksapmin number systems, using either body parts or the five conversational number words. But as we expected and in consonance with everyday practices, the body parts and

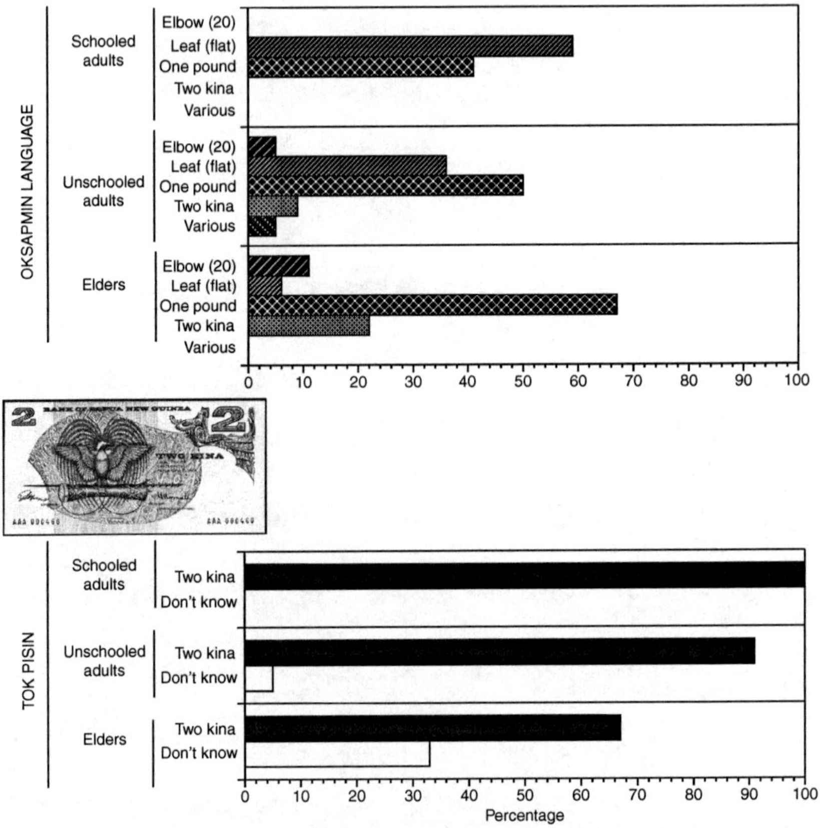

Figure 59. Percentage distributions of names used for the K2 note in the Oksapmin language and Tok Pisin by cohort (2001).

conversational numbers were not consistent with the Hindu-Arabic numerals inscribed on the tokens. Instead, the named numbers were many-to-one transformations of the inscribed numbers, the transformations reflecting a treatment of the 10t coin as a unit (the obsolete shilling equivalent) and the K2 note (the obsolete pound equivalent). Thus, the 10t coin was generally *not* referred to as shoulder (10), but instead as thumb (1) or *tit* (conversational number "one"). The 20t coin was treated as index finger (2) or *yota* (conversational number "two"), not elbow-on-the-other-side (20). The 50t coin was treated as pinky (5) or *banen* (conversational number "five") and the K1 coin (the value of ten 10t coins) was treated as shoulder (10). Recall that some even referred to the K2 note as elbow-on-the-other-side (20), the equivalent of the pound in units of shillings (or the equivalent of the K2 note in units of 10t coins.) For notes, the K2 note was commonly referred

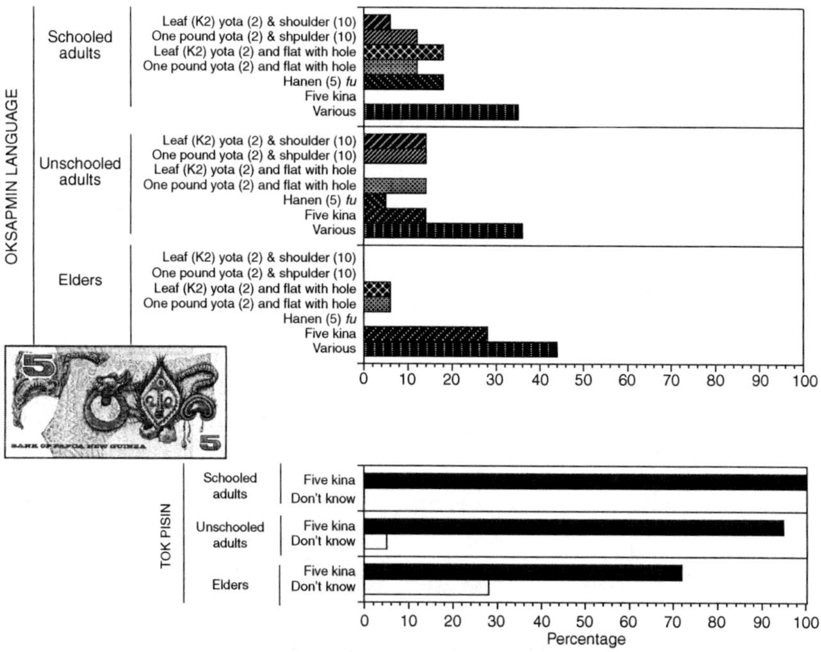

Figure 60. Percentage distributions of names used for the K5 note in the Oksapmin language and Tok Pisin by cohort (2001).

to as a pound, and the K5, K10, and K20 notes contained either explicit or tacit reference to the pound (or its kina equivalent, the K2 note).

People's names for notes in the Oksapmin language also made use of the expression *fu*. In current practice, *fu* is in an important sense a stand-in for the K2 note with a history rooted in the pound, and commonly occurs following a body-part expression or conversational number word. The use of *fu* is salient in Figure 61 and Figure 62 when referring to the K10 and K20 notes, respectively. To illustrate: *hanen fu*, a common expression for K10 across cohorts, can be understood as follows: *hanen* (5) is the conversational number word for five, and *fu* is the 2-kina equivalent; thus 5 *fus* is an expression for 10 kina.[5] On occasion, rather than stating *fu*, some elders

[5] The reader may recall that earlier I noted that *fu* is a word form traditionally exclaimed when one reaches the 27th body part – typically *fu* is an exclamation made with fists raised. More recently the word form has undergone varied shifts in meaning, with one meaning the equivalent of a pound (or K2 note), and another a doubling of the value of a body part. The sociogenetic analysis of the shifting functions of *fu* is presented in an article published several years after the fieldwork (Saxe & Esmonde, 2005), and is extended further in Chapter 6.

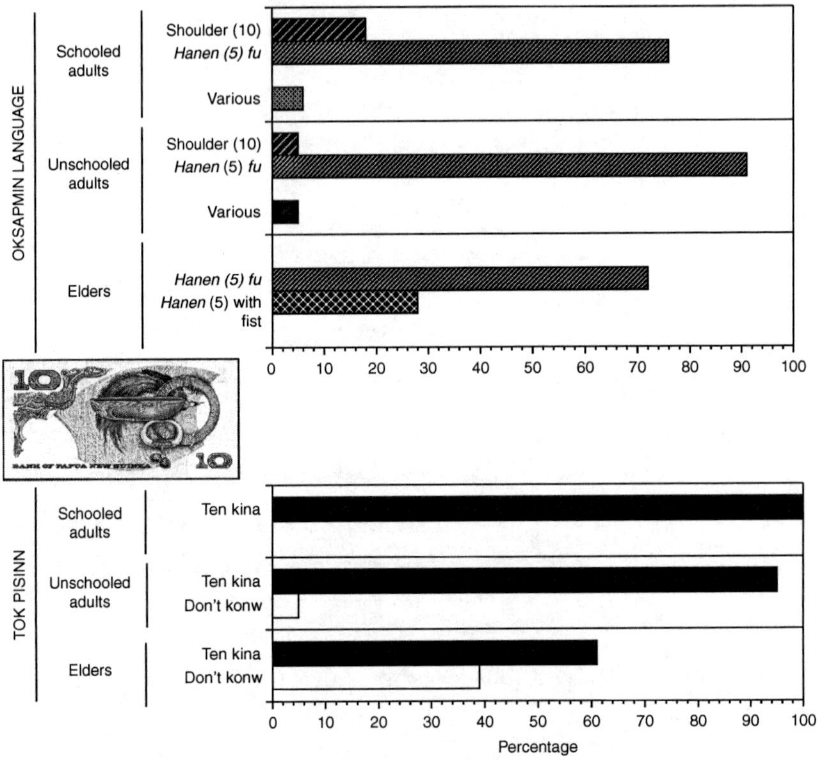

Figure 61. Percentage distributions of names used for the K10 note in the Oksapmin language and Tok Pisin by cohort (2001).

merely gestured with their fist, a gesture that accompanied the completion of a count to 27 and the explanation of *fu!* Finally, the most common response across all cohorts for the K20 note was shoulder (10), again followed with the suffix *fu*.

3. Cohort Similarities/Differences. Figures 55 through 62 reveal differences in naming practices across cohorts for currency tokens, perhaps linked to communicative networks in which people participated. One of the clearest cases of this is the K1 coin (Figure 58). Elders tended to treat the coin in terms of the body-part system (shoulder (10)), whereas schooled adults and to a slightly lesser extent unschooled adults tended to treat the coin with a trope – "flat with a hole." In the Oksapmin language, the only token for which there was a substantial degree of uniformity in usage of a single variant was the K10 note, *hanen* (5) *fu*. For this token, 70 percent, 90 percent, and 75 percent of the elders, unschooled adults, and schooled adults,

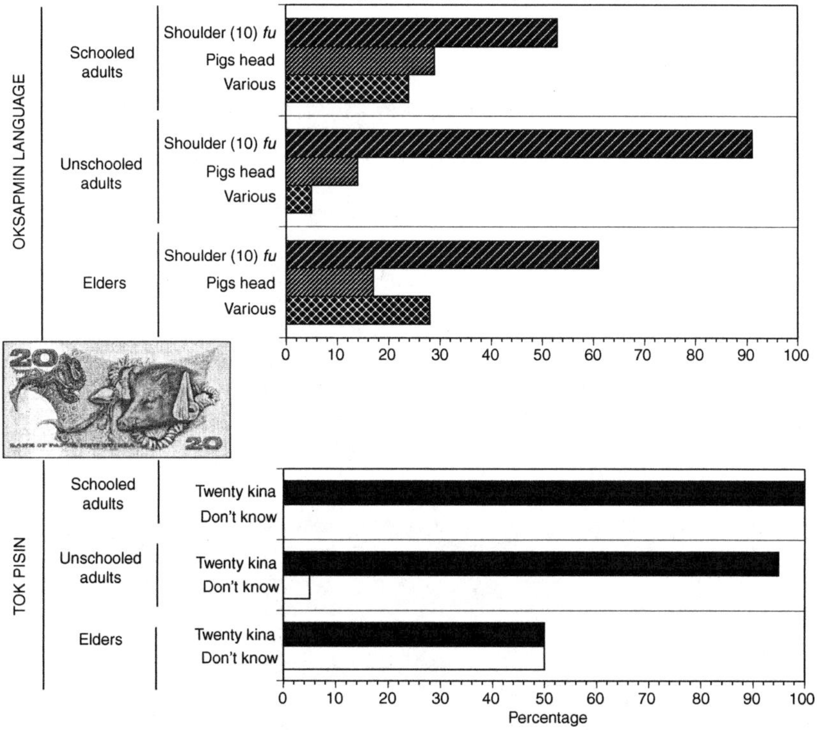

Figure 62. Percentage distributions of names used for the K20 note in the Oksapmin language and Tok Pisin by cohort (2001)

respectively, used the expression *hanen (5) fu*. Further, the elders who did not use this expression instead used a hybrid form, *hanen* (5) accompanied by a raised fist, the stereotypic gesture associated with the use of *fu*.

4. Variation in Names. Variation in names for any given token is marked across cohorts. People used various expressions for each token type in each cohort. For the K5 note, the number of variants is striking (see Figure 63), where the idiosyncratic category, "various," is the modal category for each cohort (45 percent, 35 percent, and 35 percent, for elders, unschooled adults, and schooled adults, respectively).

5. A Special Note on Hybrid Forms. People used tropes and combined elements of body parts and conversational numbers in new ways that differed from earlier uses. The results of these new uses for token reference are hybrid forms that have become or are in the process of becoming specialized registers to refer to currency tokens. The hybrid property of names for currency tokens is well manifested in the many ways that people referred

Oksapmin language	English translation
1. h^th^ta fu	Pinky (5) *fu* (kina)
2. h^th^t-hai temsi	Pinky (5), hole (1 kina coin)
3. haben tan tita haben tan tita kat-hai tita	Leaf (K2 note) leaf (k2 note), shoulder (10) (ten 1-shilling coins)
4. haben tan yota kat-hai gon tit	Leaf (K2 note) twice shoulder (10) (ten 1-shilling coins)
5. haben tan yota temsi tita	Leaf (K2 note) twice hole (1 kina coin) once
6. hanen fu	Five *fu* (kina)
7. hanen fu gon	Five *fu* (kina) all
8. hanen gona	Five all
9. hanen temsi moha	Five hole (K1 coin)
10. iratarin yota temsi	Leaf (K2 note – dialect variation) twice hole (K1 coin)
11. kat-hai hanen gona	Shoulder (10) (ten 1-shilling coins) five
12. kat-hai tita haben tan yota	Shoulder (1) (ten 1-shilling coins) leaf flat (K2 note) twice
13. kat-hai tita wan faun yota	Shoulder (10) (ten 1-shilling coins), one pound twice
14. moh yot fu mot kut	This two *fu* (K2 notes) then part-way to the next body part (meaning 1 more)
15. moh yota kat-hai	This two (K2 notes), shoulder (1) (ten 1-shilling coins)
16. moh yota moha	This two (K2 notes), and this
17. temsi tana wan faun yot tana	Hole flat (K1 coin) and one pound (K2) twice
18. wan faun moh yot tana kat-hai	One pound (K2 note) twice, shoulder (10) (ten 1-shilling coins)
19. wan faun tana wan faun tana kat-hai	One pound flat, one pound flat, shoulder (10) (ten 1-shilling coins)
20. wan faun tana wan faun tana temsi tana	One pound flat, one pound flat, hole flat (K1 coin)
21. wan faun yot temsi	One pound (K2 note) twice, hole (K1 coin)
22. wan faun yot temsi tan	One pound (K2 note) twice, hole flat (K1 coin)
23. wan faun yota kat-hai tit	One pound (K2 note) twice, shoulder (10) (ten 1-shiling coins)
24. wan faun yota moh kut	One pound (K2 note) twice, then par way to the next body part
25. yot fu hama tit-o kut	Two *fu* (K2 note) and once part way to next body part
26. yot tan hat forhan kat-hai tit-si	Twice flats (meaning K2 notes) and then shoulder (10) (ten 1-shilling coins)

Figure 63. Various ways of identifying the K5 note in Oksapmin.

to the K5 note (see Figure 63). Examples include "two flat leaves [2 pounds or 4 kina] and shoulder (10)," "two flat leaves and flat with hole," and "one pound *yota* (2) and flat with hole."

Tok Pisin: Properties of Token Names

Token naming in Tok Pisin was remarkably uniform. In general, people named a token in Tok Pisin by the numerical value of its worth (e.g., "ten," "twenty") by the unit the currency token ("toea" or "kina"). For example, when the 10t coin was presented, a participant's response was almost always "ten toea." When the K2 note was presented, the typical response was "two kina."

Though people used uniform Tok Pisin expressions for virtually all tokens, cohorts differed in their knowledge of the Tok Pisin language, though even some elders who claimed that they could not speak Tok Pisin knew some Tok Pisin token names. For the 10t coin, almost everyone in each cohort identified the 10t token as "10 toea." For the seven tokens worth more than 10 toea, 100 percent of schooled adults produced the same Tok Pisin expression for each token. Unschooled adults performed similarly, with the rate of the same Tok Pisin identifications ranged from 90 percent to 100 percent. The elders, showed less knowledge of Tok Pisin identifications: The percentage of elders who produced the same (appropriate) names for the tokens varied from about 45 percent to 90 percent with the median across tokens at 60 percent.

Population-Wide Properties of Token Names and Their Sociogenesis through Collective Practices of Exchange

This study provided a systematic documentation of names for currency tokens in the Oksapmin and Tok Pisin languages. Now I take a broader perspective, offering a sociogenetic analysis of naming practices with particular regard for the historical circumstances in which naming practices arose. Body-part and conversational numbers appeared to be used in irregular ways in the Oksapmin language, and these irregularities in usage may appear perplexing. For example, *hanen* (5) by itself means 50t, but when followed by *fu* refers to a K10 note. *Tipana* (1) by itself means a counted value of one but when used in conjunction with currency means 10t. Shoulder (10) when used with counted objects means a value of 10, though when used in conjunction with currency it means 1 kina; further, when followed by *fu*, shoulder (10) means 20 kina. Such irregularities in Oksapmin naming appear to have emerged with the rise of exchange practices. People, at different sites, used indigenous forms of representation to solve shifting problems of communication, problems made complex by the penetration of different currency systems through the community (e.g., pounds and shillings, the short-lived Australian dollar, the later issue of kina and toea). Such processes of rapid change may also account for the many variant names for tokens in the Oksapmin language. The number of variant names for the K5 note stands out in this regard. There was one variant in Tok Pisin, while there were 26 principal variants listed in Oksapmin (see Figure 63).

Why were there so many Oksapmin names for tokens? for the K5 note compared to others? Several hypotheses are plausible. One that I considered was that the K5 note had been in circulation for a shorter duration than the others, perhaps decreasing the opportunity for stabilization of names. Though reasonable, this hypothesis turns out not to be the case. All notes used in the study were distributed with the first issue of Papua New Guinea currency.

Another plausible hypothesis, for which there are corroborative data, is that people substituted the Tok Pisin form, "five kina," for an Oksapmin expression. The borrowing of the lexical form means that the Oksapmin expression may not have had an opportunity to stabilize. Support for the hypothesis comes from an analysis of elders' names for the K5 note. When elders were prompted for Tok Pisin expressions, they commonly used "five kina," unlike their fluency in Tok Pisin for other expressions.

But even if there had been an increased usage of the Tok Pisin expression for the K5 note across cohorts, one must ask, why the tremendous variation only for the K5 note? There is good reason to believe that the answer lies in the fact that the K5 note is the only note with an odd-numbered value (compared to the K2, K10, and K20 notes) and the odd number complicates its representation in the Oksapmin language. Consider that the odd number of kina (five) creates a structural problem: The root representation for notes is the pound-equivalent K2 note, or the value of 2 kina. When body parts or conversational numbers are used to modify the 2 kina, the result is that even-numbered notes (or currency values of kina) can be represented in straightforward ways (whether in the form of body parts, conversational numbers, or modifying pound equivalents), but the odd-numbered K5 note cannot. The proliferation of names for the K5 note appears linked to this property. This is evidenced by the host of variant compound expressions for the K5 note noted in Figure 63, in which we find different ways of adding the surplus value of 1 kina to the already variant ways of expressing 4 kina. Some of these ways of adding the value of 1 to the expression of 4 kina include a traditional word form, *kut*, which signals a part of something else. Here, a speaker indicates only partway to another body part (that is, an unnamed midway position in the body system). There are also ways of referring to a value of 1 kina by indexing the K1 token, like *temsi*, *temsi tana*, or ten 10t coins, like *kat–hai* (10).[6]

[6] The ascendance of the Tok Pisin expression "faiv kina" may also have slowed the winnowing down of variants for the K5 note. Indeed, as noted previously, some elders made use of the expression "five kina" both when asked for an Oksapmin name and when asked for a Tok Pisin name.

Preliminaries for a Sociogenetic Model

The propagation of token names in the Oksapmin and Tok Pisin languages would appear to have proceeded in opposite directions. Oksapmin names emerged from within Oksapmin communities and were reproduced and altered in collective practices of economic exchange. Tok Pisin names, in contrast, had already undergone a process of systematization outside the community and penetrated the community from the outside in.

The constraints on these propagation processes must have been different, for why else would properties of Oksapmin names differ from Tok Pisin names so markedly? An approach that I will elaborate subsequently distinguishes between two different models of sociogenetic propagation. The models provide a basis for understanding similarities and differences between Tok Pisin and Oksapmin token-naming practices. I refer to the two models as *local emergence*, which best fits the Oksapmin language case, and *broadcast*, which best fits the Tok Pisin language case. I sketch the basic properties of the models below in idealized form, appreciating that in reality, propagation processes display mixtures that may tend toward one model prototype or the other.

To illustrate the *local-emergence* model, let's consider Oksapmin language use and common names for tokens. Imagine the time when tokens first began to circulate in the community. During this period, people at multiple sites are engaged in interaction, referring to tokens in their communications with one another. The lack of names for these foreign objects in the Oksapmin language creates coordination problems between interlocutors as they try to orient one another to their communicative intents. Initially, solutions to these coordination problems are likely different across sites. But through time, as interlocutors move from one site to another and communicate with different people, a waxing of some ways of referring to tokens occurs based on precedent (what was successful previously) and a waning of other ways occurs. In this diachronic process, some names become privileged and people anticipate that particular communicative choices will be more successful than others, leading to some variants being used with greater frequency, and others dropping in use. Of course, the process is complex, and the production of variants and the ascendency of some to a privileged use may be relatively slow in taking shape.

To illustrate the *broadcast* model, let's consider the Tok Pisin language and the uptake of a stable set of names in the community. In the case of propagation of Tok Pisin names, some people were likely treated as having more authoritative knowledge about Tok Pisin – teachers, trade store owners or clerks, missionaries, and patrol officers – and deference was likely given to their Tok Pisin name choices. Further, among these people of social prominence, the use of Tok Pisin token names was relatively standard, the stability coming from their status as "brokers." They are brokers in the sense that they interact with institutions outside of Oksapmin in which Tok Pisin is entrenched and stable in various social institutions; when they interact with Oksapmin people, they then produce a more or less uniform broadcast of Tok Pisin names, a broadcast also consistent among the "brokers" themselves. Further, these people of social position in some sense are point sources, interacting with multiple Oksapmin people whether in classrooms, trade stores, churches, or patrol posts.

There is a lot more to be said about these two models of propagation, and I will discuss each one further in subsequent chapters. For now, I simply note that Oksapmin names for tokens such as *temsi* (hole), *kat-hai* (shoulder (10)), or *gamantan* (married), as well as Tok Pisin names, are vestiges of a history of interlocutors' solutions to communicative predicaments, and the names that people commonly use are vestiges of these solutions. As such, the names that people use may have little to do with the character of their arithmetical thinking or understanding. Indeed, we may be able to learn very little about the numerical and arithmetical thinking of people by studying the names that they use for tokens. In the next study, I turn to people's computational activity with tokens and analyze the shift in function of token names and other representational forms as individuals use them to represent currency values.

Study 5-2. Shifting Quantification Practices with Currency

Token names, whether in the form of body parts or tropes, are not only used in acts of denotative reference – to refer to tokens in communication. They also enter into computations and the representation of the value of groups of tokens. Thus, token names can become vehicles that support summations of token values, a function that goes beyond denotative reference.

To analyze people's use of token names in summations of token values, I designed tasks that would require people to sum the values of token arrays. In these tasks, Indigo and I presented a group of tokens, asking individuals how much they were worth altogether. Three kinds of conditions were used, and in each of them, people were asked to perform five separate tasks. Two condition sets contained currency tokens; one of these summed to relatively lesser values (five tasks that included values from K.40 to K2.90) and the other summed to relatively greater values (five tasks that included

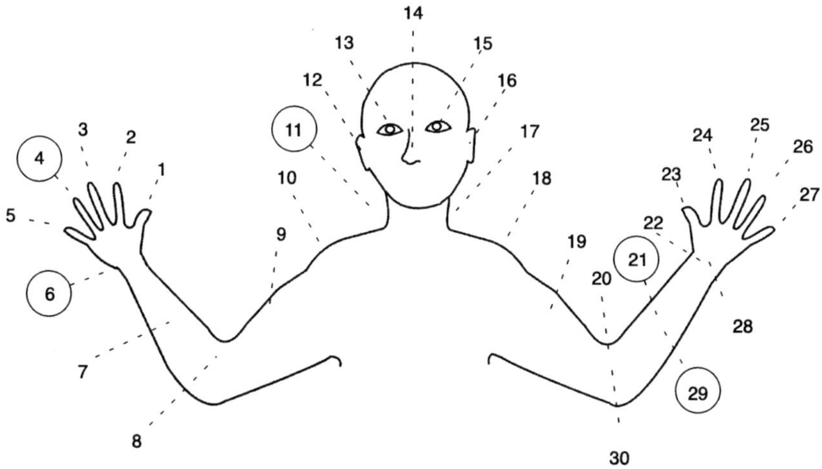

Figure 64. The body-part counting system and the selection of five root positions for summation tasks: ring finger (4), wrist (6), neck (11), forearm (21), and forearm (29).

values from K4 through K29). In the third condition set, stones rather than tokens were used in order to compare participants' solutions to the denominationally structured currency with items that had no denominational structure (five tasks that included values of from 4 stones to 29 stones).

For purposes of comparability, each condition set contained tasks with the same set of root numbers keyed to "transition points" in the body-count system – that is, 4, 6, 11, 21, and 29 (see Figure 64). These selected root numbers afforded a targeted sampling of people's use of representations in summations. The choice of values afforded an analysis of people's use of the conversational system (root number: 4); a lesser number but beyond the limit of five conversational number words (root number: 6); a value beyond the decimal structure of kina and toea (root number: 11); a value beyond the base-20 structure of Australian shillings and pounds (root number: 21); and a value beyond the end of the 27-body-part cycle (root number: 29).

The root numbers selected led to five tasks for each condition set, depicted in Figure 65. The tasks involving stones (noncurrency objects) asked individuals to determine the number of haphazardly arranged stones, each presented separately (4, 6, 11, 21, and 29 stones).

The five lesser-currency-value tasks used the same root numbers as the tasks using stones, though these values were conceptualized in the kina and toea equivalents of the target numbers, and the root numbers were

		Conditions and values		
Number	Rationale for choice	Stones	Lesser value tasks	Greater value tasks
4	Within range of the conversational number system	4	K.40	K4
6	Just above range of conversational number system	6	K.60	K6
11	First position beyond key unit (10) in the Western system	11	K1.10	K11
21	First position beyond key unit in the adapted body system (20)	21	K2.10	K21
29	Position just beyond the completion of 27 body parts	29	K2.90	K29

Figure 65. Numbers used in stones, currency-of-lesser-value, and currency-of-greater-value tasks and their rationale.

defined in relation to their shilling equivalents (or the value of 10 toea). Thus, currency tokens were used that summed to the values of K.40 (root number: 4), K.60 (root number: 6), K1.10 (root number: 11), K2.10 (root number: 21), and K2.90 (root number: 29).

The greater-currency-value tasks were again chosen using the same root numbers, though they were selected based on the kina and toea equivalents of the target number, and the root numbers were defined in relation to their pound equivalents (or the value of 2 kina). Thus, currency tokens were used that summed to the values of K4 (root number: 4), K6 (root number: 6), K11 (root number: 11), K21 (root number: 21), and K29 (root number: 29).

I expected to find, consistent with earlier results, that people would make use of the body-count system to quantify the stones, establishing one-to-one correspondences between body parts and the target objects. Further, in the currency conditions, I expected to find that most people would respect the denominational structure of tokens for kina and toea. Building upon prior observations that were corroborated in Study 5-1, I also expected people to conceptualize the multiunit-unit relations in terms of the earlier (and now obsolete) system of pounds and shillings. This would be manifest in the units that people counted (10t coins, K2 notes) and their representation of sums. I also expected to find some evidence of the body-part strategies documented in Study 4-1, though our initial piloting with some of the original tasks showed minimal use of such strategies, with many individuals engaged in covert computational activity.

As a secondary but no less important concern, I expected to find additional evidence of a growing differentiation in function of body-part names. The differentiation was from body-part names denoting body parts used to count objects and body parts used as a number system that were not inherently linked to specified body parts. I reported evidence of this emerging differentiation in earlier chapters. For instance, in cohorts with greater participation in the money economy, Study 4-1 (1980) showed that people increasingly turned external correspondences inward in their computational approaches, using double-enumeration, body-substitution, and halved-body strategies in order to keep track of additions and subtractions of terms. In this redirection, people incorporated body parts into varied kinds of microgenetic constructions, thereby using them flexibly, as when calling one body part by the name of another (body substitution). In Study 5-1, I further noted that body parts had acquired a token-naming function, a function quite independent of the use of body parts in counting. For example, even among the elders, *kat-hai* (shoulder (10)) was used to refer to a K1 coin, and *kat-hai fu* (shoulder (10) *fu*) was used to refer to a K20 note.

Methods

Participants. The same individuals who participated in the currency-identification study participated in the quantification study. Typically, interviews occurred on the floor of a house or on the ground in a location that was away from onlookers.

Quantification Tasks. I generated tasks using the five key root numbers of 4, 6, 11, 21, and 29 for the conditions of stones, currency of lesser value, and currency of greater value. In the five tasks within each of these three conditions, Indigo and I presented participants with a specified number of stones or currency tokens and asked them to determine the value of the material presented.

Procedures. The tasks and procedures were back-translated in accord with the approach described in Appendix C. All interviews were conducted orally in Tok Pisin, and a local assistant provided an oral translation into Oksapmin.

Root	Value	Tokens used	Materials
4	40t	Two 20t coins	
6	60t	Six 10t coins	
11	K1.10	Two 50t coins; one 10t coin	
21	K2.10	Three 50t coins; three 20t coins	
29	K2.90	Two K1 coins; three 20t coins; three 10t coins	

Figure 66. Lesser-value tasks: root numbers, currency value, tokens used, materials.

QUANTIFYING STONES. Participants were asked to determine the number of stones presented for each of the root numbers – 4, 6, 11, 21, and 29. The tasks were presented consecutively: first the 4-stones condition was presented, next the 6-stones task was presented, and so forth. Stones were arranged in linear arrays that varied in number of rows depending upon the number of stones used.

QUANTIFYING CURRENCY OF LESSER VALUE. Participants were asked to determine the value of the tokens presented for each of the tasks for the five target values, with tokens for each task presented consecutively – K.40, K.60, K1.10, K2.10, and K2.90. The sets of coins and notes used for each root value are shown in Figure 66.

QUANTIFYING CURRENCY OF GREATER VALUE. Participants were asked to determine the value of the tokens presented for each of the tasks for the five target values, with tokens for each task presented consecutively – K4, K6, K11, K21, and K29. The sets of coins used for each root value are shown in Figure 67.

Data Recording. During each interview, the interviewer recorded several properties of participants' solutions.

Root	Value	Tokens used	Materials
4	K4	Two K2 notes	
6	K6	Six K1 coins	
11	K11	Three K2 notes; one K5 note	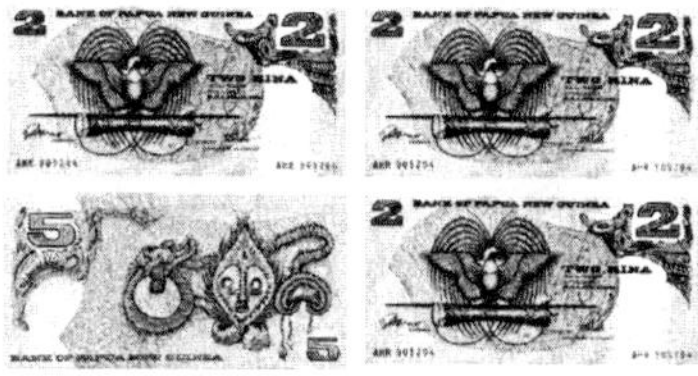
21	K21	One K20 note; one K1 coin	
29	K29	Two K10 notes; four K2 notes; one K1 coin	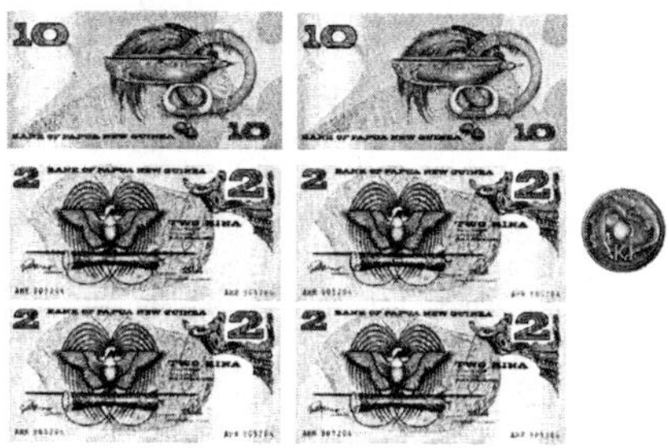

Figure 67. Greater-value tasks: root numbers, currency value, tokens used.

COUNTING STONES. For each task, in the process of a count, we recorded whether the participant made an overt gestural link between body parts and the stones being counted (as distinct from only verbally reciting a list of body-part names). For example, for four stones, individuals might touch each successive stone with their thumb (1), index finger (2), middle finger (3), and ring finger (4), respectively, as distinct from just naming the four verbal tags of thumb (1), index finger (2), middle finger (3), and ring finger (4).[7] Then, once a participant produced a representation for a set of stones, the interviewer noted whether the person (a) gestured to a body part in

[7] Individuals were assigned this code if such correspondences were included in part or all of a count.

communicating a final answer or (b) used a body-part expression in the final answer.

TASKS USING CURRENCY OF LESSER OR GREATER VALUE. In tasks with currency present, we recorded similar information to that in the stones task. This included (a) whether the individual made a gestural link between body parts and the coins or notes being counted (in either part of or all of the count); (b) whether the individual gestured to a body part in giving a final answer; or (c) made reference to body parts in the final answer. In addition, we also recorded whether we observed the individual counting the currency in units of shilling/pound equivalents – that is, counting in units of 10 toea or units of 2 kina.

As in the prior study of naming practices for tokens, Indigo and I found considerable variation in the names that participants used to represent the given quantities. In an effort to facilitate coding, we standardized the names by creating name categories that included close variants under a single descriptor. To illustrate the scheme with some English variants: we would include names like "a dollar sixty," "one dollar sixty," and "one dollar and sixty" under a single category, say "one dollar sixty."

Results

To understand participants' performance across the 15 tasks used in the study, I present a series of analyses. I begin with the accuracy of computations and proceed to qualitative features of participants' quantification activities with a focus on use of the body and token names in quantifications.

General Remarks on Accuracy. In general, most of the participants across cohorts were concerned with incorporating all of the tokens in their quantifications. A few of the elders simply named the tokens present without composing their values. Other forms of "inaccuracy" were to produce approximate roundings to a decimal number. For example, some people indicated that K2.10 toea was the 20th body part, what we took to be the equivalent of 20 *silings* (or 10t coins). Others indicated that K21 was *kat-hai fu*, translated as shoulder (10) K2 notes, or possibly shoulder (10) double. Those people (generally elders) who did not incorporate all of the tokens or whose solutions were markedly inaccurate were excluded from the analyses that focused on qualitative properties of people's quantifications. One caveat to this rule relates to properties of representations in the Oksapmin language. Often, Oksapmin representations are dependent upon context for a numerical interpretation. For example, in using the body system

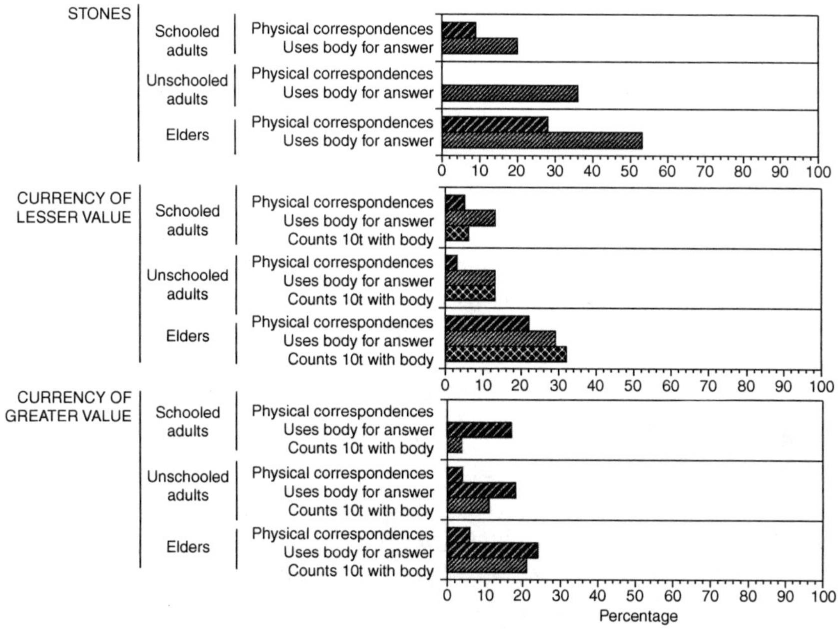

Figure 68. Uses of body and *silings* in the quantification and representation of materials in the conditions of stones, currency of lesser value, and currency of greater value.

to represent a number, an individual might drop the prefix "*tan*," which indicates the "other side of the body," when referring to, for example, *tan kat-hai* (18), or shoulder on the other side. In these cases, we recorded the lexical form used (what was literally said) and we included these people in our analyses.

Gestural Features of Strategies, Currency Units, and Representing Summations with Number Names. How were body parts used in the process of quantification? Figure 68 reveals the pattern of findings of overt features of quantifications and experimental conditions pooled over root numbers (4, 6, 11, 21, 29). The category "physical correspondence with body" was recorded when participants touched locations on the body to indicate their correspondence with objects. The category "uses body for answer" was coded when an individual gestured to a relevant body part in the production of a final answer (the body-part name and the cardinal suffix typically accompanied the gesture). The category "counts 10 toea with body" was used when participants counted the value of a coin or note in units of 10t. For example, if a participant pointed five times to a 50t coin as he or she composed its value with other values, this code was recorded. Codes were not mutually exclusive; a participant's performance could be assigned all three codes for a single task.

For most of the performances across conditions, participants used only oral reference to body parts. That said, Figure 68 reveals that when physical references to body parts were made, elders made these references with the greatest frequency, and the conditions in which the most references to body-part locations were made were those involving the noncurrency tokens, the stones. With the stones, the elders frequently gestured to body parts in their final representations; less frequently, they created a correspondence between body parts in the process of quantifications. These findings show a general decrease in use of the body across conditions from elders to schooled population, perhaps reflecting the increasing autonomy of an oral system of body-part names divorced from their reference to spatial locations on the body. Such an interpretation is consistent with Study 4-1, conducted in 1980, in which the elder cohort often appeared to treat the body more as a means of tallying objects, establishing correspondences between external objects and positions on the body.

Figure 69 contains a representation of the Oksapmin expressions used most frequently for the tasks in each of the three conditions. The figure reveals that for most of the tasks across conditions, body-part names are commonly used. In fact, body parts are the most frequently used form for all the values of stones and the lesser currency conditions up to and including K2.10. From K2.90 and beyond, the most frequently used names are typically multiunit expressions, involving counts of K2 equivalents (such as *faun* (pound), *iratarin* (leaf), *haben tan* (leaf (flat)), and *fu*) and counts of 10t coins in body parts (expressed as body parts).

An inspection of the expressions in the figure reveals that among these most frequent representations of currency sums, a majority of participants produced a compact, single-value representation, not simply a list of the names of tokens presented. For example, consider the case of K2.10, for which participants were presented with three 50t and three 20t coins. The most common expression for this condition was forearm (21). For K11, participants were presented with three K2 notes and one K5 note; the most common expressions were *hanen-fu* (K10) (trans. five double, or five K2 notes) and *temsi tana* (K1) (trans. flat with hole, or the K1 coin), thus, the English equivalent of 5 double and 1. Clearly, these representations go beyond the given, transforming currency tokens into numerical units that show little surface relation to the printed numerical values on the tokens themselves. Further, the representations of the worth of token arrays are

		STONES		
4 Stones	**6 Stones**	**11 Stones**	**21 Stones**	**29 Stones**
H^tdip-bai {29} Ring finger	*Dop-bai {44}* Wrist	*Gwer-bai {42}* Neck	*Tan bes-bai {18*}* Other side forearm	*Tan-o-bes-bai {10*}* Other side forearm
		CURRENCY, LESSER VALUE		
K.40 **2 20t coins**	**K.60** **6 10t coins**	**K.1.10** **2 50t coins, 1 10t coin**	**K2.10** **3 50t coins, 3 20t coins**	**K2.90** **2 K1 coins, 3 20t coins, 3 10t coins**
H^tdip-bai {32} Ring finger	*Dop-bai {45}* Wrist	*Gwer-bai {28}* Neck	*Tan bes-bai {9}* Other side forearm	*Wan faun tana bat forban towat-bai tit-si{13*}* One pound flat and bicep once
H^tb^ttimbas {6} Pinky, one absent	*Dopa {5}* Wrist	*Gwer {3}* Neck		
		CURRENCY, GREATER VALUE		
K4 **2 K2 notes**	**K6** **6 K1 coins**	**K11** **3 K2 notes, 1K5 note**	**K21** **1 K20 note, 1 K1 coin**	**K29** **2 K10 notes, 4 K2 notes, 1 K1 coin**
Wan faun yota {13} One pound two	*Wan faun yetir (tana) {12}* One pound three (flat)	*Hanen fu tit temsi tit {12*}* Five FU once hole once	*Kat-bai fu tita temsi tita {23*}* Shoulder FU once hole once	*Arum-bai fu bama tensi tana {3*}* Nose FU and hole flat
Haben tan yota {9} Leaf flat two	*Dop-bai {7}* Wrist	*Gwer-bai {4}* Neck	*Kat-bai fu bama kat-bai gon tit si {9*}* Shoulder FU shoulder all once	*Imab kak-si bat forban towat-bai{2}* Pig head and bicep

NOTE: Frequency of listed representation is indicated in braces {}. An asterisk (*) indicates that the listed representation is a prototype – the value in braces is the number of close variants. The use of the suffix-*bai* is used in speech to index a cardinal number meaning for a body part (as distinct from an ordinal or count meaning).

Figure 69. Most frequently used expressions to represent each of the 15 token arrays. Appendix E contains some of the auxiliary word forms (and their translations) used in Oksapmin quantitative expressions. See http://www.culturecognition.com/videos/ for video support.

numerically coherent in accord with semiotic systems that have emerged in the Oksapmin community.

A comparison between the most frequent representations used for stones and currency of lesser values is striking; Figure 69 reveals little difference. The source of the parallelism, of course, is that the 10t coin is treated as a unit. Four stones and 40t, for example, are both represented by ring finger (4). Six stones are referred to as wrist (6), and the same expression is used to represent 60t. The representations do not shift at the boundary of the K1 unit: 11 stones is represented as neck (11) and K1.10 is also represented as neck (11). The same is true for 21 stones and the value of K2.10 – each are represented as *tan bes-bai* (21). There is a noteworthy exception, however.

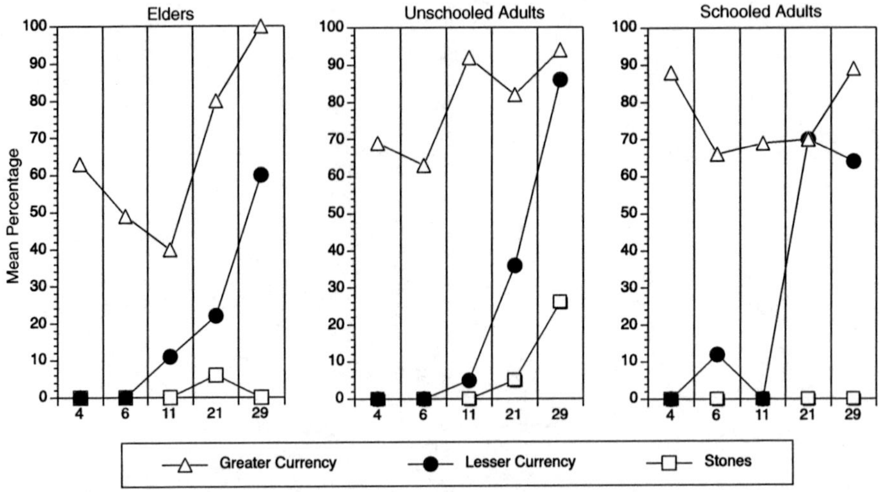

Figure 70. Use of pound-equivalent forms in final representations on the stones, lesser-currency-value, and greater-currency-value conditions by cohort.

In the case of 29 stones, people frequently looped around the 27th body part at the pinky (27th) to the wrist (28th), and ended on the forearm (29th), referring to the body part as *tan-bes-hai* (29).

For currency tokens of greater value, we find common use of pound-equivalent expressions as a means of representing the composition of tokens. The expressions include *haben* (leaf), *iratarin* (leaf), and *faun*.[8] In addition, we again find the occurrence of *fu*, as was noted in Study 5-1. I pursue a detailed analysis of *fu* in Chapter 6.

Figure 70 contains the mean proportions of use of pound-equivalent forms as a function of cohort on each of the condition sets (stones, currency of lesser value, and currency of greater value). The pound-equivalent expressions included *faun* (pound), *haben/iratarin* (leaf), and *fu*. These pound equivalents were used most frequently in the representation of greater currency values. Almost all used a pound-equivalent expression in the representation of K29; the majority of people in all cohorts used a pound-equivalent expression for the other values in the currency of greater

[8] "*Faun*," as noted previously, is the Oksapmin pronunciation for the Australian pound. The term has been well incorporated into the Oksapmin language and was still used commonly in 2001, even though pounds had not been in circulation for about 35 years. *Faun* was used in expressions like *wan faun yota* (K4) where *yota* is the conversational number word for two, or *wan faun yetir* (K6), where *yetir* is the conversational number word for three. While some participants used the term *faun* under the currency conditions, it was rarely used under the stones conditions.

value conditions. The exception was again elders, only a minority of whom used a pound-equivalent expression for the values of K6 (48 percent) and K11 (40 percent).

COUNTING STONES: INDIRECT EFFECTS LINKED TO COLLECTIVE PRACTICES OF ECONOMIC EXCHANGE. A question of interest related to sociogenetic processes was whether innovative uses of the body system linked to currency were incorporated into the general structure of the body system. For example, were innovations such as the use of 20 as a stopping point in counting shilling equivalents observable in the people's representations of the stones, occasions when the material counted had no denominational structure? Additional analysis of people's performances on the 21- and 29-stone condition tasks indicates that this was, in fact, occurring (further corroborating a finding reported in Study 4-2). Some people counted to the 20th position, and the subsequent body part was the thumb (21), traditionally the first body part in the ordinal sequence, but in this case it stood for the 21st. In such cases, people appear to have made use of the modified system.

Figure 71 shows, as a function of cohort, the percentages of people who used the 20th body part as a privileged point when counting stones. Elders, the group with the least participation in the money economy, typically used the traditional trajectory for quantifying stones. In contrast, both unschooled and schooled adults used a modified trajectory, treating the 20th body part as a point to re-cycle to the thumb (1) either in the Oksapmin language (saying thumb (1) but pointing to the forearm (21)) or in both physical location and language (pointing to the thumb (1) and saying thumb (1)). This, then, may be understood as an invisible use of the cyclic (or base) structure associated with the denominational system of currency. While historically the cyclic structure was generated through currency exchanges, now people extend the structure to a nondenominational countable (stones). It also points to a more general development in the social history of the Oksapmin body-part counting system.

CONVENTIONALIZED FORMS FOR REPRESENTING ODD-NUMBERED VALUES. Recall that in naming the K5 note in Study 5-1, people used many variants. I attributed the many variants to the odd-numbered value of the note together with the fact that expressions for K2 and beyond were rooted in the one-pound unit.[9] In the present study, in generating and naming

[9] A constraint that limits the use of body parts to refer to odd-numbered values of kina is linked to the fact that kina are represented by expressions linked to the pound. For example, one could refer to 4 kina or 6 kina using the expression *faun* (pound) modified

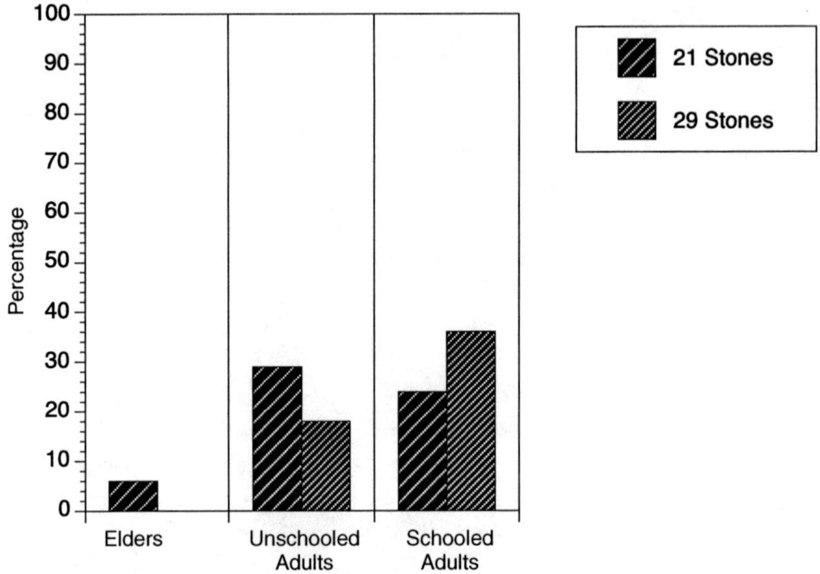

Figure 71. *Tan kir* (20) used as a base point on quantifying 21 and 29 stones.

odd-numbered values in tasks involving currency of greater value (K11, K21, and K29), people made accommodations similar to those used in the names for the K5 note. They appended even-numbered representations of kina with the use of *temsi* (hole), *kat-hai* (shoulder (10)), and *kut* (partway) in their representations.

(a) *Temsi* (trans. hole). Recall that *temsi* is one way to refer to the K1 coin. In representing odd-numbered values of kina, we find representations such as "*kat-hai* (10) *fu* (double) *tita temsi tita* (1)" for K21. In this case, *kat-hai* (shoulder (10)) modified by *fu* refers to a value of K20, and *temsi* (the K1 coin (flat with hole)) signifies a value of 1 kina.

(b) *Kat-hai* (10) (trans. shoulder (10)). Recall that *kat-hai* (10) is another way of referring to the K1 coin (or the equivalent of 10 10t coins). In representing odd-numbered values of kina, we find expressions such as "*kat-hai* (10) *fu* (double) *hama kat-hai* (10) *gon tit si*" used again to refer to K21.

by the appropriate body parts (pinky (5) or wrist (6)), but one cannot refer to 11 kina. Thus, odd-numbered values of kina require an accommodation. One might ask, why not simply refer to odd-numbered values with body-part names? The problem here is that body parts when not used as modifiers carry meanings linked to the shilling, so that expressions like "*gwer* (11)" (neck (11)) are taken to refer to values of 11 10t coins, or K1.10, rather than K11.

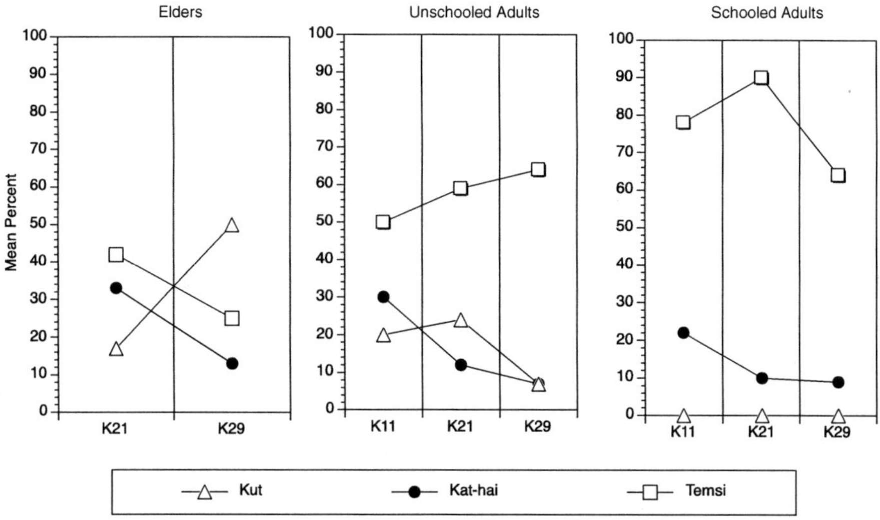

Figure 72. Accommodations to the use of *fu* in representing odd-numbered values by cohort: the use of *temsi* (hole), *kat-hai* (shoulder (10)), and *kut* (partway).

In this case, *kat-hai* (trans. shoulder (10)) in conjunction with *fu* refers to a value of K20, and *kat-hai* (trans. shoulder (10)) alone (its second occurrence in the expression) refers to 10-shilling equivalents (10 10t coins or 1 kina).

(c) *Kut* (trans., part, partway, half). Recall that *kut* has varied meanings, the most relevant here being a meaning like part or partway. Thus, to refer to K11, we find people stating "*hanen* (5) *fu* (double) *hat forhan dop* (6) *-o kut* (partway to wrist)." In this case, *hanen* (5), the conversational number word, modified by *fu* means K10. Since *hanen* is equivalent to the pinky (5) in value, the next body part is the wrist (6) or *dopa*. Partway to the wrist from the conversational number for five (the equivalent of pinky (5) in body counting) then is one more beyond 5, or the value of 1 kina. Thus, the expression translates as K11.

Figure 72 contains plots of the use of *temsi*, *kat-hai*, and *kut* as a function of cohort for those occasions in which individuals used *fu* to serve a doubling function for the greater-valued-currency condition for values of K11, K21, and K29. Because only a few elders used *fu* on the K11 task, I included only the K21 and K29 tasks for this group.

The figure shows differences in cohorts' use of linguistic devices to accommodate odd-numbered values. For example, the elders who used *fu* on the K21 and K29 tasks used all three linguistic methods, and the majority of the sample for the K29 condition used *kut*. In contrast, the schooled adults showed much less variability. Most of them used *temsi* as the method of preference and never used *kut* at all in any of the three tasks. The unschooled adults present a profile of findings midway between elders and schooled adults. Like the elders, the unschooled adults used all forms for these tasks, though as with schooled adults, with unschooled adults the use of *temsi* (hole) was dominant.

Discussion

The findings of Studies 5-1 and 5-2, coupled with those from prior chapters, provide a fascinating case of shifting relations between forms of representation and functions they serve in collective life. Over time, as people used the body system to quantify currency in collective practices related to economic exchange, the structure of the currency system became incorporated into the structure of the body-counting system. In this process, new forms for naming currency tokens were spawned. The findings of Studies 5-1 and 5-2 provide some clues to how this sociogenetic process occurred from both synchronic and diachronic perspectives. Let me summarize some principal points in this process.

1. *Introduction of pounds and shillings: Naming and counting tokens.* As pounds and shillings penetrated the area in the early 1960s, people used body-part counting to quantify tokens related to exchanges (note Officers Rogers' and Hicks' early reports, described in Chapter 3). The body-part names for tokens and tropes like "leaf" (*haben*, *iratarin*) documented in Study 5-1 may well have their roots in these early encounters. Further, as some people engaged in currency exchanges with patrol officers and later with missionaries, their use of the body to represent pounds and shillings created representations that seeded variants in the body-counting system.
2. *Trade stores.* The participation of Oksapmin men in the late 1960s and 1970s in setting up trade stores contributed to a burgeoning money economy. Their frequent exchanges using currency, in turn, led to new communication and problem-solving goals and concomitant alterations in the Oksapmin body system, such as the use of elbow (20) as a stopping point (as documented in 1978 fieldnotes, Study 4-2, and further in Study 5-2). In the alteration of the stopping point, the base structure of the pound/shilling relation came to be incorporated into the structure of body counting.

3. *Kina and toea.* With the application of the body system to kina and toea, a range of semiotic properties emerged in naming currency tokens.
   a. *Denominational units.* In the case of token naming (Study 5-1), people extended units of pounds and shillings to the representation of kina and toea in using the body system.
   b. *Body-part counting and denominational units.* The use of units of pounds and shillings applied to kina and toea carried with it an increased range of values that the body could represent. Indeed, expressing the pound as K2 and the shilling as 10t increased the numerical value of unit representations by two- and tenfold.
4. *The rise of Tok Pisin.* The increasing use of Tok Pisin in the area, along with increased relations with the Western world, was well represented in Study 5-1. The interpenetration of Tok Pisin and Oksapmin language practices was evident in the borrowing of word forms (*faun*, *siling*) and perhaps in selective substitution of Tok Pisin names for certain tokens even by monolingual Oksapmin elders, as in the expression *faiv kina*.

The transformation of semiotic systems over historical time through people's participation in shifting collective practices of exchange is a complex and remarkable process. The findings of this chapter, coupled with those of Chapters 3 and 4, provide some clues as to how this process occurred in the Oksapmin world as people made efforts to communicate with one another in collective practices of economic exchange under shifting conditions. But the specifics are far from clear. For example, among the central questions: In what way do individuals' microgenetic constructions of the sort detailed in this chapter enter into processes of sociogenesis? How did new functions arise for cultural forms documented like the recruitment of tropes to name new currency tokens? And more generally, is there further evidence of the processes of local emergence and broadcast at work in population-wide propagation processes? How would one find out? These are challenging questions that go to the crux of conceptual and methodological treatments of culture-cognition relations. I pursue these issues in the next chapter with a focus on *fu* and its shifting numerical functions in recent social history in collective practices of exchange.

6 Representational Forms, Functions, Collective Practices, and *Fu*

A Microcosm

In this chapter, I narrow but also deepen the concern to understand microgenetic and sociogenetic processes in the use of representational forms and the functions they serve. I focus on the single word form *fu*. I show that the expression presents an extraordinary microcosm, an analysis of which can illuminate the interplay between collective practices and genetic processes as they operate through a short stretch of Oksapmin social history.

The documentation and analytic focus on *fu* had a curious beginning for me. It started in the context of my 2001 reunion in Tabubil with Oksapmin friends while en route to the Oksapmin community. During the reunion, the conversation in the larger group shifted to mathematical topics, and it was in this context that I heard the word form *fu*. The expression was used in a way that puzzled me (though there should now be no surprise after reading Studies 5-1 and 5-2, conducted after these interactions). I learned that when one points to the nose (*aruma*, the 14th body part in the Oksapmin sequence) and says *arum-hai fu*, the meaning was to double the value of the nose (14), yielding a value of 28. Now this practice was perplexing at the time, especially at this point in my reentry into the community. As an apprentice to the numerical practices of Oksapmin elders in 1978, I had learned that saying *fu* with fists raised meant the completion of all 27 body parts in a count (see Figure 13). In fact, in traditional life (as detailed in Chapters 3 and 4), there were no arithmetical procedures for adding or subtracting body parts, and doubling body parts would have been an alien idea. For me, a student of mathematical cognition, this development in the use of *fu* was a remarkable turn of events.

How did *fu* come to take on a multiplicative function in Oksapmin mathematics? Was this *fu* a historical descendant of the earlier *fu* that I had learned back in 1978? Or was it a new word form, perhaps borrowed from a neighboring group? Did someone invent it as a way to represent values greater than those permitted by the indigenous number system? And if any of these processes fits the case, how did new uses of the word come to spread in the social history of Oksapmin communities? Or, was my untrained ear mishearing a sound? Perhaps the *fu* that I'd learned and the *fu* I encountered in ongoing 2001 talk were simply different words, with different sound patterns.

I put these questions aside early in the 2001 visit, having organized my effort around more general concerns to document continuities and discontinuities in Oksapmin mathematics since my 1978 and 1980 stays. But my questions about *fu* were rekindled in the course of piloting interview procedures and conducting interviews for the studies described in Chapter 5 after entering Oksapmin communities. I had the sense that many people used the expression, and the pronunciation was in fact much the same as what I had learned from my earlier visits to Oksapmin. On occasion I again would ask people about the origins of the word, but no one had much to say. My thoughts that the *fu*-double was a descendant of the early *fu* were spurred on by the vague intuition that no single person might know the genetic history of *fu* and, further, that people might not be aware of their own role in the production of historical change. That is, that a shift in function of the word may well have emerged unwittingly, without local awareness of the shift.

Keller (1994) offers a traffic jam as a simple illustration of collective human activity that arises through unintended consequences of purposeful actions: Cars are proceeding on a crowded freeway at 65 miles per hour, and the first one in a line inadvertently slows down to 60 miles per hour. The driver immediately behind, upon seeing the brake lights of the car in front, adjusts her speed, overcompensating slightly – reducing the speed not just 5 miles per hour but with an added safety margin of 5 miles per hour to 55. All of the drivers operate with the same maxim – "adjust your speed in accord with the car in front with a safety margin" – and the maxim leads to an eventual standstill by the *n*th car. Keller points out that the traffic jam is the product of individual human actions, actions that were all purposeful and concerned with not bumping into the car in front. The intentions were unrelated to the final product, however, and the makers of the jam are unaware of their contribution to it. Keller argues that a wide range of collective human products can be understood as analogous to a traffic jam – a structure emerges as epiphenomenal to the intentional actions of participants.

Of course, traffic jams are not collective systems of representation and bear little relation to them. Nonetheless, historical linguists like Keller find this general idea useful in thinking about historical change in systems that appear to be products of intentional design. How might the traffic jam metaphor be relevant to the analysis of historical shifts in Oksapmin communities?

Let's consider the case of collective representations used in Oksapmin communities related to the money economy. As recurrent collective problems gradually shift in daily life with the rising money economy, people face new kinds of coordination problems (Lewis, 1969) – problems of making one's meaning clear to one another to enable coordinated action and reference. Such coordination problems may be especially challenging when there are no previously "given" collective representations (precedents) in the language. To accomplish a communicative act, interlocutors interacting at different locations may recruit varied forms of expression to solve coordination problems for objects like new currency tokens. Some may make use of gestural forms, like pointing or reference to a body part, to refer to a currency token; others may create tropes of varied sorts. Initially, these means of communication may be idiosyncratic and vary markedly over locations and interlocutors. Over time, as interlocutors shift interactional partners and locations, people may make use of forms used successfully with previous interlocutors with new interlocutors. The process of iterative uptake of local solutions to coordination problems through time over shifting interlocutors and across different locations may have an important consequence for an analysis of the sociogenesis of form-function relations: Some solutions become recognizable forms in the community, becoming collective representations used throughout. Just as the members of Keller's traffic jam do not know of their complicity in it, no one individual may be aware of his or her contribution to the changing structure of representational forms and associated functions. Here we have systems that emerge well adapted to shifting local conditions and personal predilections, but of which no single person is conscious. Indeed, the "design" of a collective system would emerge out of the unintended consequences of local communicative activity. I will return to these ideas as I consider sociogenetic issues in mathematical cognition in relation to the dynamics of historical change, both in the case of *fu* as well as in the sketch of a general treatment in Part IV. For now, *fu* will be a test bed for ideas I later elaborate.

With this initial framing of collective practices and genetic processes and with an eye toward historical change, I move to my 2001 return to Oksapmin communities. Of particular concern is to understand relations between social history and mathematical thought.

Leading Observations

In a single day after Indigo and I had completed our interviews for Studies 5-1 and 5-2, I found, by serendipity, an important lead to help account for form-function shifts in the use of *fu*. I had occasion to chat with three people separately – an elder (of perhaps 65), an older man (in his 50s), and a middle-aged man (of about 35). In these conversations, I was concerned with understanding some of the varied properties of these people's mathematics, and in each case, the conversation included some discussion about *fu*. But this time, instead of beginning with questions about the origins of the meaning "to double" – a gambit that had led to dead ends in previous conversations – I began each conversation with a query about the man's knowledge of the position of *fu* on the body, in an effort to find evidence of the meaning and origins of this arithmetical expression. I heard three different accounts, which were important clues to a continuity of form and the shifting functions of the form. These locations are depicted in Figure 73.

The discussion with the elder corroborated my early 1978 observations. He showed the ending of the count at the 27th position with fists raised and an exclamatory *tit fu gon a!* This is a multiword expression that I had not yet heard in 2001, but one that I had learned back in 1978.[1] He also demonstrated that to count beyond 27, one loops back around from the little finger on the second hand to the wrist, a trajectory that I had learned in 1978 that is also one of the 2001 variants found in Study 4-2.[2]

The older adult used *fu*, but he used it to refer to either of two alternative positions. One meaning was similar to the one the elder had communicated and that I had learned in my own early apprenticeship in 1978: a complete round of 27, ending with the pinky. The other meaning was a puzzle for me: He said that *fu* also referred to the elbow on the other side (the 20th position). My thought, of course, was, Why the elbow (20)? Was this an idiosyncratic use to be treated as "noise" in our early efforts to piece together a genetic history? But this usage converged with another use of the elbow-on-the-other-side current among trade store owners in 1978 – as the count of 20 shillings to their pound equivalent. Was there a relation here?

The interview with the middle-aged man provided independent corroboration that the 20th body part was the referent for *fu*. Unlike the older

[1] Recall that *tit* (1) is the conversational word meaning "one," *gon* refers to "all" or "complete," and *a* is a demonstrative – like "this" or "that." Thus the expression "*tit fu gon a*" communicates "one complete (from thumb on one side to the pinky on the other)."

[2] Study 4-2 was conducted after the completion of the study exclusively on *fu* (Study 6-1) reported in this chapter.

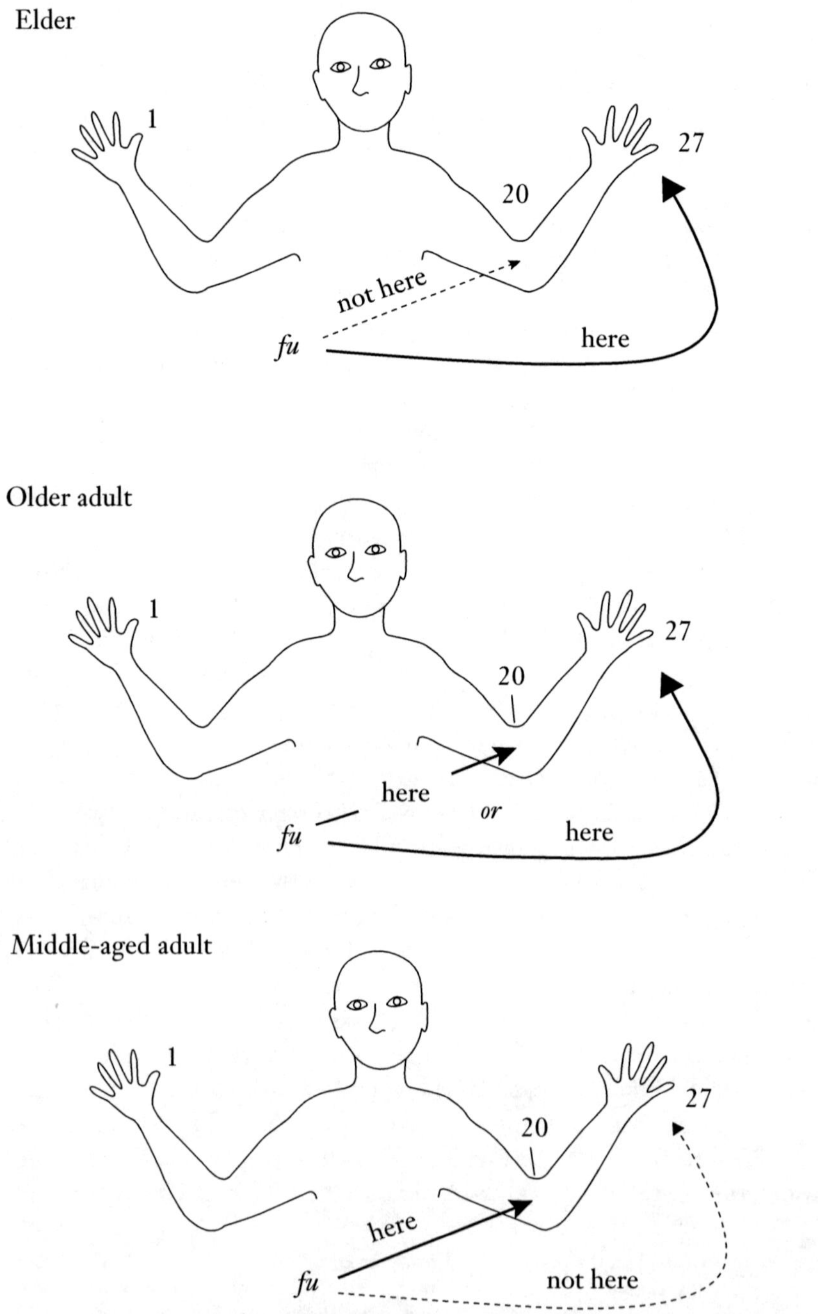

Figure 73. Three different locations of *fu* as reported by an elder, an older adult, and a middle-aged adult.

adult and the elder, however, he claimed that *fu* did not refer to the 27th at all – but to the 20th.

Were there other meanings and uses of *fu*? How general was the reference to 20? Why might it be used to mark the 20th position? How widely distributed was this usage in the population? Indigo and I had not yet developed coding schemes to understand how *fu* appeared in the many interviews that we had engaged in for Studies 5-1 and 5-2, and the procedures used in Study 4-4 had yet to be conceived. But I did wonder.

Might the truncated use of the body system (terminating at the elbow (20)) that I described in Chapter 4 have spread further through the Oksapmin communities subsequent to my last visit to Oksapmin in 1980 (see Figure 29)? Might *fu* have come to be used side-by-side with the pound (or *faun*) as a marker for the 20th body part? If so, why might this change have occurred? Such questions and reflections prompted a new set of studies that were designed to produce data that would bear on the shifting functions of *fu*. My plan was to collect systematic interview data on *fu* that would inform conjectures about form-function shifts.

Indigo and I field-tested and then back-translated a series of interview questions that could be used in cross-cohort designs, a process guided by several central questions. First, does *fu* have an everyday meaning in the Oksapmin language that is unrelated to counting that might help us understand its history and its use in quantification? I had some sense, based upon a discussion with a couple of additional elders, that there were meanings of *fu* that had fallen out of use. I wanted to pursue this lead, which might help me to understand the origins of its meaning in the context of body counting, and perhaps the shift in meaning to the 20th body part.

Second, I considered the lead about variant positions of *fu* on the body as an important clue that both the 20th position, elbow (20), and the 27th position, pinky (27), were locations for the exclamation of *fu!* How common was this knowledge? Were there other body-part positions? Did thinking about the location of *fu* vary by cohort? If so, could understanding variation help us in piecing together a social history of shifts in function?

Third, like the expression for leaf, was *fu* also used to refer to the (obsolete) pound?

Fourth, if the use of *fu* had emerged in relation to currency, to what extent was it now applied to noncurrency items. How general was its use?

Fifth, and finally, even though it was not a part of daily practices, could people use *fu* to represent greater numbers without using currency as a medium – numbers that extended considerably beyond the 27th body part?

Indigo and I thus included questions about how to represent the numbers 50 and 60, suspecting that this might be another context in which *fu* might be invoked by participants.

Study 6-1. Shifting Meanings, Locations, and Uses for *Fu*

Methods

Participants

Almost all of the participants in this study lived in the hamlets of Tekin and Divanap. This is in contrast with the participants for the 2001 studies reported in Chapter 5, who lived mostly in Daburap, Kweptanap, and Tekap. The two regions used different dialects.

Indigo and I interviewed four different groups of people. These included 15 Grade 7 and 8 students at Divanap Community School, 19 adults (in their late 20s or 30s who had completed at least Grade 6), 20 unschooled middle-aged adults (perhaps a little older than the schooled adults), and 20 unschooled older adults. Our methods for determining age were identical to those described in the Chapter 5 studies. We selected three well-known people for whom we had good estimates of age, and then asked study participants to whom they were closest in age. Again, our Oksapmin assistants helped recruit participants for us. Because of their status as members of a particular clan and village within the region, it was easier for them to recruit only members of their own clan. As a result, we traveled to central locations in different villages and tried to approach people of diverse clans in different villages. Again, as in the case with all studies with adults, we offered participants K1 as a token of our appreciation for taking the time to be interviewed.

Procedures

The interview procedures used in the study, like those in the prior studies, were first written in English, translated into the Oksapmin language and Tok Pisin, and then back-translated. We piloted the interview with local teachers and our research assistants as translators. In the course of the interview, we instructed translators not to adhere strictly to the translated material when interviewees were having difficulty understanding the intent of the questions. The process of translation served an important training function for our translators insofar as the intent of the questions became explicit.

The questions that we used to find out more about people's knowledge of and meanings for *fu* covered issues of (1) the everyday meaning of *fu*, (2) the position of *fu* on the body, (3) the currency equivalent of *fu*, and finally (4) the meaning of *fu* with reference to a noncurrency quantity, the sweet potatoes. The interview protocol is contained in Appendix F, with English, Tok Pisin, and Oksapmin translations, though I note key elements of the protocol below.

Results

The Meaning of Fu

The interviewer began by asking, "Do you know the meaning of *fu*?" and participants were asked questions to probe their understanding of the word form as it was used traditionally. People produced varied responses to our queries. Some common responses included that they did not know the meaning; others communicated that the meaning involved some sense of a complete group of plenty of things, sometimes associated with a payment of traditional valuables in an exchange. To give a sense of the varied responses that show threads of common source meanings, I provide illustrative excerpts from the corpus of interviews.

Some participants linked *fu* to a payment that was enough or complete in traditional valuables used in exchanges.

> Unschooled adult no. 1. He indicates that *fu* was a term invoked in bride price exchanges: "In the old days it was used for bride price – if plenty or enough to pay for bride, then call it *fu*."
>
> Unschooled adult no. 15 communicated a similar idea, linking *fu* to *bonang*, or traditional valuables used to pay for a bride: "If you are paying bride price, or have plenty of traditional money, this will be *fu*."
>
> Elder no. 4 dropped the reference to bride price, simply stating, "If it is plenty, call it *fu*."

Other elders provided more abbreviated responses, and perhaps were communicating the same idea.

> Elder no. 7 simply said that "*fu* is plenty of money."
>
> Elder no. 12 said that "It is a word used when counting money in Oksapmin."

Some people said that *fu* was a complete group consisting of plenty of elements that included areas other than the functional value of traditional valuables in exchanges.

> Elder no. 20 said that "In old times, *fu* refers to pandanus nuts that will be divided up. Before dividing, they are referred to as *fu*."[3]
>
> Elder no. 2 communicated a similar idea to elder no. 20. Elder no. 2 said that *fu* can mean "one clan (*moh tit fu*) and then another (separate) clan (*moh tit fu*)."

Others linked *fu* to the counting system, though they conveyed a similar sense in the identification of *fu* with a sense of "complete" (or enough traditional valuables to accomplish a function like payment for a bride).

> Elder no. 18 pointed around her body to show *tit fu*, saying, "it is plenty." Unschooled adult no. 19 said, "when you count up to plenty; like plenty of sweet potatoes, taros, like that."
>
> Elder no. 15 said it was when you counted up to the pinky (27), and schooled adult no. 19 said, "It is a word you use when counting – you count to *fu* and then you start over again."
>
> Elder no. 13 combined the counting meaning with that of the payment of traditional valuables in an exchange, "*Tit* (1) *fu* means to count all the way around your body – do it again and it is *yot* (2) *fu*, again and it is *yetir* (3) *fu*. *Fu* on its own is used to refer to a big heap of things, or a group of things that is put together. When paying bride price, you put a heap of things, if it is plenty, and it is *fu* – you say something like '*fu bilong yu*' when you give it over."

Finally, I note some idiosyncratic interpretations. These point out how widely word forms may be interpreted.

> Elder no. 14 offered an onomatopoetic interpretation, indicating "sometimes the sound you make when you blow the fire." But when probed about other meanings, she indicated that "in money (perhaps traditional valuables) the fist is *fu*."

Elder no. 14's reference to the fist takes up the stereotypic gesture of fists raised at the completion of a count, which is associated with the exclamation of "*tit fu!*" or "*tit fu gon a!*"

[3] Pandanus nuts are somewhat similar to the seeds of a sunflower, insofar as many nuts are contained in a single large flower. This may convey the sense of a complete group of plenty.

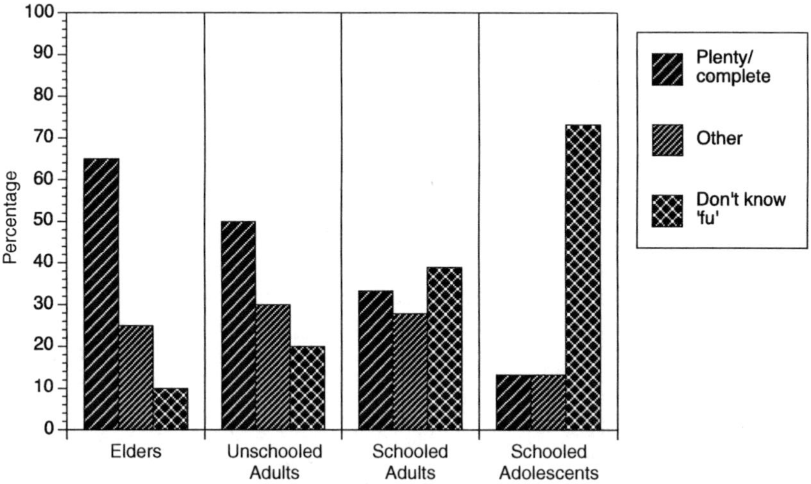

Figure 74. Percentage distribution of plenty/complete/group meaning for *fu* as a function of cohort.

Indigo and I coded people's statements about the meaning of *fu*, assigning statements to one of three categories: (a) statements in which *fu* conveyed a sense of a complete group of plenty of elements; (b) statements that did not include this strain of meaning; and (c) statements that they did not know the meaning of *fu*. Figure 74 shows the percentage distribution of individuals assigned to these categories by cohort. It shows a decrease in the number of individuals per cohort who said that *fu* referred to a complete group of plenty. Whereas most of the elders gave this meaning, most of the schooled adolescents did not know a meaning for *fu* – only 12 percent of them indicated a meaning that was related to the idea of a complete group of plenty.

Location

In the questions about the location of *fu*, Indigo and I asked whether interviewees knew a position for *fu* on the body: "If you count on your body like this (interviewer shows thumb, index finger ...) can you show me where *tit fu* is?" (See Appendix F.)

We followed up this query by asking whether they knew of other locations for *fu*. If someone gave more than one location, both of the responses were recorded and coded. Consistent with my prior conversations with the three men, the interviews revealed that both the elbow (20) and the pinky (27) were seen as privileged locations, though some people did not know of any position at all. Notably, the position chosen and knowledge of a position differed over cohorts.

Like their reflections on the meaning of *fu*, people's responses to queries about the location of *fu* on the body also elicited commentary. Certain elders limited their answers to body locations, each indicating that the count to the elbow on the other side (20) and the count to the pinky (27) are both places for *fu*.

> Unschooled adult no. 14. Now people say *tit fu* for elbow on other side (20); this is also 20 kina. But in the old times, they called pinky on other side (27) *tit fu*.
>
> Elder no. 1. Pinky on other side (27). (Do you know another place?) Only two places. In Tok Pisin, elbow on other side (20). In Tok Ples (the Oksapmin language, as represented in Tok Pisin), pinky on other side (27).
>
> Elder no. 3. Elbow on other side (20), saying elbow on other side (20) is "*tit fu gon a!*" (Do you know another place?) Counts all the way around to pinky on other side (27).

Whereas elder no. 3 indicates that both the pinky and elbow are locations for *fu*, the two others make reference to the history of *fu*. In "olden" times, unschooled adult no. 14 says, *tit fu* referred to a count to the pinky, and in "modern" times it is associated with the emergence of Tok Pisin in the community, *tit fu* is associated with the elbow (20). Elder no. 01 makes a similar distinction, identifying the pinky with the Oksapmin language and the elbow with Tok Pisin. Elder no. 3, perhaps assuming that he is communicating with a person who speaks Tok Pisin, first identifies *tit fu* with the elbow (20) and when prompted indicates that the pinky (27) is another place for *tit fu*.

I also noted some dissociation from counting meanings of *fu*.

> Schooled adult no. 6. *Tan-hatatah* (27). (Do you know another place?) No. Just one complete round to pinky (27) on the other side. Money is different. When count up to 10 that's *hanen (5) fu*. Twenty kina is shoulder (10) *fu*.

Schooled adult no. 6 indicates that the pinky (27) is the place for *fu*, but when asked about additional locations, responds, "No," and goes on to mention money: "Money is different. When count up to 10, that's *hanen* (5) *fu*. Twenty kina is *kat-hai fu* [shoulder (10) *fu*]." Schooled adult no. 6 is aware of the multiplicative function of *fu* but doesn't identify this answer with any sense of the historical source of this function.

Other people seemed to relax the constraint that when used in counting, *fu* communicated a sense of completeness or enough of something.

> Elder no. 19. Pinky on other side (27). (Do you know another place?) if plenty it will be *tit fu*, so other places on the body that shows plenty.
>
> Elder no. 7. Don't know – maybe *wan faun* or maybe *tipanah* (1). (Do you know another place?) Yes. Anywhere on the "other" side – you say *fu* because it is plenty.

For elder no. 2, like elder no. 19 and elder no. 7, the meaning for *fu* appears to drift, incorporating in new ways everyday observations of the use of *fu*.

Further drift from the pre-contact meaning of *fu* seems to draw from the stereotypic gesture of fists raised. In the following response, elder no. 2 appears to interpret *fu* with reference to a fist, the traditional gesture of raising fists at the completion of the 27th body part.

> Elder no. 2. *Tit fu!* is one fist and then the other fist, which is shoulder (10) *fu*; both fists together.

Elder no. 2 appears to identify *fu* with the traditional gestural display but mixes it with the current Tok Pisin meaning of *fu* as one element of a currency expression. Recall from Study 5-1 that *kat-hai* (10) *fu* is commonly used to index the K20 note. Elder no. 2 may be focused on the fists coupled with the shoulder (10) as equivalent to K20.

Finally, a schooled adult drifts further from the common meaning, indicating that while *fu* refers to a count to the elbow on the other side (20), it also refers to the shoulder (10th).

> Schooled adult no. 8. Elbow on other side (20). (Do you know another place?) Yes. Shoulder (10).

It is unclear whether the reference to the shoulder (10) is a residual of the common use of the expression *kat-hai* (10) *fu* to refer to K20 or whether this statement reflects the notion that 10 is a complete unit.

We conducted two frequency counts of people's statements about the location of *fu*. The first analysis included the informants' (a) spontaneous statements about the location of *fu* and (b) responses to our open-ended follow-up questions about whether there was another place in addition to the original position cited. For this analysis, response type rather than

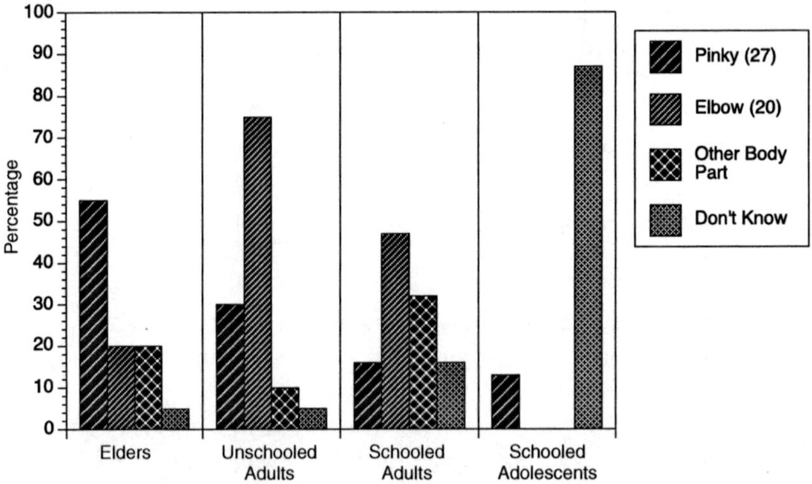

Figure 75. Percentage distribution of references to *fu* on body prior to probe questions as a function of cohort.

participants was treated as the unit. A person could claim that both pinky (27) and elbow (20) or other places were all places for *fu*. So, each response category was coded as "used" or "not used." The complete results are shown in Figure 75.

Consistent with my prior conversations with the three men, the analysis of participants' responses revealed that both the elbow (20) and the pinky (27) were privileged locations, though some people did not know of any position at all. The majority of the (unschooled) elders' responses (55 percent) indicated that the little finger (27) was the location for *fu*, with 20 percent of them indicating that elbow (20) was the location. Twenty percent of responses indicated some other body part. In contrast, for unschooled middle-aged adults, the relationship was reversed. Seventy-five percent cited the elbow (20), whereas only 20 percent cited the pinky (27). Three of these people cited both the elbow and the pinky as locations for *fu*. Only about 10 percent of the responses indicated a different body part. The pattern for schooled adults was similar to that for unschooled middle-aged adults, though there was an increase in the number of people who stated that they did not know a position or referenced a different body part. Almost all of the adolescents responded that they did not know the position of *fu*.

In the second analysis, I used a different source of convergent data to explore the same issue about the position of *fu*. Recall that if participants had mentioned only the pinky (27), only the elbow (20), or not mentioned

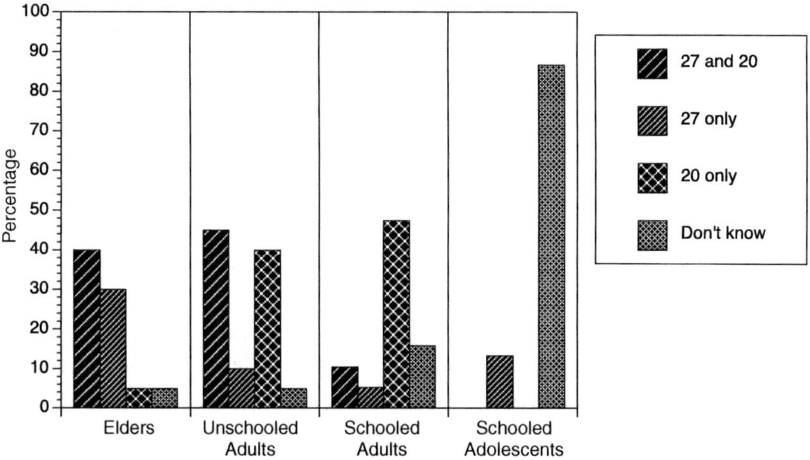

Figure 76. Percentage distribution of indications of places of *fu* on body after probe questions as a function of cohort.

either, Indigo and I used directed prompts, requesting them to consider whether using one or the other or both of the unmentioned body parts was a normative use of *fu*. Figure 76 shows the distribution of participants' responses by cohort.

The figure shows that a substantial portion of the unschooled populations – the elders and unschooled adults – indicate that *fu* can refer to two different positions on the body: the elbow (20) and the pinky (27). However, these populations also differ in claims for an exclusive position for *fu*. Whereas a substantial number of the elders (30 percent) indicate that the pinky (27) is the exclusive position, a substantial number of the unschooled adults (40 percent) indicate that the elbow (20) is the exclusive position. The schooled populations rarely indicated that *fu* had two positions on the body. Rather, a substantial number of schooled adults indicated that the elbow (20) was the exclusive position, and almost all of the schooled adolescents indicated that they did not know the position for *fu*.

Taken as a whole, these findings on the location of *fu* on the body indicate that the early conversations that pointed to the elbow (20) as the position of *fu* represented a widely used phenomenon. The use of *fu* at the elbow (20) was widespread through the middle-aged cohorts, though the link to *fu* with particular positions on the body rapidly disappeared in the younger generations sampled. The inquiry into the meaning of the Oksapmin word *fu* that I describe next provided important clues to the social history of the shift in position.

Currency Choice

To explore the way in which people translated *fu* into currency, Indigo and I presented an array of the following currency units: 1t, 2t, 5t, 10t, 20t, 50t, K1, K2, K5, K10, and K20. Participants were asked, "Show me how much money *tit fu* is."

Presented with the array of currency denominations, participants responded in varied ways. Some simply pointed to the K2 note. But other answers were also common. Some people added all of the currency together to determine its amount, and some of this addition appeared to be oriented to seeing whether the sum was "enough" to be called a *fu*. Other people pointed to a K20 note, referring to it as a *fu*, and still others pointed to several denominations, like the K20 and the K2 notes. In the following excerpts, I show the principal variations in responses.

Some focused on the meaning "plenty" in organizing their currency choice.

Elder no. 20. K2, K5, K10, K20 – I would just call them altogether *tit fu!*

Elder no. 19. If it is plenty of money, they will say *fu*.

Others justified their currency choice with reference to the K2 note as linked to elbow (20).

Elder no. 4. Selects the K2 note, counting (imaginary) shillings in the note with body parts, reaching *tan kir* (20), calling it *tit fu* and indicates that the note is called *tit fu*.

Elder no. 16. Explains that K2 note is elbow on other side (20).

Elder no. 13. Explains that 20 10t coins are inside the note, counting around to *tan kir* (20) and saying, *tit fu gona*. He explains that any value of money that has the value of 2 kina is *tit fu* – not just 20 10t coins. He also says that K2.70 could be called *tit fu*.

Unschooled adult no. 5. K2 note – shows elbow on other side (20).

Elder no. 12. Not only points to the K2 note, but also shows that the use of *fu* is not limited to the note itself, collecting the K1 coin and two 50t coins, indicating that this sum is *tit fu* as well.

Elder no. 3. Says 20 kina; illustrates a count of 10t coins to elbow (20) to indicate the 20 in the K20 note. Elder no. 3 appears to have the denominational structure incorrect, but shows the logic of selecting the K20 note, as many people did.

Elder no. 7. She indicates that she doesn't know *tit fu* – only *hanen* (5) *fu*, which she indicates is 10 kina, or "*kat-hai* (10) *fu*," which she indicates is 20 kina.

These observations provide further support for the conjecture that *fu* has shifted in function in recent Oksapmin social history – from meaning a complete group of plenty to perhaps serving as an operator that doubles the value of a body part, adding more nuance to the initial leading observations.

Some people simply identified the K2 note as *fu*, and these narratives provide information on a different variant on the meaning of *fu*, one that may well have been a progenitor of the doubling function. In the process, they defined *fu* with reference to other equivalent words that provide additional information on the meaning of *fu* in contemporary practices.

Unschooled adult no. 6. Says *fu* means the K2 note, or "*wan faun*."

Elder no. 15. Selects the K2 note. Says that *haben* (leaf) *tan* is *tit fu* and that also the K20 note is shoulder (10) *fu*.

Elder no. 2. Selects the K2 note, saying, "because it's *wan faun*, so it's one – *tit fu* means one."

In addition to identifying the K2 note with *fu*, unschooled adult no. 6 identifies the note as a "pound," whereas elder no. 15 uses another name, "flat leaf" (*hapen tan*), for the K2 note. Elder no. 2 puts the emphasis on "*tit*" in his identification, noting that he selects only *one* K2 note.

Some schooled adults appeared to incorporate Western-style base-10 structure principles in the interpretation of *fu*. In the populations sampled, this was idiosyncratic, and it may well be linked to applying a schema of Western numeration on the indigenous system.

Schooled adult no. 3. First chooses K10 note for *tit fu*. He then calls the K20 note *yot* (2) *fu* and the K30 note *yetir* (3) *fu*. Later in the interview, he says that the K20 note is *tit fu*, and that the K10 note is not *tit fu* – perhaps in a concern for consistency with the practice of identifying the K20 note as *kat-hai* (10) *fu*.

Schooled adult no. 5 (see also schooled adult no. 8). Like schooled adult no. 3, chooses K10 note as *tit fu* and indicates the K20 note is *yot fu*.

In these schooled individuals, it may be that *fu* is only rarely used, and that the press of questions is leading them to cite variants in use of *fu*. Indeed, these responses may well be efforts to make sense of only dimly known practices followed by an older generation and those who have not been to school.

To determine to what extent individuals referred to the K2 note as a *fu*, Indigo and I coded whether or not participants ever selected the K2 note as a reference for *fu* in their currency choices. Figure 77 reveals participants' responses as a function of cohort.

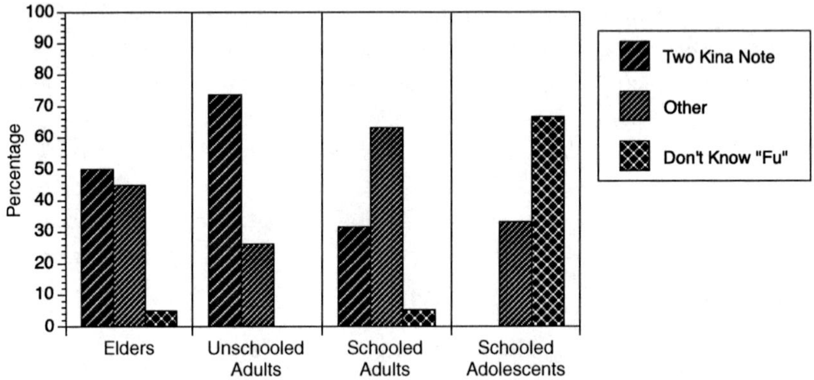

Figure 77. Percentage distribution of currency selection for *fu* as a function of cohort.

In all but the schooled adolescent cohort, some individuals knew the K2 note as a *fu*, but this identification was not common knowledge. Though most of the unschooled adults and half of the elders chose the K2 note as *fu*, only 30 percent of the schooled adults did so. Such findings distinguish *fu* from other pound equivalents, such as *haben* (leaf) or *iratarin* (leaf). None of the schooled adolescents chose this note as *fu*. Thus, knowledge of the link between contemporary use of *fu* and the value of 2 kina (or pound) is disappearing, and if it continues to be used in currency representations, its origins may well be invisible to younger generations.

Fu *Applied to Noncurrency Quantities: Sweet Potatoes*

In order to determine whether *fu* could be applied to non-currency, Indigo and I asked about whether sweet potatoes could be referred to with the word *fu*. Of those individuals who were familiar with *fu*, most indicated in the affirmative.

Most elders indicated that *fu* in the sense of plenty/complete or in the numerical sense could be used for sweet potatoes. Some used it in a qualitative sense of completeness, referring to a full string bag of sweet potatoes.

> Elder no. 5. Yes. If it goes to the other side – if it's plenty. He puts on his *bilum* (string bag) and says if it's filled, it's called "*tit fu gona*."
>
> Elder no. 19. Yes. If *bilum* inside contains plenty. So one full *bilum* is *tit fu* (very full! – with fists up and elbows out, demonstrating).

Other elders referred to a numerical value of sweet potatoes as a *fu*.

> Elder no. 3. Yes. I can count sweet potatoes to *tit fu gona* (20).
>
> Elder no. 12. Yes, *tan kir* (20) sweet potatoes.
>
> Elder no. 13. Yes, 20 or 27. He explains that in the old days, *tan hatatah* (27) was used for *tit fu*, but now because of money, *tan kir* (20) is used for *tit fu*.
>
> Elder no. 4. Yes, counting around body again to *tan hatatah* (27) and exclaims, "*tit fu gona!*"
>
> Elder no. 10. Yes, *tan hatatah* (27) sweet potatoes.

But some others said that the use of *fu* was limited to money. These were people who had been to school, with the exception of one unschooled adult.

> Schooled adult no. 15. First says yes, it would be one sweet potato, then changes mind and says you can use *tit fu* only for money.
>
> Schooled youth no. 1. No. can't say *fu* to sweet potato because sweet potato is not money.
>
> Schooled youth no. 6. No; because sweet potato is different from money.
>
> Unschooled adult no. 8. No; it is only used for money.

Figure 78 shows the wide agreement about the extension of *fu* to discrete material like sweet potatoes. Of the elders, all who knew a meaning for *fu* indicated that *fu* could be applied to sweet potatoes. It is of interest to note that in the case of its application to a noncurrency quantity, *fu* does not serve a doubling function, though for some, the reference to *fu* as 20 sweet potatoes shows the influence of the shilling–pound equivalence relation. Such results provide further corroboration with the Study 5-2 tasks involving stones, in which *fu* served to merely mark either the 20th or the 27th position in a counting activity and in a cardinal representation of number.

Representing Unusually Large Values

The questions about representing values of 50 and 60 were unsuccessful. Most people did not know, and we finally terminated the question. Further, the way that the question was posed relied on knowledge of Tok Pisin.

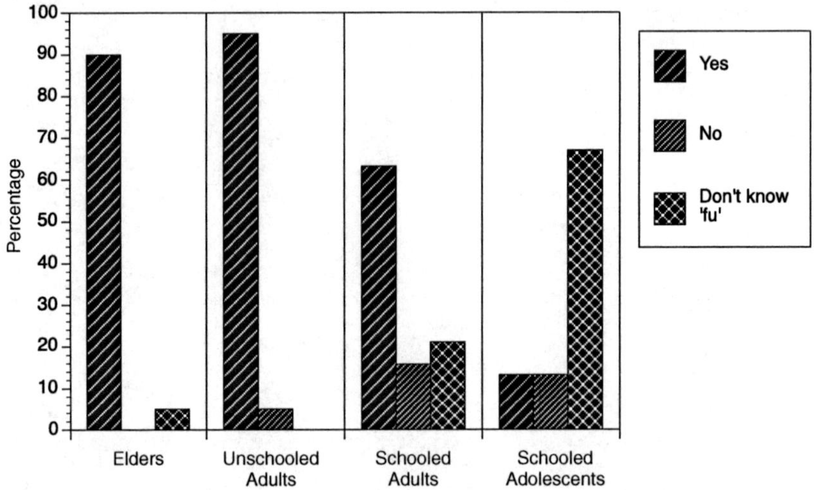

Figure 78. Percentage distribution of answers to whether or not *fu* could be used to refer to quantity of sweet potatoes as a function of cohort.

I take up this issue again in Part III, where the focus is on collective practices of schooling.

Summary

The data produced on the location, meanings, and currency choices for *fu* vary substantially over cohort. Together, the analyses reveal clear patterns as we consider the cohorts of elders through the schooled adolescents.

The majority of *elders* expressed the meaning of *fu* as something akin to a complete group of plenty, and they identified the location of *fu* at the completion of the 27th body part in a count (pinky). Half of them also indicated that *fu* was associated with the K2 note. The *unschooled adult* cohort displayed a similar profile, though many more identified the elbow (20) as the location of *fu* (75 percent), and fewer knew the complete-group-of-plenty meaning; the majority identified *fu* with the K2 note. The *schooled adult* cohort also tended to identify *fu* with the 20th body part (47 percent), and few identified the location of *fu* with the pinky (27); only a minority knew the complete-group-of-plenty meaning of *fu*, and a minority identified *fu* with the K2 note. Finally, the *schooled adolescent* cohort generally did not identify *fu* with a location on the body, nor know the complete group of plenty meaning of *fu*, nor associate *fu* with the K2 note.

The findings convey much about the shifting meanings and locations of *fu*. They not only corroborate the lead offered by our initial three observations, they provide much greater nuance and cover a much greater swath of the Oksapmin population. But they are reflections on *fu*, not analyses of *fu* as it is used in quantification practices. In the next study, I draw upon the data collected for Study 5-2 to explore the extent to which *fu* is used across cohorts, with a focus on the way particular conditions and values used in tasks may support greater or lesser usage of *fu*.

Study 6-2. Distribution of *Fu* Use in Representing Values of Currency and Numbers of Stones: Additional Analysis of Study 5-2 Data

Method

Participants and Procedures

The participants and procedures in this study were the same as in Study 5-2, since this study provides additional analyses of those data. Recall that across each of the five root numbers in each of the three conditions participants were asked to use the body system to tell us how much was presented. The three conditions included stones (task values: 4, 6, 11, 21, and 29), currency of lesser value (task values: K.40, K.60, K1.10, K2.10, and K2.90), and currency of greater value (task values: K4, K6, K11, K21, and K29). For the present analysis, I focus only upon occasions in which *fu* was used in the final representation.

Results and Discussion

Figure 79 indicates the mean percentages for which participants used *fu* as a function of cohort across the five conditions of tasks involving stones, currency of lesser value, and currency of greater value. These findings point to the broad spread of *fu* through multiple strata of Oksapmin communities. All cohorts used *fu* frequently for greater values of currency, especially when those values were at least 21 kina. Indeed, for values of 11 kina, 21 kina, and 29 kina, at least two-thirds of all middle-aged interviewees, both schooled and unschooled, used *fu* in their expressions. Among the elders, at least two-thirds used *fu* in their expressions for 21 kina and 29 kina. *Fu* was infrequently used to express lesser values or to count stones, and it was especially rare among schooled adults.

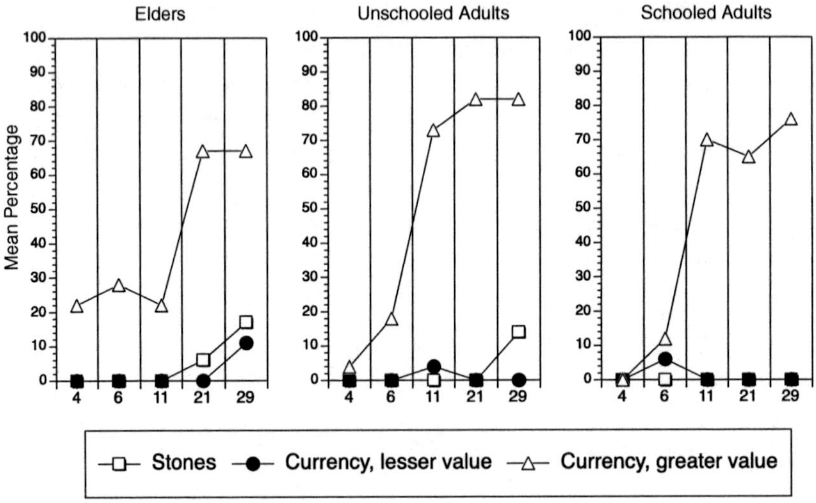

Figure 79. Use of *fu* by task, condition, and cohort.

Figure 79 provides further evidence that *fu* had come to take on the generalized meaning of a value of 2 kina (or a doubling function). Because the tokens were of various denominations across conditions, one can conclude that *fu* use was not restricted to a count of 20 10t coins or to a K2 note. To be clear on this point, consider people's summations for the task in which 21 kina was presented as one K20 note and one K1 coin. Over 60 percent of our participants used *fu* in their response, although there were no K2 notes or 10t coins present.

A Cautionary Note

In the interviews involving arithmetical compositions of currency tokens and stones, Indigo and I presented people with currency, recorded their responses, and then reconstructed the meaning of their representation – what amounts to conjectures about the microgenesis of their representations. This technique necessarily involves some educated guesswork, and leaves open lingering questions. For example, in general, the data did not allow for distinguishing between informants who used *fu* as a unit of 2 kina and those who used *fu* to double a unit of 1 kina. Even in simple examples, like using *kat-hai* (10) *fu* to represent 20 kina, there are several possible meanings. For example, the speaker could mean a count of 10 K2 units

(10 *fus*) or the speaker could mean double a count of 10 K1 units. Most likely, our respondents conceptualized the units of currency in a variety of ways, so the same response may take on different meanings in the quantitative activity of different people. For example, the schooled people that I met in Tabubil and the headmaster of a school in Oksapmin made explicit reference to doubling. But perhaps elders who made use of *fu* in which one could interpret *fu* as "double" were in fact conceptualizing *fu* as a count of 10 K2 notes.

Because of the centrality of varied ways in which *fu* might be generated in the microgenesis of quantification, I report a final study of *fu* use next, this time focusing on reading Hindu-Arabic numerals of the sort posted as prices in trade stores, with a focus on people's use of *fu* across cohorts.

Study 6-3. *Fu* Use in the Reading of Prices Written with Hindu-Arabic Numerals

In Chapter 3, I noted that many stores post the prices of goods. The originating purpose of the present study was to document the extent to which different cohorts could decode written prices. To this end, Indigo and I had presented prices that made use of the same root numbers used in Study 5-2, but requested participants to read written prices. In the course of analyzing people's translations of written prices into the Oksapmin language, I noted the use of *fu*, and it is the use of *fu* in reading written prices that is now my focus. The analysis of *fu* use in reading-prices tasks is important because it constrains conjectures about the microgenesis of *fu* that bears on the interpretation of performances on the Study 5-2 tasks. If people read Hindu-Arabic numerals without materials to sum and still used *fu* in the sense of double-a-body-part's-value, their doing so would be additional evidence supporting a doubling interpretation of *fu*. So, the purpose of the study on reading written prices took on a new function in line with concerns of this chapter about the use of *fu* in quantifications.

Methods

Participants

The same individuals who participated in Studies 5-1 and 5-2 reported in Chapter 5 participated in the present study.

Materials

Materials consisted of prices written with Hindu-Arabic numerals, similar to the posting of prices that occurs in many trade stores. We used the same eight values, each presented on a 2 inch × 2 inch piece of paper, similar to the presentation of posted price labels in many trade stores. The prices presented were the same as those used in the currency quantification tasks presented in Study 5-2: 40t, 60t, K1.10, K2.10, K2.90, K4, K6, K11, K21, and K29.

Procedures

Participants were presented with each price individually and asked to tell the interviewer the written price in the Oksapmin language. If participants were unable to read the written representation, this was noted and they were given the oral equivalent in Tok Pisin and asked to provide a translation into Oksapmin.

Results

Accuracy. All responses to the ten price presentations were transcribed and assigned to one of the following three categories: neither written nor oral price correct, only oral price correct, written price correct. Participants were then assigned three scores: (a) proportion of the 10 responses for which neither written nor oral price was correct, (b) proportion of the 10 responses for which only the oral price was correct, and (c) proportion of 10 responses for which the written price was correct. I then computed the mean proportion of each category score for each cohort. The results are contained in Figure 80.

Cohorts differed markedly in the accuracy of their translations. None of the elders read any of the 10 written prices. Some, however, could translate a Tok Pisin representation of the prices into the Oksapmin language, though this varied over price from about 11 percent (for K29) to 59 percent (for K4).

A substantial minority of the unschooled adults could read the prices. This ranged from 27 percent (for K29) to 36 percent (for 40t, 60t, K4, and K11). Most of the remaining unschooled adults could translate oral representations of the prices in Tok Pisin into written representations in the Oksapmin language. The exceptions were for values of K11 and K29, for which 36 percent and 32 percent could not read written or translate oral presentations of prices.

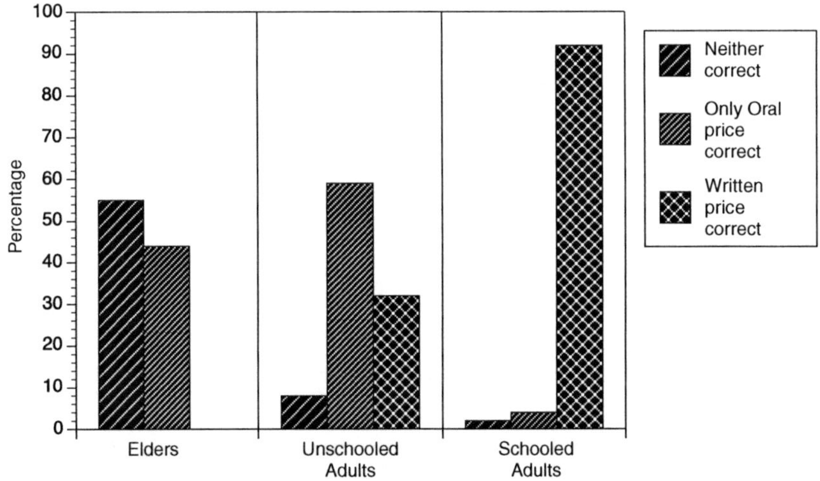

Figure 80. Mean proportions for reading/translating prices into Oksapmin by cohort on 10 written-prices tasks.

All of the schooled adults read prices between 40t and K2.90 correctly; all also read K11 correctly. For the remaining values, we found some variation, which ranged from 81 percent for K29 to 88 percent for K21.

Use of Fu

To determine the extent to which *fu* was incorporated in people's decoding of prices, I focused only on values from K4 through K29. Figure 81 contains the mean percentage of correct responses that did or did not contain the expression *fu*.

The findings in Figure 81 show that the use of *fu* varied in frequency over conditions and cohorts. The elders were less frequently correct than other groups, and for the larger value conditions, elders showed more correct responses with the use of *fu* than without (K21, 40 percent correct usage with *fu;* K29, 28 percent correct usage with *fu*). The majority of unschooled and schooled adults read prices correctly on the larger value tasks, and most of them used *fu* when they did read prices correctly.

These results are impressive in that they show that people, particularly the unschooled and schooled adults, use *fu* as a way to express large values of kina without engaging in a process of repeated addition of values. Still, these figures need to be interpreted cautiously. Alternative interpretations are that some people may simply learn identifications of particular verbal

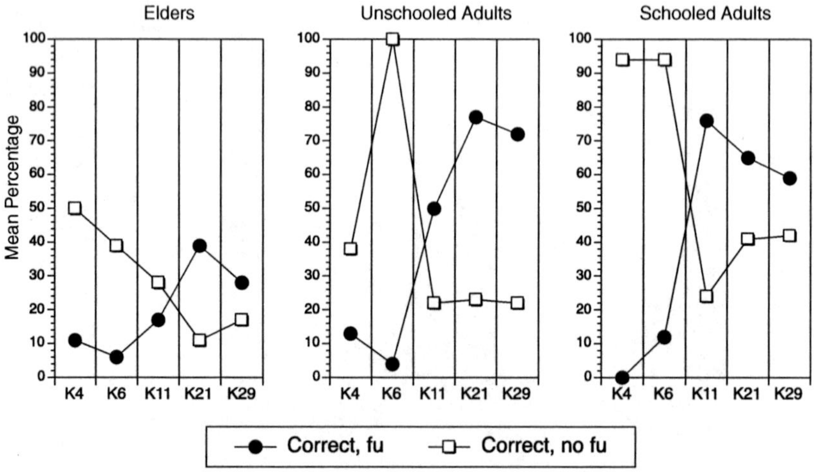

Figure 81. Percentages of *fu* use in written-prices tasks by cohort.

representations as single expressions – for example, that *hanen-fu* simply means K10 without any computation implied whatsoever.

Shifting Forms and Functions of *Fu* in Collective Practices of Exchange: Coordinating Synchronic and Diachronic Analyses

It should be clear at this point that there is an abundance of data to address the diachronic question about whether the exclamatory, fists-raised *fu!* that I was taught as a part of traditional counting in 1978 bears a genetic relation to the *fu* that serves a doubling function for many in 2001. My task now is to coordinate the elements of the case in a single analysis. Figure 82 incorporates the timeline previously used to illustrate the emerging differentiation between subsistence and cash economies (see Figure 15, Chapter 3), but I now use the figure to present an analysis of the shifting functions of *fu* over recent Oksapmin social history.

In the figure, I use diagonal right-moving arrows to represent new functions of *fu* and their roots in prior functions. The figure also contains two types of horizontal lines, each of which varies in thickness. The solid lines show rough approximations of shifting proportions of Oksapmin people using *fu* to serve an associated function over historical time (greater thickness signifies greater proportion of Oksapmin people using *fu* to serve the respective function). Similarly, the dashed lines depict related developments that are important for understanding the genetic history of *fu*,

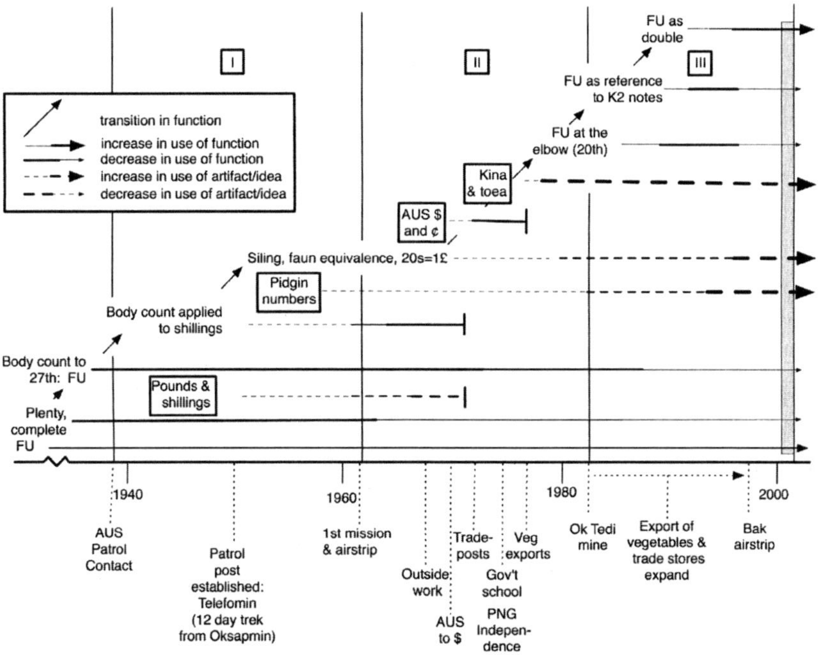

Figure 82. Shifting form-function relations of *fu* represented in synchronic time (vertical organization) and diachronic time (horizontal organization).

such as the use of currency systems and Tok Pisin. Finally, the vertically elongated, gray rectangle at the right is used to represent the time interval during the 2001 research visit to Oksapmin and the multilayered functions for which *fu* was used. I begin the analysis with the region marked by the elongated rectangle in the figure, a cross section of *fu* use in 2001, and then review the emergence of varied functions and meanings of *fu* over a short span of social history to the left of the rectangle.

In the early summer of 2001, the expression *fu* was used by people as a suffix that followed a body-part expression. My Oksapmin friends and acquaintances in Tabubil indicated that *fu* was used to indicate that a listener was to take double the value of the specified body part, as in *kat-hai* (10) *fu* (shoulder-cardinal double), which meant 20 or 20 kina. Indigo and I later found that this was true in the Oksapmin area as well. Had I understood more about current practices and knowledge at that time, we would have found that other segments of the Oksapmin population used *fu* to serve other functions, and that these functions were distributed unevenly through Oksapmin cohorts. The functions include the use of *fu* to refer to the K2 note, to mark the completion of 27 body parts, to mark the completion of 20 body parts, as well as its traditional meaning of a complete group of plenty. Through my varied empirical projects over the course of the three stints of fieldwork, I came to understand that the variety of functions that Indigo and I documented in 2001 had their genesis in prior historical periods. The forms of expression in which *fu* occurred are linked in complex ways to the functions that *fu* served over historical time. Notably, Indigo and I found no single person who was knowledgeable about what I now regard as the transitional links between these forms and the varied functions of *fu*.

Now, let's take a step back and consider the roots of *fu* use and transitional links informed by studies of this and prior chapters. In this analysis, let's consider the periods delineated in Figure 82, beginning at the left with a focus on the way form-function shifts are interwoven with changing collective practices of economic exchange over historical time.

Pre-contact Period

I noted earlier that archeology in the Telefomin area documents human activity 17,000 years ago (see timeline contained in Figure 82). But the long history of human habitation in this area is for the most part unknown and not recoverable. We do know that Oksapmin used two number systems before Western contact: the conversational number words (*tit* (1), *yota* (2), *yetir* (3), *yota yota* (4), and *hanen* (5)) and the 27-body-part counting system. How these two systems originated is subject to speculative analysis, based largely on the kinds of cross-group considerations I presented earlier.

There may have been considerable heterogeneity in representational practices in prehistory. One example of this heterogeneity is the existence of different methods for counting numbers higher than 27. As Study 4-2 documented, in 2001 unschooled adults and elders reported there were two different ways to do so, both of which have some support as pre-contact methods for counting. Many elders in 2001 (and most people in 1978) reported that to count beyond the 27th body part, one would move on a trajectory that proceeded from the little finger (27) and then loop back to the wrist (28th). In this method of counting, we find no recursive structure. In contrast, other elders and unschooled adults in 2001 reported a method of counting values beyond 27 that, after reaching the 27th position, began again at the thumb (1), suggesting a cyclic structure, one marked by *fu!*. But I found no reports of this method in 1978 or 1980. So, it is unclear whether these two methods coexisted early on.

Shift from Fu *as a Complete Group to the Completion of a Count to the 27th Body Part*

During pre-contact times – at least as far back as elders in 2001 could remember – *fu* was used in ordinary language to connote a complex of meanings involving "plenty," "group," and "complete." A complete count of 27 body parts became known as a more specialized *tit fu!*, in an expression that unites the 27-body-part counting system with the conversational number word *tit*. This shift, from the use of *fu* as a term in ordinary language to its use as a part of counting, constitutes a shift in function of the word form.

Post-contact: Period I

Body Count Applied to Shillings and Pounds

Sometime during the first post-contact period, Australian shillings and pounds were brought into the area by patrol officers and missionaries (see timeline contained in Figure 82). Tokens for currency may well have been targets of counting by some Oksapmin people. If they did count tokens, it is likely that the body system would have been applied without respecting the tokens' denominations. Even as late as 2001, trade store owners and clerks reported that some of their elderly customers were unfamiliar with the equivalence relations between currency denominations; those customers believed that more coins meant more money, regardless of the value of each coin. The story about the trade store owner who accepted only 10t coins (see n. 32, Chapter 4) hints at the difficulties people must have had in quantifying the relationships between the various currency denominations. In itself, the activity of counting shillings and pounds was not a shift in function of body counting, but perhaps foreshadowed a subsequent shift in which the 20th body part, as a representation of 20 shillings, became treated as a value equated with the value of a pound.

Introduction of Tok Pisin/English Numbers and Terms for Currency

Increasing contact with early patrol officers and missionaries introduced Tok Pisin and English to the area, providing a new language for number. During this early period only a small portion of the Oksapmin population interacted with missionaries and patrol officers, and it is likely that few of them became fluent in the new languages through this early contact. Indeed, in 1978, Tok Pisin or English was spoken by a minority of Oksapmin people, primarily those who attended school (where they learned English) or had worked outside of the area on plantations (where they encountered Tok Pisin).

Post-contact – Period II

Pound-Shilling Equivalence and the Elbow (20) as a Pound

During Period II, with the early emergence of local trade stores, some parts of the Oksapmin population engaged in currency exchange at different sites (see timeline contained in Figure 82). Some people came to equate a count of 20 shillings with a pound, with the 20th body part taking on a privileged position in this translation. Because there were an increasing number of trade stores established during this era, the practice of equating the 20th body part with the pound likely emerged at one or a small number of sites. This practice was reproduced and adjusted to local circumstances, perhaps "snowballing" through networks of interlocutors across an increasing number of sites. Of course, such knowledge was distributed unevenly in Oksapmin communities, with *fu* as the 20th body part becoming more commonly known among that portion of the population who engaged in economic exchange with currency.

From Australian Pounds and Shillings to Australian Dollars and Cents to Papua New Guinea's Kina and Toea

In the late 1960s, the currency used in the Oksapmin region changed from Australian pounds and shillings to Australian dollars and cents, and then again in the late 1970s it changed to Papua New Guinea kina and toea. Although the currency system changed, the denominations of bills and coins closely matched each other, and it is probable that some Oksapmin terms for currency stayed the same. For example, the words *faun* (pound) and *siling* (shilling) were still commonly used in 2001 to denote K2 notes and 10t coins. Over time, with roots in prehistory, a complex system of language for currency evolved that combined body-part number words, conversational number words, and ordinary Oksapmin words that are used to name currency.

Post-contact – Period III

With the striking influx of currency and increased exchange during period III, varied linguistic devices emerged to represent quantity in everyday activities, some based on the Oksapmin language and the body system, and some on Tok Pisin and English (see timeline contained in Figure 82). *Fu* is one of the varied devices used in the Oksapmin language.

Fu *as the Elbow (20th Body Part) and K2 Note*

The association between the 20th body part and the pound was likely a precondition for *fu* to take on a new meaning in quantification. The *faun* (pound) marked a complete count of (plenty) *silings* (shillings) at the elbow (20). *Fu* had been used to mark a complete count of 27 body parts, but with the use of shillings and pounds as countable objects, "complete" took on a new meaning. The shift from the 27th to the 20th body part for a *fu* appears to have grown out of a shift in sense of a "complete" number of shillings. In this way, it appears that *fu* took on both the meaning of elbow (20), as a complete count of the body parts that constituted a *faun*, and the meaning of *faun* itself, as a complete group of *silings*. The treatment of *fu* as a pound equivalent and *fu* as the K2 note equivalent may have occurred in a number of ways. One way might be linked to the common 20:1 exchange ratio between shillings and pounds, on the one hand, and between 10t coins and K2 notes, on the other. The diagram contained in Figure 83 shows the dual 20:1 ratio with *fu* as signaling that the pound and the K2 note each represent a complete group.

Fu *as Double the Value of a Body Part*

Using *fu* as the completion of a count of 20 *silings* (or 10t coins) became a useful way to refer to larger values of currency. The generalization of *fu* to refer to a value of 2 kina may have occurred relatively quickly in the transition. This meaning has taken root and spread so that in 2001 when some people represent greater values of kina in the Oksapmin language, they use *fu* with the meaning of double the value of a body part. When they count noncurrency units, such as stones, *fu* has maintained its prior meaning for some people as the completion of the last body part in the Oksapmin counting system.

Multiple Functions of Fu *in 2001*

Studies in Chapter 5 and the present chapter illustrate well the various functions of *fu* in 2001. Some of the different ways that people, in effect, double the value of a body part associated with the use of *fu* may differ from one another in subtle ways. For many unschooled people, the currency system

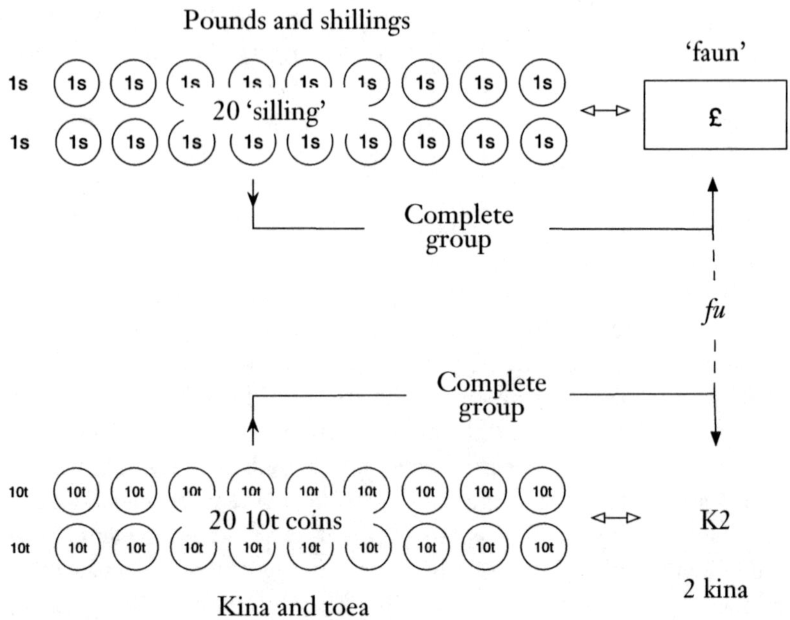

Figure 83. *Fu* as a mediating link between equivalence relations in the system of pounds and shillings and the system of kina and toea.

may be conceptualized as a multiunit system with 2 kina and 10 toea as the principal units: A value of say, 14 kina is conceptualized as 7 pounds or *bes-hai* (7) *fu* (shoulder (7)-*fu*). In this approach, *fu* is not conceived as performing a doubling function but may be more like a summation of seven K2 notes (14 kina). However, others may consider *fu* as a multiplier. This orientation may lead people to interpret expressions like *bes-hai fu* as 7 kina times 2, yielding 14 kina. For still others, particularly those in the mining area of Tabubil, *fu* appeared to mean simply to double the value of a body part (independent of the currency system). Of course, any one individual may cognize *fu* in all three ways along with more traditional uses, like to refer to a complete group of plenty or the completion of a count.

Collective Practices and the Micro-, Socio-, and Ontogenesis of Form-Function Relations

In their critical remarks on treatments of culture reviewed in Chapter 1, Sperber and Hirschfeld argued against the received view that cultures are homogeneous, well bounded, and coherent. They argued that such conceptualizations point only to fictions, and that there are no "cultures" per se, only cultural phenomena such as artifacts and sign forms, norms, social organizations, and values. According to Sperber and Hirschfeld, these phenomena are the products of "population-wide distribution processes" that vary in their properties and concentrations through communities. The process-oriented aspects of their formulation resonate well with my argument that individual action in relation to collective activity is the locus of both the reproduction and the alteration of cultural phenomena, whether collective practices of daily life or cultural forms of representation. The conceptualization provides leverage for understanding the findings I have reported on uniformity and variation during any single period of time (synchronic analysis) as well as continuity and discontinuity over historical time (diachronic analysis). Consider a brief review of findings from Chapters 3 through 6 with both synchronic and diachronic lenses.

Chapter 3 reviewed continuities and discontinuities in the social organization of collective practices of exchange over recent history, from trading partners in prehistory (without Western coinage) to the proliferation of trade stores. Through time, the collective problems that emerged for interlocutors shifted, as new forms of exchange with Western coinage were introduced, as currencies shifted, and as the amount of money and goods in circulation changed. Through this period, people were recruiting forms of representation to communicate about and accomplish varied communicative and problem-solving functions, as they reproduced and altered the organization of exchanges as they participated in collective practices.

In Chapters 4 through 6, I reported on studies of the body system form and functions that it served in the shifting collective practices of exchange. Study 4-1 revealed variant arithmetical functions used to solve arithmetic problems in 1980; for some people, the function of body counting was limited to external correspondences between body parts and objects, whereas for others, the function of body counting was extended to the generation of internal relations between body parts to keep track of arithmetic compositions and decompositions. Study 4-2, conducted in 2001, documented multiple variants for treating the ending of a count, some ending with the thumb (23) to the little finger (27), the form I first learned in 1978. Others ended from the little finger (23) to the thumb (27), the ending that a neighboring group, the Bimin, reportedly used in 1978. Study 4-3 documented variant language forms for counting and quantification in economic exchange; these forms included quantifiers in Tok Pisin and Oksapmin that were linked to age and amount of schooling. Finally, Study 4-4 revealed that the ways in which exchanges occurred were distributed differently through the population, with exchanges by elders taking on different properties than exchanges by younger, schooled populations.

Chapter 5 shifted the focus to naming practices for currency tokens and their use in arithmetic compositions. Study 5-1 showed that, in the Oksapmin language, people used a range of names to index each common token, and for the K5 note, the list of names was extraordinarily long; in contrast, there was a striking lack of variations for Tok Pisin. There were different patterns of coherence and incoherence in use of naming practices in different cohorts. People often mixed shilling/pound units with kina/toea units, resulting in some paradoxes. For some, numerical word forms in everyday use took on opposing meanings; for example, when one person called the 10t coin "one" (thumb (1) or *tit* (1)) and the K1 note, "ten" (shoulder (10)). Another example, this time in the Tok Pisin language, is calling the 10t coin "ten" and the K1 coin "one kina." To an outsider, these contrasting values may appear incoherent; to an insider, the forms work to communicate intended meanings.

The Chapter 6 focus on *fu* provides some insight into the micro-, onto-, and sociogenetic processes that lead to both uniformities and variants in form-function relations over historical time. In microgenetic constructions, people made use of word forms like *fu* to communicate intended meanings, anticipating breaches as they adjusted their uses to local circumstances. Such public uses of *fu* supported sociogenetic processes – people's reproduction and alteration of the function of *fu* keyed to shifting collective problems. Historical shifts in function that result from the iterative process of reproduction and alteration of uses of *fu* include the use of *fu* to refer to (a) a complete group of plenty, (b) all the 27 body parts of a count, (c) the attenuated position (20) on the body as a complete group of shillings, and (d) a doubling of the numerical value that a body part signifies.

As a group, the studies on form-function relations linked to collective practices of economic exchange reveal, in some striking ways, the interplay between cultural and cognitive processes in the cultural history of mathematical ideas – in momentary interactions between interlocutors and in communities over historical time. In the next part of the volume, I extend the conceptual framework but now with a focus on a very different set of collective practices – those linked to Western-style schooling.

Part III

Schooling

In prior work on cognition and schooling in traditional societies like Oksapmin, investigators have often asked whether schooling influences cognitive development in any fundamental way. To address the question, investigators have taken advantage of the rare opportunity to contrast schooled and nonschooled populations living in the same traditional communities using Western measures of cognitive development. Foci of study have included a wide range of general cognitive abilities, including perception, memory, concept formation, and logical problem solving (for reviews, see Cole & Hatano, 2007; Rogoff, 1981).[1] Though the comparative approach has led to fairly consistent findings of associations between schooling and cognitive development, the approach is subject to critique. One critique is methodological, questioning the extent to which the use of standard measures can produce valid and reliable estimates of cognition in traditional societies. Another critique is conceptual, probing the appropriateness of treating schooling as a packaged independent variable that determines the cognition of individuals.

My work on schooling takes a very different approach from the standard comparative treatments of schooling and cognitive development. Like economic exchange, collective practices in the classroom are understood as jointly constituted by actors, whether adults or students. Rather than conceptualizing schooling as a direct cause of cognitive development of students, my focus is on the dynamics of the reproduction and alteration of

[1] Researchers have suggested a range of possible mechanisms for the documented effects. These include the introduction of explicit verbal instruction out of the context of the object of instruction (Bruner, Olver, & Greenfield, 1966; Vygotsky, 1986), the emphasis in school on the search for general rules (Cole et al., 1971), and the introduction of specific skills taught in school such as literacy or numeracy (Olson, 1976).

forms of numerical representation and functions they serve as students and teachers participate in collective practices of classroom life.

As with my work on economic exchange, my concerns in the study of schooling and Oksapmin mathematics are with continuities and discontinuities in the reproduction and alteration of form-function relations over a stretch of Oksapmin history. Chapter 7 provides a historical sketch of the emergence of Western-style schooling in Oksapmin communities. During a 40-year stretch, some traditional practices for educating youth have waned and schooling practices have shifted in some marked ways. Chapters 8 through 11 contain studies drawn from three periods of fieldwork, in 1978, 1980, and 2001. Chapter 8 focuses on evidence of form-function shifts in unschooled Oksapmin children's body counting in 1978 and 1980. Chapter 9 shifts the focus to children in school during the same era and the way they have developed uses of the body form to serve new functions linked to collective practices of classroom life. In Chapter 10, I examine shifts in classroom practices associated with the 1998 school reforms and ways that new school practices support shifts in form-function relations involving both the body system and Tok Pisin. In Chapter 11, I close with a discussion of the way new collective practices mandated by school reform have had unintended consequences; my focus is on new alterations in the body-counting system and Oksapmin word forms for naming currency tokens.

An overarching theme in Part III is that, as in the case of economic exchange (Part II), representational forms and functions are reproduced and altered as the unwitting consequences of people working to accomplish local goals in communicative and problem-solving activity. My analyses of form-function relations in Part III builds on my prior treatment of microgenetic, sociogenetic, and ontogenetic processes, but now with a focus on students and teachers. As in the studies of economic exchange, I use varied research designs and empirical techniques across the chapters to support both synchronic and diachronic analyses. Figure 84 provides an overview of these sources by chapter and study.

Study/period number, (Date), Title	Empirical techniques	Data sources/ designs	Principal forms & functions of interest: Forms	Functions
Chapter 7. Collective Practices of Schooling in Oksapmin: A Brief History				
Period 1. Prehistory through the Baptist Mission station (1962)	Review of archival/ scholarly reports	Archival/ scholarly reports	Not applicable: No Western schooling	
Period 2. The Bible school	Informants	Informants	Not applicable: No Western schooling (in math)	
Period 3. The Community Schools	Review of archival records; ethnography	Archival, ethnographic	Body parts, Western mathematics	Counting, arithmetic, communication
Period 4. The school reforms	Review of archival records; ethnography	Archival, ethnographic	Body parts, Western mathematics	Counting, arithmetic, communication
Chapter 8. Unschooled Children's Developing Uses of the Body System				
Study 8-1 (1978). Counting to compare and reproduce sets	Task-based interviews	Cross-sectional age comparisons	Body parts, stones	Compare and reproduce discontinuous quantities
Study 8-2 (1978). Comparing numerical values of body parts	Task-based interviews	Cross-sectional age comparisons	Body parts	Compare body part value
Study 8-3 (1980). Measuring with the body	Task-based interviews	Cross-sectional age comparisons	Body parts, string bags	Measurement
Chapter 9. Children's Adaptations of the Body System in School: An Unintended Consequence of Post-Colonial Schooling				
Study 9-1 (1980). Observations in classrooms	Classroom observations	Cross-cohort grade level comparisons	Body parts, written arithmetic problems	Solving arithmetic problems in written format
Study 9-2 (1980). Arithmetic with the body: Unintended consequences	Task-based interviews	Cross-cohort grade level comparisons	Body parts, arithmetic problems	Solving arithmetic problems
Chapter 10. About Twenty Years Later: Schooling and Number				
Study 10-1 (2001). Third grade school mathematics	Classroom observations	Observation	Number in three languages; written forms	Solving arithmetic problems
Study 10-2 (2001). Quantification practices known by Grade 3 children	Task-based interviews	Within group contrasts	Body parts, Oks. & Pisin quantifiers, written number	Counting, arithmetic with currency, reading prices
Chapter 11. Teachers and Students as (Unintentional) Agents of Change				
Study 11-1 (2001). Observations of classroom instruction	Classroom observations	Observation (case studies)	Written number, quantifications in the Oks. language	Counting, addition, subtraction, multiplication
Study 11-2 (2001). Elementary lesson plans	Interviews about lesson plans	Interviews (case studies)	Numerical inscriptions, Oks., Pisin number, instructional artifacts	Counting, addition, subtraction, multiplication

Figure 84. Research designs and empirical techniques used in chapters on schooling.

7 A Brief History

Collective Practices of Schooling in Oksapmin

Schooling's growth in the Oksapmin world was dramatic in the period from the mid-1960s through 2001. Evidence of growth is reflected in the number of schools in the area and in student enrollment capacity of the schools. As depicted in Figure 85, starting from the initial Bible school in the mid-1960s, which only a few adults attended, schools have proliferated over the years. Further, today many schools are specialized: Elementary schools serve younger children, community schools serve older children, and after the end of my last field trip in 2001, a high school was instituted in the Oksapmin community (2005).

Figure 85 partitions the penetration of Western schooling in the Oksapmin world into four principal periods. Period 1 occupies the time just prior to the introduction of Western schooling. Period 2 marks the introduction of a Bible school for adults that accompanied the Baptist mission in Tekin and extends to the establishment of the first community school. Period 3 is a relatively long stretch of time that includes the rise of the community school during the colonial and postcolonial periods, and ends with the introduction of a wave of elementary schools linked to the 1998 reforms. Period 4 begins with the 1998 reforms.

No Western Schooling (Before 1962)

The pre-Western period represented in Figure 66 extends from Oksapmin prehistory through the founding of the Bible school at the Tekin station in the early 1960s.[1] To understand the alien institution of Western schooling

[1] Weeks reports that before the mission opened in Tekin, about a dozen people from Oksapmin attended a mission school in Telefomin, a trek of about 80 kilometers across rugged, mountainous territory (see Weeks, 1981b).

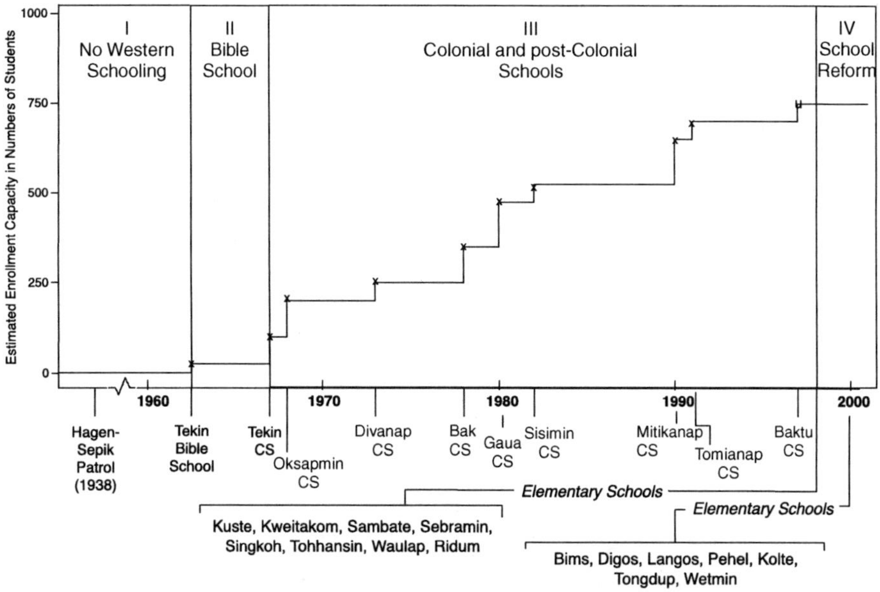

Figure 85. Community school and elementary school openings in the Oksapmin area and shifting enrollment capacity of schools. The figure was produced using information provided by Chris Mamsi, Elementary Coordinator, Sandaun Provincial Administration, Telefomin District Administration, Telefomin, Sandaun Province. "CS" is an abbreviation for community school.

from an Oksapmin perspective, it will be useful to consider the way that Oksapmin adults educated their youth prior to Western contact.

Traditionally, Oksapmin used varied forms of education. One form common to Mountain-Ok communities was initiation, a process through which important knowledge from one generation was taken up by another (Barth, 1987). Initiations were specific to males. Females had a milder marking of a passage to womanhood. For boys, initiations were physically challenging, and they occurred in stages during which secrets of Oksapmin society were communicated. Some information on these practices can be found in Weeks (1981d), Boram (1980), Brutti (2000), and Barth (1987). The documentation of Oksapmin initiations is weak, in part because they were secret and communicated cultural secrets. Boram (1980, pp. 364–369) describes initiation cycles in Oksapmin that included the passing down of knowledge and skills required for taro gardening as well as hunting and family management. Weeks (1981b, p. 99) reports the range of knowledge that was passed down during the long cycles of initiations as boys advanced through the protracted process, as well as aspects of the process:

> Rituals related to hunting, fishing, health, and family management were also taught. Many of the secrets reserved for men were taught. These included a secret language, names, songs, houses for spirits, foods, secret stones, bones and skulls of the ancestors, and most important, the secrets of the initiation itself.
>
> Through the process of the initiation youth were introduced to the privileges of adult males – hunting marsupials, eating marsupials with salt, the use of a bilum [string bag] with feathers, the special topknot on the head, playing the Kundu [specialized drum], but most important the right to sit with other men and discuss issues.
>
> The phases of the initiation are carried out with a great sense of drama and surprise. Each new building is introduced in a special sequence [buildings are way stations for different stages of the initiation]. The use of pig grease and natural red colours is done in such a way as to startle. Initiands are shocked when a fire is made on their chest, or they are led into a pool of water while being told they are going to be castrated with bamboo knives.

Prior to the introduction of Western schooling in Oksapmin, informal education also occurred between fathers and sons as well as between mothers and daughters (Weeks, 1981b). At around five years of age, boys and girls both began active learning of skills important for participation in community life. Boram (1980, pp. 267–268) provides a chart of 29 areas, how they were learned, by whom, and at what age. Boys, for example, helped their fathers cut down trees, clear bush, gather firewood, find edible insects, and raise pigs; they also began to play with toy bows. Girls helped their mothers retrieve water and attend to younger siblings, gather firewood, cook, and help with weeding gardens; they also learned food-gathering activities like hunting for tadpoles, frogs, insects, and rats. Girls also learned how to make string from the bark of a tree, the material used for making string bags (described in Study 8-3).

Weeks (1981b, p. 97) provides the following description of the kinds of knowledge acquired by older children:

> At eight or nine, boys are introduced to more complicated tasks like making rope from vines, harvesting crops, hunting birds, making bark cloth, and carrying loads. Girls at this age can now care for babies, make grass skirts and bilums (string bags), plant seeds, and join in the above tasks of boys like harvesting crops, making bark cloth, and transporting produce.

> At puberty youth join in much heavier tasks like house building; the girls gathering leaves for the roof, the boys poles and bark for the walls and floor. The girls must build their own menstrual huts. Now the boys go into the mountains with their fathers to plant karuka nut trees (pandanus). They are big enough to make garden fences, to learn to snare wild game, and to go on trips to exchange goods.

In 1977–1978, Weeks organized an interview study in which elders were asked about traditional skills that were disappearing, skills once used but no longer taught to developing children. The elders described skills related to tools, like using stone axes, files, and traditional knives. Weeks also reported the lack of education about varied practices that he describes as dominant in the pre-contact period but fading or absent in 1977:

> Examples of these would be rain making, traditional childbirth techniques, ritual garden magic for taro and *kaukau* [sweet potato], love songs to win affections, and medicines to protect men from the poison in a woman's vagina. The art of poisoning, they said is dying out. The Taman (witch) eating another's soul is going. Divination to tell why someone has died and who killed them is not being taught. In the past every man and woman could learn Sanguma from his or her parents. It was used to kill only within the clan, relatives, brothers and sisters. Also dying out are the skills of making and using traditional medicines, collecting plants for medicines, collecting plants for poisons, and how to make natural dyes. (Weeks, 1981b, p. 97)

The knowledge forms and educative processes prior to the entry of Western life provide a revealing backdrop for the alien practices of Western schooling and the knowledge forms promoted.

The Bible School (~1962)

The early missionaries to the Oksapmin area introduced the first Western-style schooling. As noted previously, the Baptist mission was established in Tekin in the early 1960s. The missionary recruited students from local hamlets for Bible study. I have not found much documentation of this era, except that a key concern was to translate the Bible into Oksapmin, and to this end, turn the oral Oksapmin language into a written language. Further, by engaging a small group of Oksapmin in Bible study, missionaries hoped that the evangelical concern to spread God's word could be accomplished by Oksapmin people themselves. I suspect that instruction in this first school in the Oksapmin community was not related to number, but I have no information about this.

Colonial and Postcolonial Community Schools (1967)

The first wave of government-sponsored schools came to Oksapmin when Papua New Guinea was a territory of Australia. The first community school was established in Tekin in 1967, and other schools followed. The teachers came from other regions of Papua New Guinea.[2] They were unfamiliar with Oksapmin language and practices. English was the primary language of instruction, though teachers also used some Tok Pisin in noninstructional contexts.

Schooling was colonial in style, with a focus on Western school skills. The official school rules did not permit children to speak the Oksapmin language while on school grounds (Gearhart, 1980).[3] Teachers themselves, with some exceptions, knew little about the Oksapmin community and did not understand the Oksapmin language or local knowledge systems (including number). Because of language differences, teachers did not communicate with parents and tended to stay in their positions in Oksapmin for only one-year stints. Indeed, they viewed teaching in the Oksapmin region as an undesirable assignment and they often felt not adequately compensated.

Weeks (1981b) collected systematic data from teachers on schooling during this era. Weeks' focal questions were on delivery of schooling to the Oksapmin community from the points of view of teachers. Weeks wrote:

> Most of the teachers feel isolated. They sense that they have been dumped in Oksapmin. They feel neglected and forgotten. One Headmaster complained of writing eight letters to Vanimo in eight months and having had replies to none of them. They complained of a lack of supplies and support, that the inspector came infrequently and never on schedule. Some teachers have actually had to go to Vanimo to get supplies for their school (and would argue their absence from the school was justified). The teachers complain of a lack of support from parents, that few show up for work days. In turn the pupils are forced to do most of the construction and maintenance work for the schools. The large age range of pupils makes this possible, as one Grade 1 can range from 8 to 15, there are enough big pupils to do the heavy work....

[2] Clifford Boram, an early Oksapmin teacher who eventually earned a degree in anthropology, was an exception (Boram, 1980).

[3] These rules were variably enforced.

> On community relations they complain not only of the lack of parental support, but of the difficulties in communicating where few of the adults speak Tok Pisin and where the teachers do not know Tok Ples [the Oksapmin language]. Attempts to work through the children or translators they have found unsatisfactory. (1981b, p. 131)

It was during this era that I conducted fieldwork in Oksapmin (during 1978 and 1980), and during the latter period, Maryl Gearhart studied classroom practices at Divanap Community School (Gearhart, 1980), also a site of my own work.

At Divanap Community School, the curriculum consisted of six grade levels (Grades 1–6), although not every grade was taught during any given year. The math curriculum was Western-based and included lessons on counting (with the Western system) and arithmetic (using concrete manipulatives). Emphasis was on drill and practice of arithmetical facts and memorization of tables, often implemented as classroom chorus drills. The focus of the curriculum was academic – school subjects included English, math, science, health education, social studies, community life, and physical education. Syllabi and curricular materials were created by the National Department of Education; the school's intended function was to prepare students to take on leadership roles in the nation, and school curricula included little that was relevant to everyday community life in a remote area like Oksapmin. Figure 86 contains a 1980 photograph taken of a morning assembly at Divanap Community School.

The School Reforms (1998)

The last wave of schooling shown in Figure 85 was ushered in with the national educational reforms introduced in 1998 (for a discussion of Papua New Guinea education policy, see Litteral, 2000). Reacting to the prior colonial system, national leaders wanted to reform the schools, both to preserve traditional ways and to prepare citizens to participate in community decision making as well as further training for jobs in rural and urban contexts (PNG Dept. of Education, 2003, p. 4). Governmental reforms created government-run vernacular schools (called elementary schools) for young children. Oksapmin elementary schools taught young students initially in the Oksapmin language, and the language of instruction gradually bridged to English by the end of their third year. Elementary education was linked closely to community activities. It was intended to build upon the skills and knowledge the students brought to the school from their own language and

Figure 86. Morning assembly at school in Oksapmin area (1980).

community and to engage the community in the life of the school. After three years of instruction, students transitioned to community schools for Grades 3 through 8.

In 2001, Oksapmin children and adolescents could progress through three levels of schooling – elementary schools, community schools, and high schools. Elementary schools included three grades – Preparatory (Prep), Elementary-1 (E-1), and Elementary-2 (E-2). During these three years, the language of instruction was the Oksapmin language, bridging to English by the end of E-2. Subsequently, children attended the primary grades (community schools), years three through eight. During this period, the language of instruction was primarily English, with some Oksapmin and Tok Pisin. In high school (Grades 9–12), English was the language of instruction. Up until 2005, students had to take up residence in other parts of Papua New Guinea to attend high school. A high school was established in Tekin in 2005.

The contrast between the reform period and the early colonial period of schooling is noteworthy. In the reform period, many native Oksapmin teachers taught in schools located in or near the hamlets in which they grew up. All teachers knew the Oksapmin language and knew the traditional and out-of-school collective practices that were a part of children's everyday lives. Teachers were also involved in community life.

8 Unschooled Children's Developing Uses of the Body System

Prior to the introduction of schooling in Oksapmin history, what were the developmental patterns in children's acquisition of the body system to serve numerical functions? Of course, I could not answer this question directly. Schools had been introduced prior to my first visits to Oksapmin. But during my early visits, I was able to interview school-age children who had not attended school with a focus on the use of the body for counting. I report on these studies in this chapter. From my contemporary perspective, the studies serve two important functions. First, they provide information on developmental processes in unschooled children's knowledge of the body form to serve numerical and measurement functions during a period in Oksapmin history when unschooled children populated Oksapmin communities, and the body system was a principal form for numerical representations. Second, they provide clues about the knowledge of body counting that children who attend school may draw upon to interpret instruction in school.

In Study 8-1, my focus is on the form-function shifts in children's use of their bodies instrumentally to compare and reproduce arrays of objects. The study is comparable with work that I had done previously in the United States on young children's use of number words to serve as intermediaries in comparing and reproducing quantities (Saxe, 1977, 1979). In Oksapmin, I wanted to leverage this work to explore whether Oksapmin children would show a shift from use of nonnumerical functions of body parts in naming activities to numerical functions. The shift would be precursory to the progression from external correspondences to the varied internal correspondence strategies associated with arithmetic previously documented in Study 4-1 on economic exchange (and documented in schooled children in Chapter 9). In Study 8-2, I ask related questions about the shift to numerical functions. My focus is on whether the spatial organization of the system around the body leads to system-specific challenges for developing Oksapmin children, problems that children reared in communities with only number word systems do not face (for instance, children growing up in other parts of Papua New Guinea or the United States). In Study 8-3, my focus turns to the Oksapmin practice of using the body in linear measurement – to measure string bags. Across studies, but especially in Studies 8-2 and 8-3, my purpose was to explore the nature of form-function shifts using the body that might be specific to Oksapmin communities, with a focus on counting and measurement.

Study 8-1. Counting to Compare and Reproduce Sets

My work in the United States on early number development set the stage for this study on body counting in Oksapmin. In my studies in the United States, I had presented children as young as three years of age with numerical tasks that involved the numerical comparison of arrays as well as the numerical reproduction of an array with an available set of objects (Saxe, 1977, 1979; Saxe, Guberman, & Gearhart, 1987a). I found that younger middle- and working-class children in my sample could use number words when asked to compare or reproduce groups of objects, but they did not use number words instrumentally – as an intermediary to serve numerical comparison or reproduction functions. For example, for numerical comparisons, these young children would count one array or two arrays continuously as if they were one and then offer a judgment, rather than using their counting to mediate a judgment. Sometimes, when set sizes were reduced to small numbers, the children's use of counting to serve intermediary functions for number was somewhat more frequent.[1] Older children did use counting to mediate their numerical comparisons and reproductions of groups. The shift in function of count words in the United States thus led to the present study. In Oksapmin, I asked, Do young Oksapmin children show knowledge of the body form but not use the form as an intermediary to support numerical comparisons and reproductions? If so, what is the character of the progression? To explore the questions, I adapted tasks to materials and conditions in the Oksapmin world, systematically varying levels of set size in task designs.

[1] As I noted in Chapter 1, the finding on set-size effects is common to research on early number development with a range of number tasks (e.g., Gelman & Gallistel, 1978; Piaget, 1952). Small set sizes may have properties that support children's developing numerical capabilities, and subsequent work confirms this (Starkey, Spelke, & Gelman, 1990; Wynn, 1992; Xu & Arriaga, 2007).

Methods

Participants

The 1978 sample included 28 children, including thirteen 5- to 9-year-olds, six 10-year-olds, and nine 11- to 16-year-olds. Many children had no schooling at all; others had nominal experience. Although the ages of some of the children were determined through missionary birth records, others were estimates based upon visual comparisons of children of known ages with children of unknown ages. Since many of these ages were only approximations, a procedure to be described later that would allow for additional confidence in age group comparisons was used. The children were considerably older than the sample in the United States. There were two reasons for this. First, I tried interviewing younger children, but my alien status proved a formidable barrier in this effort, and I abandoned this tack. Second, I found that the age norms for the use of counting as an intermediary appeared to be later in Oksapmin, so the age distribution that I sampled appeared adequate to assess age-related patterns in shifting functions of body counting.

Materials

All materials used in the tasks were local. For the comparison tasks, the materials used were small stones and small pieces of bamboo. The materials for the reproduction tasks were small pellets of clay and small stones.

Procedures

For both comparison and reproduction tasks, I used three levels of set-size conditions. The sets were presented to children in a constant order: small (three and four elements), intermediate (six and seven elements), and large (eight and nine elements). I assumed that this order would permit maximum success, providing children with the simpler tasks first. Although my intention was to make use of all of the set sizes with all of the children I interviewed, this was not possible – some of the younger children were very shy and reluctant to continue after one or more tasks.

Comparison Tasks. In each of the three conditions, children were shown parallel rows of stones and sticks, with the longer row containing the fewer number. Children were asked whether there was the same number of sticks and stones or whether there was a greater number of stones or a greater number of sticks. If children did not count spontaneously, their judgment was recorded and they were asked probe questions to encourage use of counting (e.g., count and show).

Reproduction Tasks. Children were required to reproduce three-, six-, and nine-element collections, one set size for each task. In each case, a set of elements was placed behind the children and they were told that they should put out the same number of elements on the floor in front of them. If children did not count spontaneously to solve the task, they were again asked probe questions to encourage counting (e.g., count and show).

As in my work in the United States (Saxe, 1977, 1979), I coded performances for two properties: whether children counted to produce a numerical comparison or reproduction as well as their success in doing so.

Results

I compared children in two age groups on the reproduction tasks: children 9 years and younger and children 11 years and older. Children who were classified as 10 years old were excluded from these age group comparisons because of lack of confidence in age estimates.

For the small-set-size conditions, children's performances even without overt counting were sometimes correct, and probes sometimes elicited premediational approaches (like counting only one array for comparison tasks). Thus, children's reproductions and comparative judgments often appeared to be based on their analysis of number, but one that did not make use of body-part counting, a finding consistent with the literature on children's accurate judgments when small sets are used (often referred to as *subitizing*). These results are also consistent with the Gordon (2004) and Frank (2008) studies and with the Pirahã study cited earlier.

Figure 87 shows the distribution of children's correct reproductions and comparisons on these small-set-size conditions. Fisher exact tests revealed a significant effect for age on the comparison task, $p = .02$, and that the age trend approached significance on the reproduction task, $p = .06$. That said, a majority of the youngest group achieved accurate reproductions and just under a majority achieved accurate comparisons.

Children's performance on the intermediate- and large-set conditions had a different character than on the small sets. All Oksapmin children counted on the intermediate and large set tasks, though sometimes the interviewer elicited counting by probe questions. Children's performance was evaluated for whether they used counting as a basis for their numerical comparisons and reproductions.

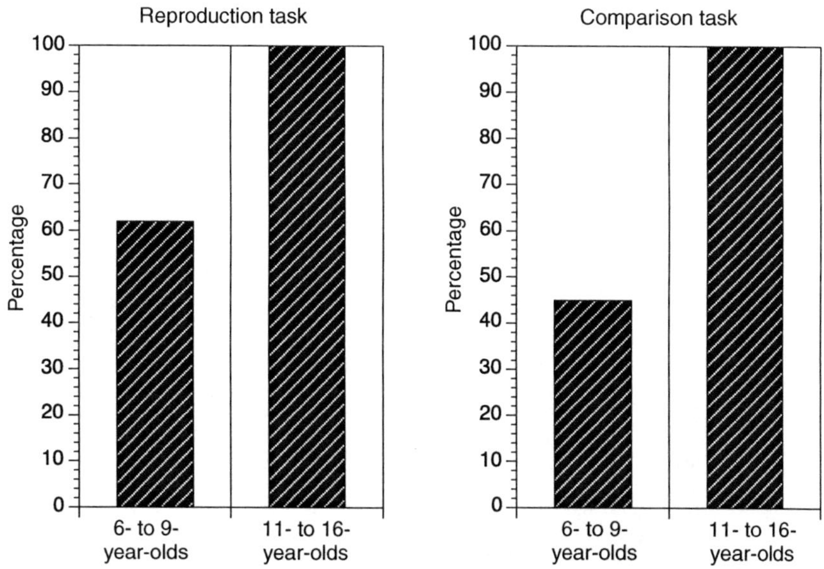

Figure 87. Percentage distributions of children's accurate reproductions and comparisons for small set sizes by age group.

Like U.S. children on similar tasks, some Oksapmin children did not use their counting to serve numerical functions. For example, in their comparisons, these children would count only a single array and offer a comparative judgment, or count two arrays continuously as if they were one and offer a comparative judgment. In contrast, those children that used counting to serve comparison or reproduction functions would often count both arrays and judge that the array that was assigned the body part further in the sequence was a greater number. For reproductions, I found that some children would count the model and then count an equivalent number for a copy, or put out an approximate copy and then count both model and copy, adjusting the copy until an appropriate number was reached.

Figure 88 contains the frequency distribution of children's use of counting to mediate reproductions and comparisons of number on the intermediate- and large-set-size conditions. As expected, young Oksapmin children tended not to use their counting to mediate their solutions to these tasks. The older age group did use counting to mediate the performances. Fisher's exact test revealed that these age trends were significant for each of the conditions (reproduction task set size six, $p = .01$, set size nine, $p = .02$; comparison tasks, set size six, $p = .001$, set size nine, $p = .001$). A comparison of children's performances as a function of set size six versus

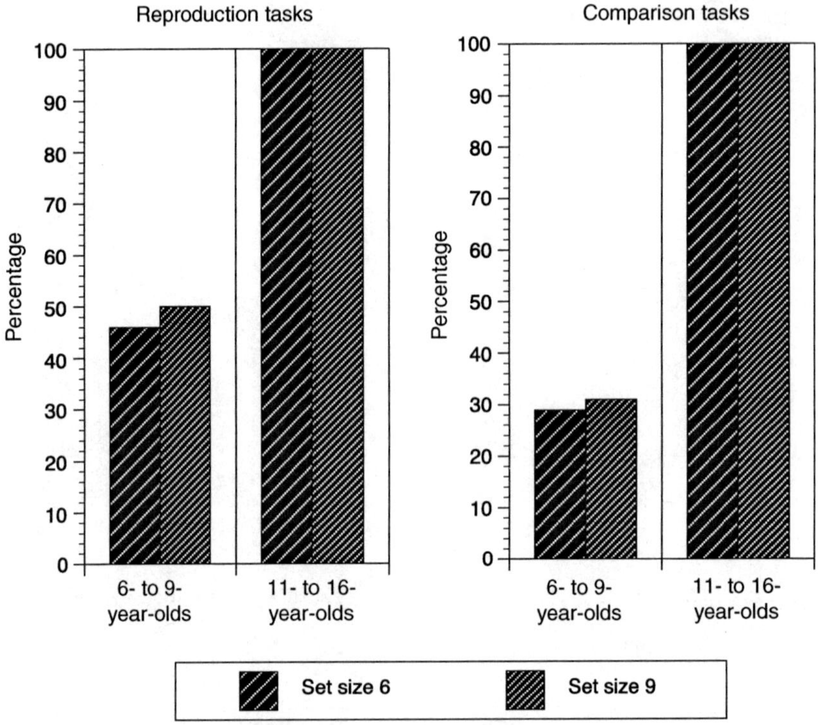

Figure 88. Percentage distribution of children's use of counting to mediate the reproduction and comparison of intermediate and large set sizes by age group.

set size nine in Figure 88 revealed no meaningful differences (Sigel 1956), though children's performances appeared depressed on the set size six and nine conditions relative to the small-set-size condition displayed in Figure 87. The reproduction task appeared to elicit greater use of counting as an intermediary, perhaps because there was no conflict between the length and number of two arrays (a potential conflict built into the design of the comparison tasks).

The age-related shifts in children's use of the body form to serve comparison and reproduction functions are clear in a general shift to the use of the body as an intermediary. What was the logic of the younger children's performances on these tasks? One approach to understanding it is to view their use of body part counting as similar to a naming activity. An analogy would be to name items in a list or one form of nominal correspondences (like Jane, Mary, and Sue, or altogether like my sister and her friends). Consistent with this view, sometimes children used body part names for individual elements that were enumerated without treating the last body part in their enumeration as a cardinal representation. This use is similar to findings reported in studies with children in the United States with English number words (Gelman & Gallistel, 1978; Schaeffer, Eggleston, & Scott, 1974). Other times, children used a slew of body-part names for a group of objects, again similar to findings reported in the United States (Fuson, 1988). The findings in the United States on similar comparison and reproduction tasks reveal parallel findings (Saxe, 1977, 1979). The findings also point to shifts in function of number words in early problem-solving activities. Younger children use body parts to serve nonnumerical functions often seen in children's playful activities or social activities with others. Older children use body parts to serve numerical functions as children construct numerical correspondences between body parts and objects when solving elementary comparison and reproduction problems.[2]

In sum, the findings of the present study are in accord with findings on form-function shifts in counting to mediate numerical comparisons and reproductions in the United States. At the same time, they do not capture developments that may well be specific to the Oksapmin body system, the focus of the next two studies.

Study 8-2. Comparing Numerical Values of Body Parts

From a developmental perspective, Oksapmin children who are developing the use of the body to serve an intermediary function may face challenges that require cognitive coordinations not required by their Western peers. For example, with the Oksapmin body form, the symmetric spatial organization may lead to confusions about symmetrical body parts – while an elbow on one side of the body is just like an elbow on the other side, in the body-counting system "elbow" means two different numbers. Similarly, the spatial trajectory of counting could also lead to system-specific confusion. For example, does the right elbow mean the same number whether one starts from the left thumb or from the right thumb? Such potential conflicts are depicted in Figure 89.

The purpose of the present study was to explore whether some children make numerical comparisons based upon the similarity of appearance of body parts. I designed two kinds of tasks to explore this question. The first

[2] There are counter-positions to this interpretation that favor nativist interpretations of structures of one-to-one correspondence (see, for example, Bever et al., 1968; Mehler & Bever, 1967; but also Piaget, 1968). Some of these positions were reviewed in Chapter 1.

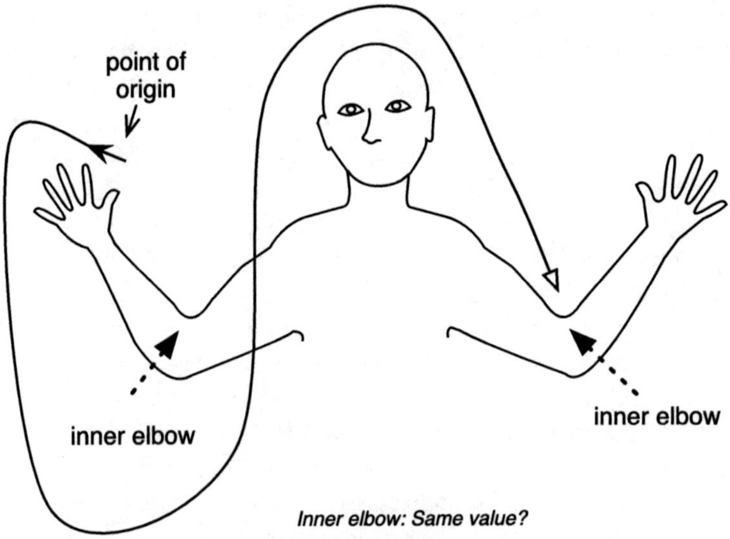

Inner elbow: Same value?

(a) Symmetrical body parts with common origin point

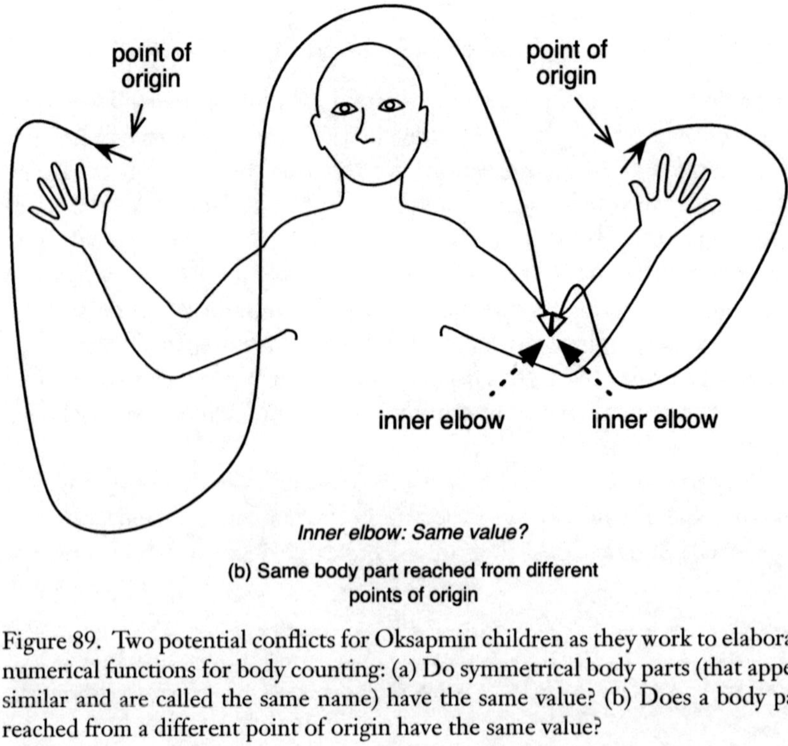

Inner elbow: Same value?

(b) Same body part reached from different points of origin

Figure 89. Two potential conflicts for Oksapmin children as they work to elaborate numerical functions for body counting: (a) Do symmetrical body parts (that appear similar and are called the same name) have the same value? (b) Does a body part reached from a different point of origin have the same value?

kind was used to determine whether children judged *symmetrical body parts* reached from the same point of origin as the same numerical value. The second kind was used to determine whether children would judge the *same body part* reached from different points of origin as the same value. In each case, the tension would be between the physical appearance of body parts and the order of occurrence of a body part in an enumeration.

Methods

I piloted procedures for both the symmetrical body-part comparison tasks and the point-of-origin tasks, modifying each to support clarity. To support the interpretation of children's responses, I included some control conditions. For the symmetrical-body-parts task, I included comparisons involving both symmetrical body parts as well as asymmetrical body parts. For the point-of-origin tasks, I included the nose (midpoint) as well as other body parts. As in the previous studies, I used a back-translation procedure to ensure the accuracy of the translation.

Participants

I interviewed 52 children, who ranged in age from about 7 to 16 years and were members of the same community as those described in Study 8-1. As in Study 8-1, children's ages were determined from missionary records and by comparing children of known age with children of unknown age. Due to the probability of error in some of these age estimates, when age-group comparisons were made, children who were classified in the middle age group (10–11 years) were omitted from statistical analyses, and only the 7–9 and 12–16 age groups were compared. But when children's profiles were evaluated across asymmetrical and symmetrical body-part-task questions, all participants were included in the analyses since age was not used as a variable in these comparisons.

Procedures

Children were asked to demonstrate how to count and were excluded from the study if they did not enumerate body parts in the conventionally defined sequence. Children who passed the introductory assessment (most did) were then interviewed about the asymmetrical/symmetrical body-parts task and then the point-of-origin tasks.

Producing Numerical Comparisons between Asymmetrical and Symmetrical Body Parts. Children were interviewed individually. They were told a story about a man who was counting sweet potatoes in his garden. Children were told that one day a man counted to this place on his body (the interviewer and then the translator made a sweeping gesture from the children's right thumb to a specified body part on the right side of their bodies). They were then told that another day the man was counting sweet potatoes and counted to here (the interviewer and then the translator made another sweeping gesture from the children's right thumb to a specified body part on the left side of the children's bodies). Children were then asked if the man counted to the same amount of sweet potatoes (pointing to the two body parts) or different amounts. If children responded "different amounts," they were asked which was the bigger amount. Six pairs of body parts were indicated. Three of these were symmetrical pairs (namely, (a) right ear (12)/left ear (16), (b) right shoulder (10)/left shoulder (18), (c) right biceps (9)/left biceps (19)), and three were asymmetrical pairs (namely, (a) right wrist (6)/left ear (16), (b) right elbow (8)/left wrist (22), (c) right side of neck (11)/left biceps (19)). In the administration of the comparisons, symmetrical and asymmetrical tasks were alternated.

Point of Origin in Body-Part Comparisons. Children were interviewed individually. They were told stories that led them to compare the numerical values of body parts reached from different points of origin. In the introduction to one story, children were told that, although people in their village count from one side to the other (gesturing), in a village over the mountains, people count the opposite way (gesturing). We then asked the child to demonstrate the way in which people count in their village and the way in which people count in the other village.[3] If a child confused these instructions, the introduction was repeated. Once the child understood the introduction, we told the next part of the story. "One day a man from [*name of child's village*] was counting sweet potatoes in his garden and counted to this place [*interviewer gestures from the right side of the child's body to one of three specified body parts*]. The man from the village over the mountains was also counting sweet potatoes in his garden and counted to this place [*the interviewer gestures from the left side of the child's body, to the same one of the three body parts specified above*]." The interviewer then asked the child whether the two men counted to the same number of sweet potatoes or

[3] Note that this story was untrue, but that did not appear to faze children; most of them seemed to enjoy the story.

whether they counted to a different number of sweet potatoes. If a child claimed that they counted to different numbers, the interviewer asked who counted to the bigger number.

The three body parts that the child was questioned about were the wrist (6), shoulder (10), and nose (14). In order for a child to achieve a correct score on the wrist or shoulder questions, he or she had to respond that the two individuals counted to different numbers and to correctly identify the greater numerical value. The question about the nose was used as an additional comprehension check. Two children responded that the nose represented different numbers and were excluded from the analysis. (Note that regardless of the side of origin, the nose (14) is the midpoint and will always be the same count away from the right thumb and the left thumb.)

Results and Discussion

Asymmetrical/Symmetrical Body-Part Comparisons

Figure 90 contains a frequency distribution of children's levels of response for each age group on the asymmetrical and symmetrical body-part comparison questions. Comparisons of the 7- to 9- and 12- to 16-year-old age groups revealed that there were significant age differences in performance, both on the asymmetrical-body-part question (Kolmogorov-Smirnov two-sample test, chi-square ($N = 52$, $df = 2$) = 11.21, $p < .01$), and on the symmetrical-body-part questions (Kolmogorov-Smirnov two-sample test, chi-square ($N = 52$, $df = 2$) = 6.85, $p < .05$). An analysis of children's profiles across the asymmetrical and symmetrical questions also yielded expected differences. Children consistently achieved more correct responses on the asymmetrical as opposed to the symmetrical body-part comparisons (sign test, $N = 25$, $\times = 0$, $p < .001$).

Point-of-Origin Tasks

Figure 91 contains a frequency distribution of the number of correct responses at the two age levels on the point-of-origin task. Consistent with the responses on the previous task, young children tended to use physical similarity of body parts as a relevant criterion for numerical evaluation, and, as age increased, this criterion shifted to the ordinal position of a body part in an enumeration (Kolmogorov-Smirnov two-sample test, chi-square ($N = 52$, $df = 2$) = 7.90, $p < .02$).

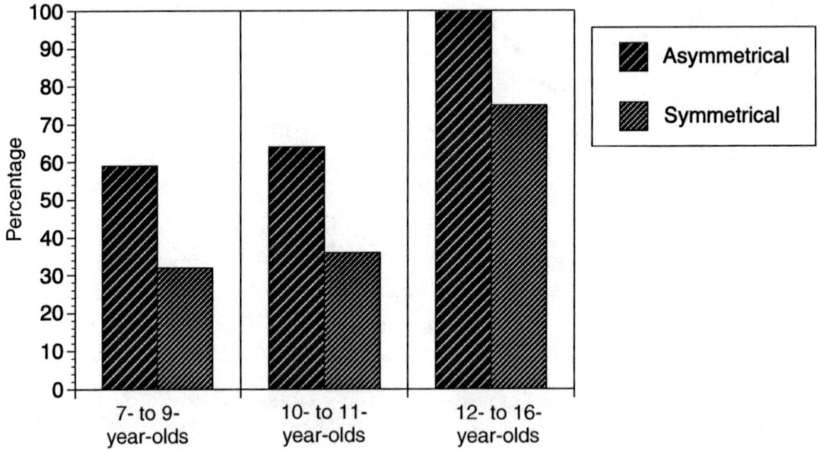

Figure 90. Percentage of children achieving at least two of three items correct on the symmetrical and asymmetrical body-parts tasks.

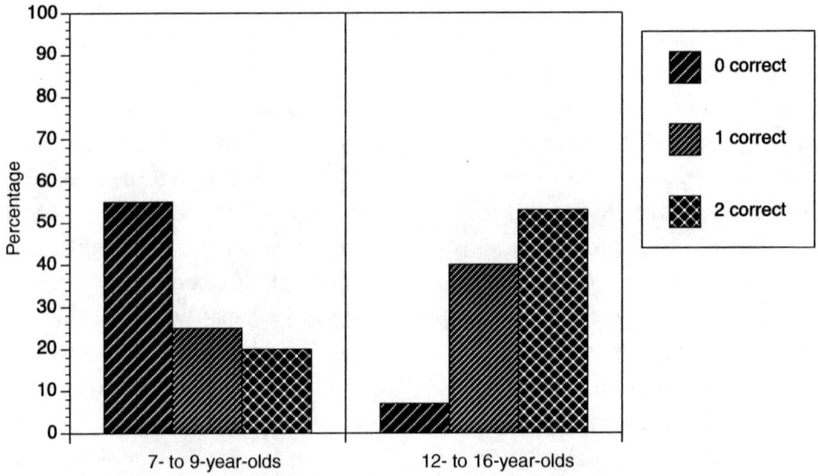

Figure 91. Percentage distributions of items correct on the point-of-origin task as a function of age group.

The findings presented in this study show that, with age, children increasingly base their cardinal judgments on ordinal position in an enumeration instead of figural features of body parts. When comparing the values of symmetrical- or asymmetrical-body-part pairs that were reached from a common point of origin (right thumb), the younger children incorrectly identified symmetrical body parts as equivalent in value. Furthermore, when they did correctly identify asymmetrical body parts as not equivalent, they did not regularly identify which of the two represented the larger value. Older children produced accurate comparisons, regardless of the physical similarity of the body parts. The point-of-origin task revealed a parallel sequence. Younger children considered the same body part reached from two different points of origin (left or right thumbs) to represent the same value. In contrast, children in the older age group considered the value of a body part to be based upon its ordinal position in an enumeration.

Study 8-3. Measuring with the Body

In Oksapmin, and throughout the Mountain-Ok region, a pervasive artifact is the string bag or *bilum*. The bags are made from a string harvested from a local bark and then created by weaving; bags may also be decorated with bird feathers. Oksapmin men and women as well as boys and girls use these bags. Children are given a small bag to wear on the shoulder or around the neck at about 5 years of age, and from that age on they are rarely seen without one.

The bags are produced in a number of sizes, shapes, and netting patterns from small bags (10 to 30 cm deep) for tobacco or personal effects to larger bags (40 to 50 cm deep) for garden produce or to carry infants and small children. Bags are also used as adornments in ceremonial dancing or used for storing ritual objects. Figure 92 shows bags worn by adults and children in 1978. During my fieldtrips in 1978 and 1980, bags were a standard part of dress, worn by men, women, and children. Figures 18 through 20, presented earlier, are photographs taken in 1938 by the Hagen-Sepik patrol; they show Oksapmin men wearing bags during the period of first Western contact.

Study 8-3 explores developmental changes in the way in which Oksapmin children and adults use the body system as a measure of a bag's depth. People measure the string bags in the course of production and to evaluate the finished product. To produce a measurement, they put their hands into a bag and allow the body of the bag to drape along the extended arms (see Figures 93). The point at which the rim of the bag falls on the arms is then reported as its length. The sizes of bags are referred to as "knuckles," "wrist (6)," "forearm (7)," "inner elbow (8)," "biceps (9)," or "shoulder (10)." With

Figure 92. String bags are worn by men, women, and children. Photographs, 1978, 1980.

the exception of the knuckles, the same body parts used in the numeration system are used to measure length. A particularly interesting feature of this system is that, unlike the conventions for measurement in the West, here units are not equivalent across individuals, since people's arms vary

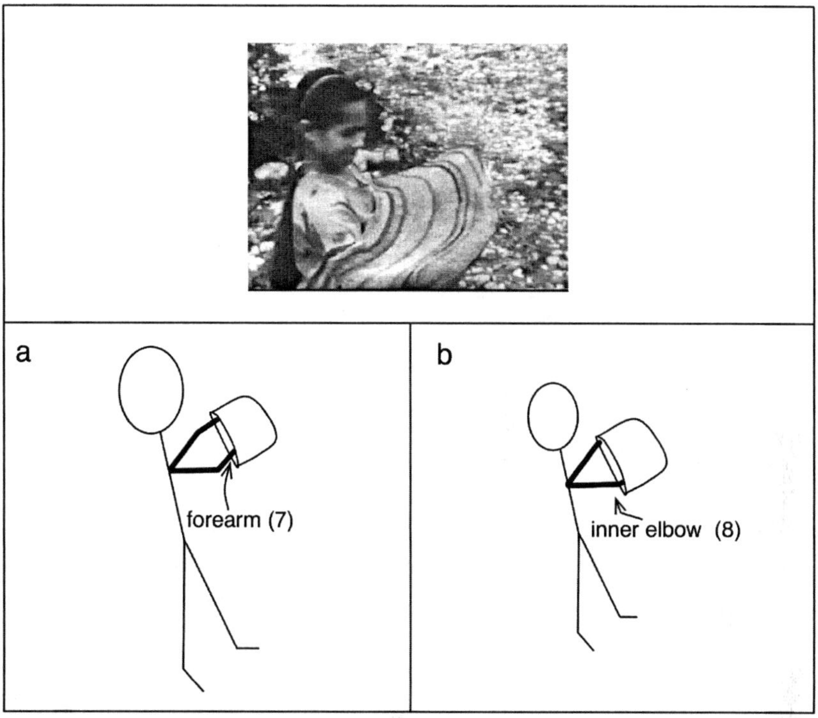

Figure 93. Top: Photograph of an adolescent measuring a string bag. Bottom: Sketches of an adult (a) and a child (b) inserting their arms into the same string bag to measure. (a) An adult inserts her arms into the bag, and the fingertips-to-rim measure is the forearm (7); (b) a child inserts his arms into the same bag, and the fingertips-to-rim measure is the inner elbow (8).

in length from one individual to another. The Oksapmin convention is widely known throughout the community; women who do not make bags (approximately half of them), men, and older children all have equal facility in using the basic technique.

To investigate developmental changes in children's use of the body as a form for measurement, I devised two task types. The first was designed to assess people's use of the conventional measuring procedure to compare the lengths of two string bags. In this task, the interviewer separated two string bags from one another and collapsed them, so that their depths were not visible; then the participant was asked to compare the bags' depths. The participant could measure each bag's depth with his or her own body (but could not compare bags visually). The second task type was devised to assess participants' understanding of the necessity to compensate for differing lengths of people's arms when comparing measurements produced by individuals with obviously different arm lengths.

Methods

Participants

Participants included 23 unschooled children (mean age = 8.4 years, SD = 1.0 years) and 30 unschooled adults (mean age = 29.5 years, SD = 7.0 years), with approximately the same number of males and females in each group. Participants' ages were based upon visual estimates.

Procedures

Prior to the interview, participants were asked to demonstrate how to measure string bags. The two children who could show how to measure were not included in the study.

Comparisons of Length Using the Body-Part Measure. The materials used for this study included three string bags that were about 30 cm, 35 cm, and 40 cm in length from the bottom to the rim, respectively. Two stories were used.

1. "Two Divanap [*name of village*] women made string bags. One woman came from one side of the path, and one woman came from the other side of the path. This is the string bag of the woman from the other side of the path. [*A string bag is held up and given to the participant.*] Put it on your arms and see what length it is and remember it. [*After the participant measured the bag, it was folded and put to one side of the participant.*] Here is the string bag from the woman from the other side of the path. [*A second string bag that is different in depth from the first string bag is handed to the participant.*] Put it on your arms and take a look. [*After the participant measured the bag, it was removed, folded, and put on the other side of the participant.*] Now, of these two string bags, is one longer or are they the same length? How do you know? [*If participant says that they are different lengths:*] Which is longer? How do you know? How long is it?"
2. The story was then repeated, substituting another village name and a different pair of string bags.

Necessity of Equivalent Units. This task involved comparing the measurements of two individuals using the body-part system. Interviewer questioned the participant about each of the following four stories to assess his or her understanding of the necessity of equivalent units when comparing measurements produced by two individuals.

1. *Daughter makes bag for father; string bag size measures forearm* (7) *on daughter*. "There was a little girl. Just a small little girl [*gesture made by interviewer indicating small size of girl*]. This little girl made a string bag for her tall father [*gesture made by interviewer indicating the size of the father*]. She made the string bag and then put it on her arms and saw that it was as long as the forearms [*teller gestures to forearm* (7)]. Then she put the strap on and gave it to her father. When her father put it on his arms, did it come to the forearm (7) also, or not? How do you know? To where did it come? How do you know?"
2. *Daughter makes bag for father; string bag size = inner elbow* (8) *on daughter*. The participant was told the same story as in no. 1, except the string bag size was changed to the inner elbow (8) on the daughter.
3. *Mother makes bag for son; string bag size = forearm* (7) *on mother*. "A mother, fully grown [*teller gestures*], made a string bag for her little son, this high [*teller gestures*]. The mother measured it and it came to the forearm (7) [*teller gestures*]. Then she put the strap on and gave it to her son. When the little son put it on his arm, did it come to the forearm (7), or not? How do you know? Where did it come to? How do you know?"
4. *Mother makes bag for son; bag size = inner elbow (8) on the mother*. The participant was told the same story as in no. 3, except the string bag size was changed to inner elbow (8) on the mother.

Results

Comparisons of Length Using the Body System. Figure 94 shows the number of correct responses a participant produced in the comparisons-of-length task. In order for a response to be coded as correct, the participant had to state correctly which was the bigger string bag and explain this judgment by reference to his or her previous measurement (e.g., "This one is longer because it came up to my inner elbow (8) and that one only came up to my forearm (7)"). In general, participants performed at a high level on this task. There was a developmental difference between children and adults (pooling 0 and 1 responses to obtain higher expected values, chi-square (N = 51, df = 2) = 16.3, $p < .001$); however, the majority of participants achieved both comparisons correctly.

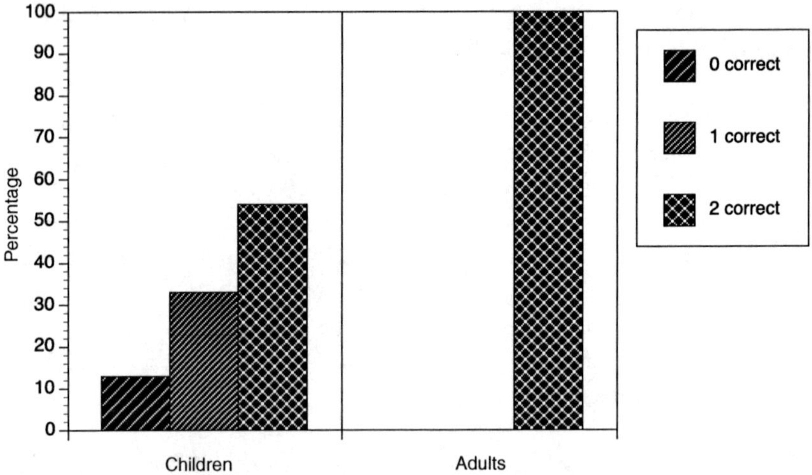

Figure 94. Number of correct responses to string bag comparison task for unschooled children and adults.

Necessity of Equivalent Units. Figure 95 shows the number of stories in which each participant correctly predicted that the string bag would reach a different body part on the recipient's arm and correctly predicted the direction of difference. Adults responded correctly to more stories than children did (chi-square (N = 51, df = 2) = 21.7, $p < .001$). The majority of unschooled children believed that the string bag would come up to the same place on the recipient as on the donor on all the stories, whereas the majority of the adults indicated that there would be a discrepancy and predicted the correct direction of the discrepancy on at least three of the four stories.

A number of responses to the equivalent-unit stories that were recorded as "incorrect" may be more properly termed intermediate responses. In these cases, a participant would respond that when the little girl gave the bag to her father it would reach a higher point on the father's body because he had long arms (in fact, the rim should come up to a lower point because he has long arms). That is to say, the participant correctly predicted that the string bag would reach a different place on the recipient's arm but made an incorrect judgment about the relation between the length of the arm and the point to which the bag would reach when measured. In these responses, the participants demonstrated, through their justifications, that they recognized the problem of the lack of equivalent units, but they did not adequately coordinate the inverse relation between the bag length and arm length for people of different sizes.

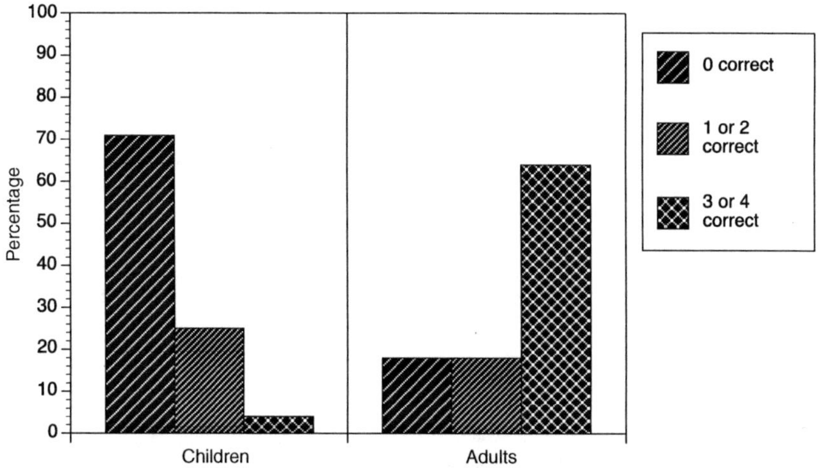

Figure 95. Number of correct responses to equivalent-units tasks for unschooled children and adults.

Discussion

The findings of the study revealed a range of microgenetic constructions in the way children and adults used the body form to serve a host of functions related to measurement. Although young children knew the conventional body parts and could place string bags on their arms to show how to measure them, they frequently did not use the body-part form to serve the normative measurement functions. On the comparison-of-length tasks, children did not show knowledge of the intermediary function of the body as a measurement form. They failed to produce correct evaluations on the string-bag-comparison tasks. On the necessity-of-equivalent-units tasks, most children did not coordinate inverse relations between the bag length and arm length enabling them to correctly identify differences between measurements using the child's and adult's arm lengths. By adulthood, all Oksapmin participants demonstrated an understanding of the transitive use of the body as an intermediary on the comparison tasks, and the large majority showed an understanding of compensatory relations in adjusting for unequal units.

Of particular interest are the findings of form-function shifts on the equivalent-unit tasks. The findings support a developmental progression specific to the use of the Oksapmin body-part system. At first, children treat the value of body parts as representative of length relations in an absolute sense: Regardless of the size of an individual, a body part functions as a representation of an absolute length. Later, children begin to differentiate between measurements produced by people of different sizes but have difficulty conceptually coordinating the inverse relations between arm length and the distance the string bag will reach on the body. Thus, though some of the participants recognized that the body-part measurements would not be the same on the children and the adults, they believed that the adults would measure the string bags to a higher value since their arms were longer. Finally, by adulthood, participants both differentiated between body-part measurements produced by people of different sizes and achieved a coordination of this inverse relation between string bag length and point on the arm that the string bag would reach. In these performances, adults' use of body parts took on the function of a standard measure when adjusted by considerations of who was measuring.

Like the studies with discontinuous quantities (Studies 8-1 and 8-2), the findings here reveal a shift in function of body parts in the ontogenesis of quantification uninfluenced by schooling. Early in the developmental process, Oksapmin children use conventionally defined points on the body in measurement-like activities, but body parts are not used to serve canonical measurement functions. Only later, presumably as children engage with emerging goals related to measurement in daily activities, do they begin to tailor their use of the cultural form to achieve solutions to measurement problems that involve metric relations for linear quantities. This process consists of conceptually organizing the procedures into a coherent system of units. The fact that Oksapmin elaborate understandings of equivalent units despite the fact that equivalent units are not an inherent feature of their system points to the autonomy of the development of cognition about measurement from the particular cultural forms used in measurement. This is an idea to which I will return in Part IV.

9 Children's Adaptations of the Body System in School in 1980

An Unintended Consequence of Postcolonial Schooling

The studies described in this chapter explore the way children's participation in postcolonial schooling in 1980 led them to construct new arithmetical functions for the body-part system. The studies have a notable beginning. During my visit to an Oksapmin community school math classroom in Divanap, I thought that I noted children using body-part counting as they were solving arithmetical problems. After class, I queried teachers about children's gestures.

Teachers didn't have much to say about children's body-part gestures, other than that they also noticed them on occasion. There was good reason for this. As I noted earlier, the teachers in this era were recent recruits from other coastal regions of Papua New Guinea. They had little understanding of the Oksapmin community and its practices; those who I talked with had no knowledge that people in this part of Papua New Guinea traditionally use body-counting systems.

So, I let teachers know that people in this and the larger Mountain-Ok region use body-counting systems. Some found that curious, but that knowledge did not appear to influence their practice or their thinking about mathematics instruction. Indeed, it seemed to be one fact among the multitude of foreign elements in the Oksapmin world. In the foreign context, teachers had plenty to manage in their own lives as well as the running and maintenance of a bush school.

The School

The postcolonial era community school was largely constructed from local materials, and there was no electricity or plumbing. In 1980, the school curriculum consisted of six grade levels (1 through 6), although not every grade was taught during any given year. During the period of the present study, Grades 2, 3, 4, and 6 were in session. As noted earlier, the emphasis was on the drill and practice of arithmetical facts and memorization of tables, often implemented as classroom chorus drills. The postcolonial school rules (which were national policy, but variably followed) were that students could not use the vernacular on school grounds (which would include body counting) and that all instruction must be in English. A 1980 photograph of the school (at the time of the study) is contained in Figure 86.

The Studies

With the help of Maryl Gearhart, who was engaged in a separate project for the Ministry of Education, and Tom Moylan, who was involved in a follow-up study related to his dissertation project in linguistic anthropology, I conducted the first of two studies.[1] Study 9-1 was observational, conducted in classrooms at three grade levels, Grades 2, 4, and 6. The key question was whether a substantial number of children were using the body system in classroom mathematics activities (or whether my initial observation was more of a fluke occurrence). As the findings presented shortly reveal, there was considerable overt use of the body system, though overt use varied by grade level.

The goal of Study 9-2 was to investigate whether participation in the collective practices of classroom life was leading children to construct novel body-part functions. At school, often children were engaged with arithmetical problems without countable objects available, and thus in an important sense they were in predicaments similar to those of unschooled adults participating in economic exchanges during this era. Would schooled children who were just acquiring English as a second language, but who knew the body system, generate arithmetical functions for the body similar to those of adults participating in the money economy? My interest was to explore this possibility through interviews with students about arithmetic tasks.

Study 9-1. Observations in Classrooms

To organize the study in ways to observe body counting in classroom activities, I focused on student test taking during class. I expected that in a class arithmetic test, students might draw on their body system as a resource

[1] In 1980, Tom Moylan, Maryl Gearhart, and I returned as a group to Oksapmin.

to help them solve problems. I collaborated with teachers in the design of the test administration so that it afforded use of an observation protocol, and I developed some standard observational procedures. Grade 2, 4, and 6 teachers administered the test in their classrooms; during test taking, Maryl, Tom, and I recorded signs of student body counting through a standardized procedure detailed in the following section.

Methods

Participants

The participants included all children enrolled in three classrooms – Grade 2 classroom (16 boys and 9 girls), Grade 4 classroom (19 boys and 5 girls), and Grade 6 classroom (13 boys and 9 girls). The ages of the children were estimated from missionary records. The mean ages in years for each respective grade level were 10.7 years (sd = 2.0), 11.1 years (sd = 2.4), and 13.8 years (sd = 1.3).

Procedure

Each grade level was administered the arithmetic test in each of their respective classrooms. I designed the test in a way that would provide sufficient challenge to students at multiple grade levels; at the same time, the procedure of test administration and the arrangement of order of test items would allow coders to record evidence of body counting on a given problem type. In this procedure, each coder would focus on single tables of students, with about five students at a table, in a procedure that would allow every student to be included as they worked on the test (see Figure 97).

The test consisted of 20 addition and subtraction problems (five sets, *a* through *e*, of four problems each) presented orally and on a blackboard. To provide a range of problem types, each set of four problems included four varieties: Two problems contained only single-digit terms ("easy problems"), and two contained at least one two-digit term ("difficult problems"); in addition, two problems were always addition and two were always subtraction. The problems were administered in the order shown in Figure 96.

For each of the successive sets of four problems, two coders observed, at a distance, separate groups of five children seated at a common table. The task of coders was to determine whether any child in the group used the body system while solving each of the four problems. For each of the four

Problem Set	Arithmetical Problems (in order of presentation: left to right)			
a	16 + 5 = ?	4 – 2 = ?	22 – 6 = ?	3 + 2 = ?
b	4 + 3 = ?	14 – 9 = ?	6 – 4 = ?	18 + 9 = ?
c	16 – 8 = ?	5 + 2 = ?	5 – 3 = ?	13 + 6 = ?
d	6 – 3 = ?	17 + 8 = ?	7 – 3 = ?	8 + 17 = ?
e	16 – 4 = ?	5 + 3 = ?	18 – 9 = ?	2 + 5 = ?

Figure 96. Five arithmetic problem sets used in classroom arithmetic test.

observation problems, children were credited with using the body system if they pointed to any series of conventionally defined body parts during the solution procedure. With each new problem set, coders shifted their focus to a new table of students, enabling matched levels of problem difficulty for each observational wave.

Results

Two analyses were conducted. The first analyzed whether grade level was associated with children's use of the body-part system during the test. The second analyzed the relation between children's performance on the test and their use of the body-part system during the test.

Children's Use of the Body-Part System

Figure 97 contains a photograph of the Grade 4 class during the test. The two children at the rear of the table are using the body system. I tabulated the distribution of children observed using the body-part system as a function of grade level. The majority of Grade 2 children (63 percent) were observed using the system at least once during the four-problem set; at Grade 4, only about one-half of the children used the system (52 percent); by Grade 6, only a small minority (10 percent) of children used the system (chi-square $(2, N = 67) = 8.24, p < .05$).

Relation between Overt Use of the Body System and Arithmetical Achievement

Children's arithmetic achievement as measured by the number correct on the test revealed the expected effect for grade level ($F(2, 67) = 22.45$,

Figure 97. Grade 4 children making use of the body system during an arithmetic test in 1980.

$p < .001$). Correlations between use of the body system and arithmetic test score were computed for the two grade levels for which there was adequate variation in use of the body system (Grades 2 and 4). There was no evidence that overt use of the body system and arithmetic achievement were correlated (Grade 2: $r\ (24) = -.10, p = \text{ns}$; Grade 4: $r\ (24) = -.05, p = \text{ns}$).

These findings show that many Oksapmin children were in fact using their indigenous body-part system overtly to help them solve problems in the Western school curriculum, albeit with varying degrees of success. Some, and perhaps the less successful, may have been using the system in ways akin to the global-enumeration strategy documented in Study 4-1 with elders (a study also conducted in 1980); others may have been generating new functions for the body system in ways that paralleled unschooled adults who were participating in the money economy. The emergence of new functions in relation to the collective practices of classroom life would be an unintended consequence of school instruction, an institution that devalued use of the vernacular and traditional knowledge. To explore the character of children's use of the body system, I conducted a second study. I created interview methods to elicit children's arithmetical problem solving, making clear to them that it was all right to use their bodies during the interviews.

Study 9-2. Arithmetic with the Body: Unintended Consequences

The study was conducted in two phases. During the first phase, I interviewed children individually about their solution strategies to single- and double-digit arithmetic problems involving values within the limits of the conventional body-part system. During the second phase, I then re-interviewed children who used the indigenous system to solve any of the first four problems; in this second phase, I used more difficult problems with numbers that extended beyond the conventional limits of the indigenous body system.

Methods

Participants

The same children who participated in the observational study (Study 9-1) participated in the present study. In addition, 22 nonschooled adolescents were interviewed using the same problems. A comparison between the nonschooled adolescents' and the Grade 6 adolescents' performances served as a further control: In this comparison, age was a constant and children's school experience was varied (no schooling vs. 6 years of schooling). None of the nonschooled adolescents had been outside of the Oksapmin area, and in general these individuals had only minor participation in the money economy.

Procedure

I interviewed all children individually, and all interviews were conducted in the Oksapmin language with the aid of a translator. The interview procedures were back-translated to ensure the accuracy of the translation.

Phase 1 Problems. I presented children with four arithmetic problems in a narrative story one at a time: 5 – 3 = ?, 5 + 3 = ?, 16 – 7 = ?, 14 + 7 = ? For the addition problems, the narrative was about a person who had the number of pigs specified by the first term of a problem and received more pigs – the number of the second term in the problem. Children were then asked how many pigs there were altogether. For the subtraction problems, children were told that they had the number of pigs specified by the first term of the subtraction problem and then killed the number of pigs in the second term. Children were then asked how many pigs were left. If children did not exhibit a body-counting strategy, they were asked probe questions such as, "Can you show me how you figured that out by counting on your body?"

Phase 2 Problems. In the second phase of the study, a subgroup of children who participated in Phase 2 were asked to solve additional problems. All of the Phase 2 problems contained integers that were greater values than those within the limits of the traditional counting system. These problems were 43 – 7 = ?, 41 – 6 = ?, 9 + 43 = ?, and 12 + 34 = ?.

Results

Children's solutions to the arithmetical problems were transcribed, and a coding scheme was developed for categorizing strategy types. All body-part strategies were coded as to whether children used the strategy spontaneously or whether in response to a probe question. The strategies varied markedly in their effectiveness.

Miscellaneous/covert strategies showed no overt evidence that the child used a systematic procedure to keep track of the addition or subtraction of values. I noted two subtypes of miscellaneous strategies. In the first, one that was virtually the same as the global-enumeration strategy used by some Oksapmin adults (Study 4-1), the child first named body parts to the value of one term of the problem and then proceeded to add or subtract the other term with body-part names. For instance, to perform 14 + 7, a child might name body parts to the nose (14) and then add the second term, begin with the eye (15) and proceed to the ear (16), and so on. With this type of solution strategy, the child showed no overt means of keeping track of the addition or subtraction of the second term. In the second subtype, the child produced a body-part answer without an overt strategy: He or she either guessed (or estimated) or produced a single enumeration of body parts that ended with an answer. Although miscellaneous/covert approaches provided no overt demonstration of a way to keep track of the addition or subtraction of correspondence relations, it is possible that children whose performances were categorized as miscellaneous in fact used a covert and adequate strategy to achieve a solution. To extract information about the possibility that children did use covert strategies, children's miscellaneous/covert approaches were divided into two types: those that yielded an accurate result and those that yielded an inaccurate result. An accurate solution achieved through a miscellaneous strategy may very well have been accompanied by either a covert means of keeping track of the addition or subtraction of values or a memorized solution.

Double-enumeration strategies consisted of overt procedures in which individuals used a method of establishing correspondence relations between the body parts that constituted the two terms of the problem, a method that permitted the individual to keep track of the addition or subtraction of terms. For instance, to perform 14 + 7, the individual might first represent the term "14" as "nose." The addition of 7 would be achieved by establishing correspondences between the following pairs of body parts: the forearm (7) and the eye (15), the wrist (6) and the ear (16), the little finger (5) and the neck (17), the ring finger (4) and the shoulder (18), the middle finger (3) and the biceps (19), the index finger (2) and the inner elbow (20), and the thumb (1) and the forearm (21), which was the solution. Alternatively, the addition of 7 could be achieved by establishing body-part correspondences beginning with the thumb (1) and the eye (15) and ending with the forearm (7) and the forearm on the other side (21) through a similar process. The strategy is the same as that used by some adults in Study 4-1. Figure 98 shows a fourth grader using a double-enumeration strategy as he was in the process of subtracting 16 – 7. Shown in the photographs are the correspondences (midway through the solution) between (a) nose (14) and pinky (5), (b) eye (13) and ring finger (4), (c) ear (12) and middle finger (3), (d) neck (11) and index finger (2), and (e) shoulder (10) and thumb (1). The next body part is the biceps (9), which was his solution.

Body-part substitution strategies were similar to double-enumeration strategies except that the procedure for keeping track of the second term no longer consisted of establishing physical correspondences by pointing to body-part pairs. Now, children constructed a correspondence by calling one body part by the name of another body part in the addition or subtraction of values. Thus, in the problem noted above, the eye (15) would be called the forearm (7), the ear (16) would be called the wrist (6), and so forth. Like the double-enumeration strategy, the strategy is the same as that used by some adults in Study 4-1.

Bilingual-substitution strategies occurred only on the Phase 2 problems, in which the terms were greater than could be easily represented in the indigenous body system. Children using this strategy would represent the larger term of the problem in English and then use the indigenous body system to keep track of the addition or subtraction of the second value. For instance, in one variant of this approach to solve the problem 12 + 34 = ?, an individual would begin with the term 34, stating the English number word, "thirty-four." To keep track of the addition of 12, the individual would point to the body parts, starting at the neck (12) and proceeding to the thumb (1), counting the iteration of body parts with English number

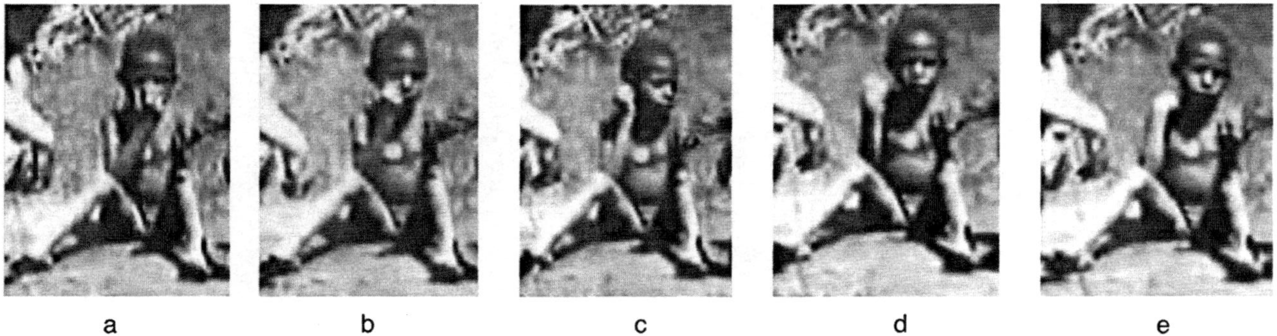

Figure 98. A child using a double-enumeration procedure, creating a one-to-one correspondence between two body-part series as he keeps track of the subtraction of one term from another. (Photographs extracted from video found at: http://www.culturecognition.com/fourth-grader-solving-16-7-body-system/.)

words from 35 (beginning with the neck (12)) through 46 (ending with the thumb (1)), the correct answer.

Children's Solution to Phase 1 Problems

Children were assigned the most advanced strategy they generated for each problem, and two analyses were conducted on these strategy codes. In the first, to understand whether participation in classroom practices led children to more sophisticated body strategies, I analyzed strategy type as a function of grade level. In the second, to determine whether participation in classroom practices was a principal determinant of strategy development, I contrasted Grade 6 students' strategies with those of their nonschooled, same-aged peers.

Figure 99 shows the percentage distribution of children's strategies at each grade level for each arithmetic problem. On the simplest single-digit problem (such as 5 – 3 = ?), virtually all of the schooled children at each grade level achieved an accurate solution by a miscellaneous approach. Although children showed no overt means of keeping track of the subtraction of 3 from 5, the accuracy of the children's answers indicates that they could keep track of the subtraction without generating an overt strategy, a finding that has some resonance with the studies on early arithmetical competence (Klein & Starkey, 1988; Pica et al., 2004; Wynn, 1992). Although the majority of nonschooled adolescents did achieve an accurate solution to this problem, they were less accurate than were their Grade 6 peers (Fisher's exact test, $p < .02$).

On the more difficult single-digit problem (3 + 5 = ?), a minority of children demonstrated advanced body-part strategies (double-enumeration or body-part substitution). Although no difference as a function of grade level existed for usage of advanced body-part strategies, a comparison of the sixth-grade students with the nonschooled adolescents revealed that sixth graders (35 percent) more frequently demonstrated the use of advanced body-part strategies than did the nonschooled adolescents (5 percent), (chi-square ($df = 1, N = 45$) = 5.91, $p < .02$).

The pattern of findings for the problems involving greater numbers differed from those for the single-digit problems. For these problems, I found an increase in children's use of advanced body-part strategies over grade: 16 – 7 = ? (chi-square ($df = 2, N = 70$) = 11.22, $p < .01$); 14 + 7 = ? (chi-square ($df = 2, N = 68$) = 10.02, $p < .01$). Only a minority of Grade 2 children (31 percent, 22 percent) demonstrated advanced body-part procedures on these problems, whereas a majority of Grade 4 (71 percent, 61 percent) and

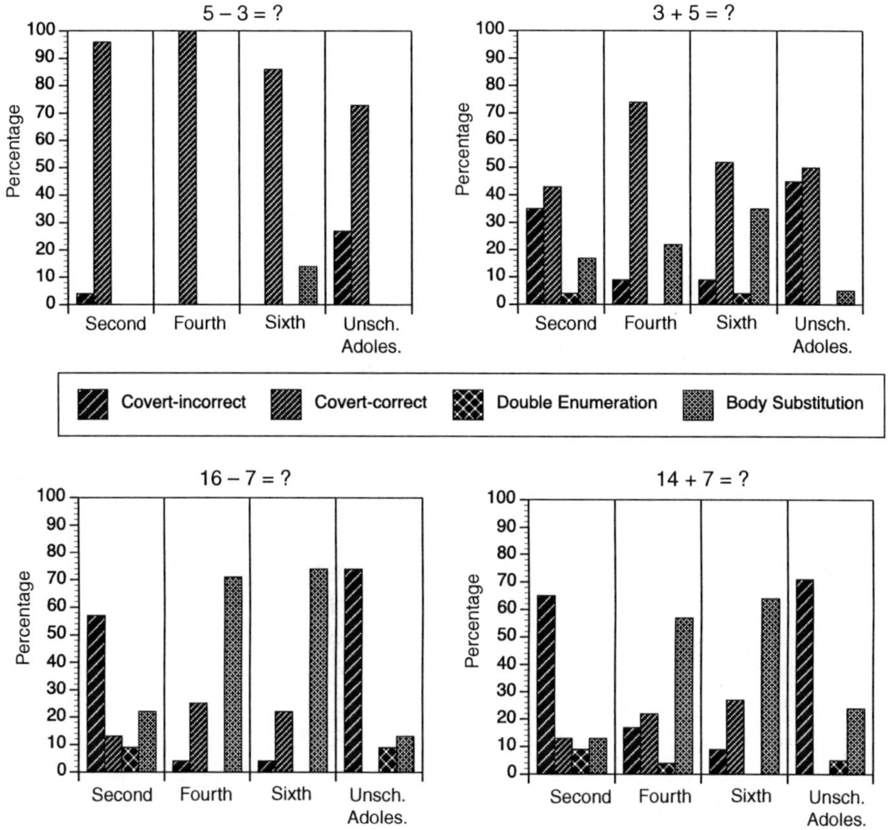

Figure 99. Percentage of children's strategy types as a function of grade level for Phase 1 problems.

Grade 6 (74 percent, 64 percent) children demonstrated advanced body-part procedures on these problems. Comparisons between the Grade 6 children and their nonschooled peers showed a similar effect for schooling: For each problem, only a minority of nonschooled children (22 percent, 29 percent) demonstrated an advanced body-part strategy compared with the majority of the Grade 6 children: 16 – 7 = ? (chi-square ($df = 1, N = 46$) = 9.83, $p < .01$); 14 + 7 = ? (chi-square ($df = 1, N = 43$) = 3.99, $p < .05$).

Children's Solutions to Phase 2 Problems

To examine whether Oksapmin children used their body-part computational system to solve problems involving numbers that extended beyond

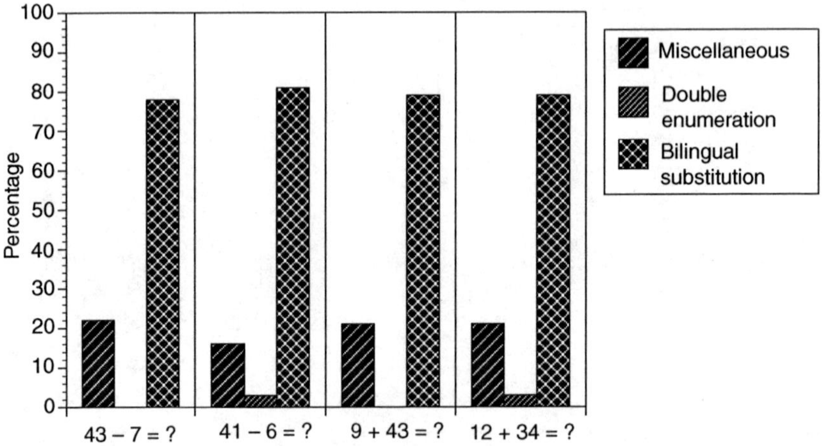

Figure 100. Percentage of children's strategy types for Phase 2 problems.

the conventional limits of the body system, I administered the four Phase 2 problems to children who had demonstrated advanced body-part procedures on any of the Phase 1 problems (largely children in Grades 4 and 6). Figure 100 shows the percentage distribution of children's solution strategies.

The strategy distributions for each of the Phase 2 extension problems were very similar. For all problems, a majority of the children used Western numeration to represent the larger value of the problem and used the indigenous system as an intermediary to keep track of the addition or subtraction of elements, generally using a bilingual-substitution strategy.

Discussion

To achieve an accurate solution to an addition or subtraction problem using body counting, children must construct a method of keeping track of the arithmetical composition or decomposition of numerical terms. When the cardinal numbers are greater than a small set size like three or four, keeping track presents an intellectual challenge. In Chapter 4, I documented the emergence of new keeping-track functions of the body system that were leading to a shift from external to internal correspondences related to collective practices of economic exchange in 1980. The focus of the present studies was to consider whether similar processes were occurring related to collective practices associated with schooling.

Like adults engaged with economic exchange, some children in school produced microgenetic constructions that extended the body system to serve keeping-track functions (Study 9-1). These novel developments were revealed through a series of findings. (a) Casual observations revealed that some children appeared to be using their body-count system during math lessons. (b) The observational study during the arithmetic test corroborated that Oksapmin students used their body system in classrooms, though its overt use declined over grade. (c) The interview studies (Phase 1 problems) revealed that, with increasing grade, children used body parts to serve internal keeping-track functions as they used the system to solve arithmetical problems. (d) The contrast with the age-matched controls revealed that the development of higher-order functions for the body system was specific to schooling, and not a development that occurred normally for unschooled children. (e) Oksapmin students in Grade 4 and Grade 6 generated approaches to solving computational problems that extended the bounds of the body system through construction of hybrid solutions to problems in which numerical terms had greater values (Phase 2 extension problems). Particularly noteworthy is that, like the internal correspondences that had emerged in adults' constructions associated with economic exchange, these ontogenetic processes developed without the facilitation of direct instruction. Indeed, teachers were unaware of students' productive and inventive use of the body-counting system to solve problems in the classroom.[2]

The microgenetic constructions I documented in children's arithmetical solutions provide insight into the ontogenetic and sociogenetic trajectories in functions of the body system. From an ontogenetic perspective, with increasing engagement with arithmetical problems indexed by grade level, children extend their knowledge of external correspondences between body parts and objects to internal correspondences between body parts and body parts to serve keeping-track functions in additions and subtractions.

[2] The progressive coordination of correspondence relations documented with Oksapmin children has similarities to that documented with Western populations of children (Carpenter & Moser, 1982; Carpenter, Moser, & Bebout, 1988; Fuson, 1982, 1988; Steffe et al., 1983), although the Oksapmin case also displays some differences. For instance, the spatial organization of the body system and its nonbase structure lead Oksapmin children to develop strategies not observed in Western children, such as the semiotic move of calling one body part by the name of another body part in the service of keeping track (referred to above as body substitution).

From a sociogenetic perspective, Oksapmin schoolchildren's use of the body system in classroom practices reveals both continuities and discontinuities in people's use of the system in Oksapmin social history. Children were reproducing the body form as they were solving arithmetical problems, supporting continuity over social history of the Oksapmin number system. At the same time, children were producing variants in the body form, inadvertently altering the use of the system to serve new functions. In this process, variant functions that the children produced may have become objects of reflection by their peers. Uptake of variants may have been a source of the widespread use of internal correspondences by schoolchildren. It is noteworthy that the process of creating new uses of the body system seemed natural and unremarkable to the children.

How durable or lasting were these body-part arithmetical innovations documented in 1980? It could be that in the multigrade community school the innovations would become common practice, crossing generations of students. But such is not the necessary outcome. Indeed, as conditions change, whether the representational systems available, or the character of schooling, or shifts in values about appropriate behavior, we may find that the ontogenetic trajectories of learning documented here have been replaced by trajectories of other sorts. The next chapter reports on studies conducted at the same school, but 21 years later, with a concern for the character of mathematics lessons, representational forms supported, and children's use of the body and Tok Pisin forms to solve arithmetical and currency problems.

10 Some 20 Years Later

Schooling and Number

In this chapter and the next, I shift my focus from the 1978/1980 postcolonial period to studies conducted early in the reform period in 2001. The site for the studies was the same school that was the focus of Chapter 9, and its appearance was similar. But the shift in time was of some consequence for classroom life. Unlike their 1980 counterparts, the 2001 teachers were all from Oksapmin, with first-hand knowledge of local practices. Their local knowledge supported the implementation of the 1998 reforms, in which the central government mandated that instruction should bridge community practices and Western knowledge.

The community school buildings of 2001 resembled the schools of 1978/1980. The photograph in Figure 101, taken in 2001, is of a typical classroom – a structure built of woven walls and a tin roof in which students sat on long benches at multistudent desks made from indigenous materials. Now the floor was covered in movable, smooth, white, small stones taken from the nearby river, and the students used these stones as calculation aids to solve addition and subtraction problems. As in 1978 and 1980, schools had no electricity or running water. But now lighting for a good number of classrooms came from sections of corrugated plastic roofing, which allowed sunlight in but at the same time prevented rain from entering. Just as in 1978/1980, the adults and students in the community were responsible for building and maintaining the school buildings and the teachers' houses, an issue that caused tension between the school staff and neighboring adults.

In line with the educational reforms, the curricular emphasis had changed. In 1980, as noted in Chapter 9, the curriculum focused on teaching Western mathematics using English. In 2001, there was much greater interest in and national support for using language flexibly to engage student understanding and to link what occurred in the classroom with

Figure 101. Typical classroom at a community school in 2001.

out-of-school activities in more substantive ways. For instance, in 2001, in the Grade 3 classroom that was the focus of the current study, the teacher moved fluidly between English, Tok Pisin, and the Oksapmin language, as she saw necessary. In the community schools, academic subjects were similar to those of the 1980s – arts, community living, English, mathematics, health, physical education, and environmental studies (science). But there was also specific instruction in vernacular language to help students continue to develop their Oksapmin language skills; this was a practice unheard of in the community schools of 1980. Unlike their 1980 counterparts, teachers in 2001 were involved in community life, attending local churches, and were frequent participants in the soccer and volleyball games that took place after school on the grounds.

Another noteworthy shift between the first community schools and schooling of 2001 was in the number of students. With new hiring at the mine in Tabubil, there was more money in circulation (that could be used for school fees). Further, as indicated in Figure 85, there were a greater number of schools in the area in 2001, meaning a shorter distance for small children to walk to get to their classes. Thus, the proportion of children attending school to Grade 6 or even Grade 8 had grown. In fact, the increasing number of students led to overcrowding – the third grade teacher whose classroom was the focus of Study 10-1 had 91 students, many more than her classroom could handle. To accommodate the students, the school split the students into two different half-day classes. No substitute teachers were available, so when the teacher was called away, classes were cancelled.

The two school studies that are described in this chapter varied in their methods and focus. Study 10-1 was observational and conducted in a Grade 3 classroom. The study documented ways in which students' mathematics was supported in the classroom and the way the classroom environment emerged in relation to children's display of mathematics. Of particular concern was instruction related to arithmetic and currency, with attention to whether students, like their counterparts in 1980, were making use of their body-counting system (and other representational forms) as they made sense of arithmetic problems.

The purpose of Study 10-2 was to gather interview data on children's use of representational forms for token naming and arithmetical problem solving. In the methods, I drew from tasks that Indigo and I used in the Part II studies with adults (Chapters 4 and 5). Interviewing the Grade 3 students supported an analysis of similarities and differences in the new generation's use of forms of representation for number and currency; the interviews would also allow for comparisons with adults in the community.

Study 10-1. Grade 3 School Mathematics in 2001

In this study, Indigo and I focused on two 30-minute math lessons in the third grade classroom of Ms. N, a community schoolteacher (though a good number of additional lessons were observed). During the lecture portion of the two targeted lessons, we sat at the back of the classroom, taking fieldnotes and observing. The fieldnotes on which the following descriptions are based were taken by Indigo. Indigo also took the lead in framing their presentation, hence the shift to the first-person plural in the narrative. The two lessons on which we chose to focus are related to earlier chapters in the book. One lesson was on arithmetic, though not using the Oksapmin body system but its Western counterpoint – Hindu-Arabic algorithms for subtraction. The other was on arithmetical representations of and computations with currency, a topic that was the focus of a number of chapters on economic exchange. As the lessons unfold through Indigo's fieldnotes, the contrasts between in-school and out-of-school practices in these domains will be apparent.

During the lessons we observed, Indigo and I (and on some days, Indigo alone) walked around the classroom, taking a closer look at what students were doing and observing the strategies they employed. While students worked individually or in pairs, we also took photographs of classroom curriculum materials and the chalkboard writing at the end of the lesson. We also engaged in considerable preparatory work and follow-up work on the analysis of these lessons, talking informally with teachers at the school.

The Teacher and Her Classroom

A typical Oksapmin classroom, Ms. N's room consisted of half of a school building with thin woven walls, dirt floor, and a tin roof. At the front of the room, a dusty and weathered black chalkboard was fixed to the wall; it was divided into three sections, labeled English, Mathematics, and Combined Subjects. The other three walls contained many bright posters provided by local and national governments; the posters advised students about proper dental hygiene, and provided maps of the country, spelling lists, drawings of Papua New Guinea cultural artifacts, and the like. Ms. N's desk stood at the front, and the rest of the room was filled with long tables and benches where the students sat – usually girls on one side of the room, boys on the other. The other half of the building was not in use, containing only a few discarded desks and a chalkboard fixed to the wall.

Ms. N was a tall, thin, friendly woman, a regular at the volleyball games held on school grounds every evening. She had grown up in the nearby hamlets, attended school and then teacher training outside of the Oksapmin area, and had returned to teach in the community schools. She had been teaching at this particular school for several years.

Two Mathematics Lessons

The two lessons we describe below cover a range of topics, some of which have been focal concerns in prior chapters – numerical representations and arithmetic. Of particular focus in the lessons was two-digit subtraction (including a borrowing and carrying algorithm), names for currency tokens in English, and addition using currency. After each lesson, we present an analysis of the knowledge domains that were treated in the lessons, with a concern for forms of representation supported, the contexts in which those forms were used, and the functions for which teacher and students adapted the forms to serve in communication and problem-solving activity. Of particular concern are the ways in which schooling supports the reproduction

$$\begin{array}{r} 54 \\ -26 \\ \hline \end{array}$$

Figure 102. A subtraction problem that Ms. N writes on the blackboard in vertical format, 54 – 26 = ?

and alteration of the mathematical form-function relations in communication and problem solving related to number and reasoning.

Subtraction: Algorithms and Counting

It was mid-morning when we entered the classroom, and Ms. N had written a series of single- and double-digit subtraction problems on the chalkboard. To introduce the problems, she begins by telling the class that they will be working on subtraction that day. She goes on to describe the day's activities using both English and Tok Pisin. The previous day the students had been working on subtraction problems, and in the following passage, she reminds students of their prior work, and guides them through one problem.

> Ms. N reminds the class that when they previously talked about place value, they had talked about the "units column and the…" The class responds: "tens column." ... Ms. N nods. She reminds them that sometimes when they are subtracting, they have to borrow from the tens column to the units column.

She turns to the first example, written on the board, which is shown in Figure 102.

> Ms. N points to the units column in the subtraction problem and asks what to do. She speaks in English, Oksapmin, and Tok Pisin to explain that when you try to take 6 from 4 in the units place, "it cannot."
>
> She then borrows from the 5 in the tens column, writing a small "1" beside the 4 in the units column. She asks the students to take 6 from 14. Audible are several students trying to solve this – one child whispers "seven," another whispers "twelve."
>
> Ms. N rewords the question. She asks how many stones would be left if you started with 14 and then took away 6. A student tells her the answer and she writes 8 on the board. As a class they finish the example under

> Ms. N's leadership – at each stage along the way she asks the students questions, like "where do I put the one that I borrowed," or "what is 4 minus 2." (I.E., fieldnotes, July 26, 2001, pp. 2–3)

In this excerpt, we see teacher and students working with a canonical cultural form in Western school arithmetic – a Hindu-Arabic algorithm for subtraction. The students began with the subtraction of one number from the other in the units column. She anticipated students' errors, pointing out that 6 could not be subtracted from 4, so one has to "borrow" from the tens column. She proceeded through the algorithm, asking students to tell her what to do at various stages of the problem. Her questions to the students focused on the steps of the subtraction problem, where to write the numbers involved, and whether or not a particular subtraction could be performed.

Ms. N moved on to solve, with help from the students, one more problem as a further exercise in using the algorithmic form, and then assigned eight similar problems for independent work or work in pairs. As they worked, Ms. N walked around the room, correcting student work and giving help where it was needed. Once again, much of the classroom talk was in English, especially when students discussed the subtraction algorithm. Though the numbers were given in English, all three languages were used during the lesson.

Observations of students in pairs or triplets revealed some of their strategies. Although many students used covert strategies, some made tick marks on their papers and counted them in English; others counted fingers and toes; still others used stones from the classroom floor as counters. In contrast to the 1980 observations, no students were observed using the traditional body-counting system as a means of solving the problems.

After circulating about the room, Ms. N called the class back together to go over a particular problem, 62 – 15 = ?, the first assigned problem in which borrowing was required. Having noted that many students were having trouble with the problem, she modeled the solution using stones from the classroom floor. Ms. N instructed students to collect stones as counters, and modeled subtraction by throwing away stones and counting how many stones were left. She did this using 12 stones for the units column, and then again using 5 stones to represent the 5 in the tens column. The next fieldnotes excerpt provides a narrative description of the episode.

> First Ms. N asks students what is the answer if you take 5 from 2. "If you said 3, then yours is wrong," she says. "Check your answer," she tells them, and there is a rustling as students turn back to their notebooks to check. They whisper to each other.

> ... Ms. N tells the students to each pick up 12 stones from the floor. (The floor, except underneath the desks themselves, is covered in white stones.) The students do so, with a lot of whispered counting in English as they gather the stones – "one, two, three, four, ..." Then Ms. N asks them to take 5 stones away from the piles on their desks. I hear the sound of many stones dropping to the floor. Ms. N asks them to count how many stones are left. D [a student] touches her fingers to the stones on her desk, counting "four six seven" very quickly. A student supplies the answer 7 to Ms. N, who writes it down on the board. Next Ms. N says they need to take "1 from 5, get 5 stones." The students do this, mostly just taking 5 of the stones already on their desks. Ms. N tells them to throw one stone on the floor. Again, audible is the sound of many stones hitting the floor. Ms. N asks them how many are left and someone tells her that there are 4 left. She writes this on the board. (I.E., fieldnotes, July 26, 2001, p. 4)

After this exercise, Ms. N had students continue with their individual or group work on the chalkboard problems. She continued checking student work and talking with individual students.

At about the time when recess normally begins, Ms. N asked students to put their pencils down. She shifted the class's focus to the correction of the exercise. As she proceeded, step-by-step, to solve each exercise problem on the board, she asked for student input. For example, at one point, she asked students the answer to 6 minus 8 in a multidigit subtraction, focusing on the units column. Several students said "eight." She again had students take stones from the classroom floor. Once all students had six stones in front of them, she asked them to take away eight stones. One student said "it cannot," prompting Ms. N to go through the procedure once again for borrowing and carrying. At about 10:10, Ms. N finished correcting the problems and sent the students outside for their mid-morning break.

Forms and Functions for Subtraction. This lesson contrasts remarkably with students' activity in 1980, when teachers in the schools were unfamiliar with traditional practices and students were forbidden to speak Oksapmin language while on the school grounds. In 1980, although their teachers offered no support to students with the body system, students used the system to solve addition and subtraction problems. Twenty years later, there was little evidence that students used the system in arithmetical problem solving.

Why was this the case? One contributing factor may be the different relationship between teachers and students in 1978/1980 and 2001. Recall that in 1978/80, many children knew the body system, a finding revealed in the studies conducted in 1978 and 1980 with unschooled children (Studies

8-1, 8-2, and 8-3). When monolingual Oksapmin children went to school then, they were faced with both new mathematical challenges and teachers who had little understanding of their interpretations of problems. These children were in a predicament: They needed to make sense of numerical problems in an alien language with an alien number system. It is likely that many initially interpreted arithmetical problems in body-system terms, a representational form they had some understanding of, adapting the body system to represent and solve those problems.

A related contributing factor may be that in 2001, the children's knowledge of the body system was more limited. In 2001 children could count more accurately and to higher numbers in Tok Pisin than they could in Oksapmin. (As will be reported in the next study (Study 10-2), this conjecture will receive some corroboration.) Contributing to limited knowledge are negative values about the system by some schooled individuals. For example, as the studies reported in Chapter 4 indicate, schooled adults in Oksapmin, from adolescents to middle-aged people, were more likely to use English quantifiers in trade stores than they were to use Oksapmin quantifiers. And, in conversations with Indigo and me, at least some adolescents and young adults made evaluative statements disparaging the indigenous mathematics. These third graders had no doubt participated in trade store transactions alongside their older peers, and observed the use of English. It may have been the case that in this community knowing English mathematics was the mark of an educated person and was therefore desirable. If this were so, even young children might be likely to use English for mathematical purposes whenever possible.

Currency Tokens and Exchanges: A Context for Arithmetic

When we arrived at the classroom, the day's math exercises were already written on the chalkboard; the heading written on the board was "Money," followed by two example problems. The teacher intended to go over the example problems with the class. In addition, she had written six exercise problems for students to work on in pairs and individually (see Figure 103).

The teacher began the lesson by making connections for students between their activities outside of school and the subject matter targeted in the lesson. Ms. N, for example, asked students to tell her how much it would cost to buy a packet of biscuits (that is, crackers or cookies) at a trade store close to the school (biscuits are a common and relatively inexpensive item in trade stores). One student responded and others agreed. The

– Money –

Examples

1. Which coins will you use to buy a tin of fish that costs K1.65?

2. What notes and coins can be used to buy a pair of shorts that costs K7.87?

Exercises

1. Which coins make up 18t?

2. Which notes make up K18?

3. Which notes and coins make up K21.55?

4. What coins can be used to buy a ruler that costs 77t?

5. What notes can be used to buy a pot that costs K36?

6. What notes and coins can be used to buy a plane ticket from Tekin to tabubil that costs K139.97?

Figure 103. Currency exercises.

children seemed familiar with the store and the prices of common items. The following passage is transcribed from Indigo's fieldnotes:

> Ms. N changes topic, moving to a coin naming activity.
>
> She asks "what are the coins that we use, here in Papua New Guinea?" Many children raise their hands and she calls on them one by one. The first child answers "wan kina," and another child answers "fifty toea." Ms. N continues to call on children until they have listed all the coins, and then she asks them about "the notes that we use." Pretty soon the students have listed all the common units of currency. (I.E., fieldnotes, July 27, 2001, p. 2)

It is noteworthy that not only were students able to name all of the locally used coins and notes in this lesson, they were even able to name coins that are very rarely (if ever) used in the Oksapmin valleys – the 1t, 2t, and 5t coins – names that they had previously reviewed in school. The children's answers throughout this discussion were all in English.

After the naming activity, Ms. N turned to the first of the examples written on the board: "Which coins will you use to buy a tin of fish that costs K1.65?" The first child that she called on to answer this question said "one one-kina, three twenty-toea, one five-toea." Ms. N wrote this answer on the board as "K1 + 20t + 20t + 20t + 5t." Ms. N then asked if anyone had a different solution to this problem. Several children provided different combinations of coins and notes, and Ms. N kept a record on the chalkboard.

Ms. N did the same with the second problem example. She stated the problem aloud and called on several students, who provided a variety of answers. She wrote only one of the possible solutions on the board. With no further discussion, Ms. N asked the students to complete the six chalkboard exercises.

> She tells them to get out their exercise books, and says "do not write the questions write only the ..." She pauses and several children say "answers" in response. "Your answers should be like this in your exercise book," she says, and points to some of the examples that she wrote on the board. (Fieldnotes, July 27, 2001, p. 4)

As they worked, Ms. N circulated, checking their work, making corrections, and helping as necessary.

Once again, we observed students whispering to each other as they bent over their books, collaborating on solutions. We observed students counting on their fingers and counting aloud, but we did not observe students using the body-count system. We also did not notice them using stones from the classroom floor or other manipulatives.

After allowing the students some time to work alone or in groups on these problems, Ms. N led a class discussion through solutions to all of the problems. Typically, she would ask a question in English and then explain the meaning of the question in the Oksapmin language. Students would answer in English. Sometimes, when students appeared to be struggling to answer, she would repeat her question or explanation in Tok Pisin or in the Oksapmin language. The following fieldnote excerpt illustrates Ms. N's strategy of calling on several students to get alternative solutions to the problem.

> Ms. N moves on to the 3rd problem. She reads it in English ("Which notes and coins make up K21.55?") and points out that in this one she is asking about "notes and coins." The first student that she calls on for an answer is a girl. The girl's answer is K20 + K1 + 50t + 5t. Ms. N asks the class, "is that correct?" "Yesss" I hear, from all parts of the room.
>
> Another girl's hand is raised, a girl at the very front of the room. Ms. N says her name and the girl gives an alternative answer: "two ten-kinas plus one-kina plus fifty-toea plus five-toea."

> Ms. N writes down what she says, reading it aloud afterwards and adding as she goes. [Her repetition of the students' answer is something like: "Ten kina plus ten kina is? Twenty kina. Plus one kina is? Twenty-one kina. Plus fifty toea is? Twenty-one kina fifty toea. Plus five toea is? Twenty one kina fifty five toea."]
>
> Ms. N then asks, "Or? How many five kina notes" can you use to get K20? She chooses a boy who answers "four." Ms. N then asks how many 50t coins [do you need] to get one kina. Someone answers "two ." Ms. N also shows that 20 + 20 + 10 = 50 and 2 + 2 + 1 = 5. She uses these to give another alternative answer. (Fieldnotes, July 27, 2001, p. 7)

After progressing through the exercises, Ms. N wrapped up the 30-minute lesson. She once again made the connection between the classroom activities around currency and children's activities in trade stores. She reminded them that in a store, one can pay for things in many different ways and may not always use exact change, perhaps a hint that the problems they did in class were not complete representations of out-of-school activity.

The photograph in Figure 104 represents the chalkboard at the end of the day, with Ms. N's transcriptions of the students' solutions to the problems. The first line of text is one answer to example 2. The next four lines of text are answers to example 1. At the end, Ms. N wrote the final exercise of the day – the payment for a trip from Tekin to Tabubil – along with one solution to that exercise.

Forms and functions for currency. A teacher instructing a classroom of students in ways of using representational forms to name tokens and produce arithmetical compositions in an important sense is a *broadcast*. The teacher occupies a social position of authority, serving as a broker of knowledge, using particular representational forms to serve numerical functions. In Ms. N's teaching, like other teachers, she supported just one correct way to name currency tokens in the English language, using terms like "one toea," or "five kina." In contrast, outside of school, there is no central broadcast, and there are a variety of ways to name currency tokens. For example, in trade stores, one might hear the term "five kina" being used to describe a K5 note, but one might also hear terms like "*wan faun yota* (2) *temsi tita*" (meaning 2 one-pound notes and one K1 coin). Indeed, one should recall that when Indigo and I interviewed 57 adults and asked them to tell us the name for a K5 note, we obtained 26 principal variants (and 48 different responses)!

Figure 104. Problems involving computation with currency in a lesson for a Grade 3 class at a community school in 2001.

One interesting clash in the broadcasts of the classroom in naming currency tokens and in the social history of naming tokens in economic exchanges outside of school is what is taken as the arithmetic currency unit in talk and arithmetic. In Ms. N's classroom, unit values of less than 10 toea are commonly referred to and represented in computation problems. Recall, for example, the photograph of the blackboard in Figure 104 with its representations of 1t, 2t, 5t, and K139.97. (With very rare exceptions, the government's issue of the 1t, 2t, and 5t coins had not reached the Oksapmin area at the time, and hence children were unfamiliar with these tokens and token values in their out-of-school lives.) In the Oksapmin language, there were no stable names for these tokens – for most, they did not exist.[1] Indeed, recall that the 10t coin (the shilling equivalent), the smallest unit known, was referred to as thumb (1) or *tit* (1). The introduction of a currency unit of one toea in the classroom posed new coordination

[1] In local trade stores, owners never priced in units below 10t. That said, the 1t, 2t, and 5t denominations were readily available and frequently used in Tabubil, the mining town where increasing numbers of Oksapmin people were employed, and where items were priced to the toea.

problems that clashed with the history of units outside the classroom. I take up this issue in Study 10-2, an interview study, which Indigo and I conducted with the students in Ms. N's class.

Study 10-2. Quantification Practices Known by Grade 3 Children in 2001

In this study, Indigo and I individually interviewed Ms. N's third-grade students with two purposes in mind. The first was to further understand cross-generational differences in the Oksapmin community in 2001 related to representational forms and functions for number and currency. To this end, some of the procedures we used were the same as those used in Part II in studies of number (Study 4-2) and currency (Study 5-1, Study 5-2). But we also introduced some modifications to the interview in order to capture better the classroom world of the children. For example, as just noted, the 1t, 2t, and 5t coins were not yet in circulation in the Oksapmin community itself (or only rarely so), and communicating about currency values of 1 through 9 toea (like the values of K2.04 or K3.97) was not a part of talk in the Oksapmin world. So, unlike in the studies reported in Chapter 5, we included the 1t, 2t, and 5t coins, asking the children to name the tokens in both Tok Pisin and the Oksapmin language. Further, since third graders did not have access to great amounts of currency, we did not include the K5, K10, and K20 notes that were used with adult participants.

The second purpose was to confirm that Ms. N's students differed from their 1980 counterparts in how they used the body system in arithmetical problem solving. Based upon the classroom observations reported in Study 10-1, I expected that we would not observe body computational strategies due to multiple converging factors. In Ms. N's classroom, students had been supported in their use of stones to solve arithmetical problems, a potential alternative to the body for classroom arithmetic. Another, and related, factor might be that their knowledge of body counting was quite limited.

Methods

Participants

Indigo and I conducted all interviews with the support of a translator. The children in the class had different schooling histories prior to third grade. About half had attended English-language community schools and about half had attended the recently introduced Oksapmin language elementary schools. However, the two groups did not differ on most of our assessments, so we consequently pooled children in most of our analyses, and we indicated these different backgrounds only when we did find differences. Thirty-one students participated in the study, with about equal numbers of boys and girls. The children were familiar with Indigo and me through discussions in their classrooms as well as through informal talk on school grounds.

Procedures

The interview procedures consisted of several parts. Most of the interview questions were identical to questions asked of adults in the studies described in Part II.

Introduction. Following some initial warm-up questions about demographics, Indigo and I explained that we were interested in understanding how mathematics has changed over the last 20 years in Oksapmin communities, expressing the hope that the children would help us.

The Structure of Body Counting in the Oksapmin Language. The interviewer asked the child to count in the Oksapmin language, using the procedures described in Study 4-2 with older populations. When children stopped enumerating body parts, they were asked if people count further than the part to which they had counted. They were then asked to demonstrate. We expected that some children would stop at the 20th place, given the innovations in body counting. To probe children's own thinking about the system's structure, Indigo and I asked if children would show the body part that was one more than the 20th body part. For similar reasons, if children counted to the 27th body part, they were asked if they would show the number that was one more than the 27th.

Counting in English/Tok Pisin. The children were then asked to count in English. If they stopped before reaching 41, they were asked if they could count higher than that. Errors children made and how high they managed to count were recorded (up to a maximum of 41).

Currency Identification: Oksapmin and Tok Pisin (Comparable to Study 5-1). Children were presented with currency tokens, one by one, in the following order: 1t, 2t, 5t, 10t, 20t, 50t, K1, K2. In each presentation, we asked what the token was called in the Oksapmin language. The procedure was repeated with a request to indicate what each token was called in Tok Pisin.

Computing with Tokens of Currency: The Oksapmin Language and Tok Pisin (Comparable to Study 5-2). Children were presented with two problems, each of which required determining the total value of a set of coins (drawn from Study 5-2). The two problems included (a) six 10t coins (total: K.60), and then (b) three 50t coins and three 20t coins (total K2.10). They were presented in the same order to each child. After each problem, if the child did not engage in an overt procedure to determine the value of the set, the interviewer asked, "Can you show me how you did that?" Children were first presented with the two problems and asked to indicate the value of each set of coins in Oksapmin. The same procedure was then repeated in Tok Pisin.

Reading Written Prices: The Oksapmin Language and Tok Pisin (Comparable to Study 5-2). Children were presented with two written prices (separately) that they might see in a trade store – 60t and K2.10. Children were asked to look at each price and then to tell us the price in the Oksapmin language and then read the price in Tok Pisin (English).

Additional Procedures. Note that in addition to the procedures described below, Indigo and I asked students questions that were used in prior studies both about the meaning of *fu* (Study 6-1) and about arithmetic in Oksapmin (Study 9-2). But only on rare occasions did they use body-part strategies documented in the 1980 study for solving arithmetical problems, and few had any knowledge of *fu* – its location or its meaning. So, these procedures and associated results are not presented.

Results

Counting in English and with the Body System. All children counted fluently in English, and all were stopped at 41. One child, however, made an error at 9, and another made an error at 21.

When displaying the body system, all but two children reached the 27th body part (one could not go beyond wrist (6), and another not beyond forearm on the other side (21)). A number of children made errors in their counts. Thirteen of the 31 children (42 percent) made errors in their count to 27 (including the two children who did not reach the 27th body part).

All children were queried about the body part that was next after the 27th. Nineteen children said that one could not count further than 27, or that they did not know. Only one child indicated looping up to the wrist, what I took as the traditional procedure in 1978. Four indicated that one should begin the count again at the thumb. Seven indicated reversing the count by counting the pinky (two children) or the ring finger (five children).[2]

In general, for numbers greater than 10, children appeared to be more proficient at counting in English than in Oksapmin. This finding stands in contrast to the performances in Study 4-2: (a) Elders were much more proficient with body counting than at counting in Tok Pisin and (b) schooled adults were equally proficient in both.

Currency Identification in the Oksapmin Language and Tok Pisin

In this section, I consider properties of students' names for the currency tokens in the Oksapmin and Tok Pisin languages and their distributional patterns. To code children's production of names, Indigo and I adapted the coding scheme used for naming tokens in Study 5-1, supplementing the codes with additional ones. Codes were added when at least three children offered a common response for a name unrepresented in the categories from Study 5-1 (otherwise the response was coded as "various"). Within each category, there were minor variations in student responses, which were collapsed into a single representation. For example, for the 1t coin, a response of *sup* (extreme form of something) and *sup tan* (extreme form of flat (coin)) would both be coded as *sup*.[3]

1t and 2t Coins. Figure 105a and Figure 105b display the percentage distributions of names used for the 1t and 2t coins, respectively. In each figure, the upper distribution is for Oksapmin names and the lower distribution is for Tok Pisin names.

Consider the Oksapmin names first for the 1t and 2t coins, the upper distributions in Figure 105a and Figure 105b. Like their adult counterparts, some children made reference to coins using number, though the units used

[2] These results must be interpreted cautiously. As Indigo and I conducted our interviews in an empty classroom in the school, children gathered outside to wait their turn. Several children coached each other in Oksapmin counting in preparation for the interview! Thus, children may be less familiar with the body system than their performances indicate.

[3] Some of the students in this study had been to elementary school, and as the next chapter reveals, in elementary classrooms, some children may have been instructed that *sup* (or *sup tan*) is the Oksapmin word for the 1t coin (see Chapter 11, Figure 119). Since the coin is not in circulation in the community, early school experience may be the source of the name for some.

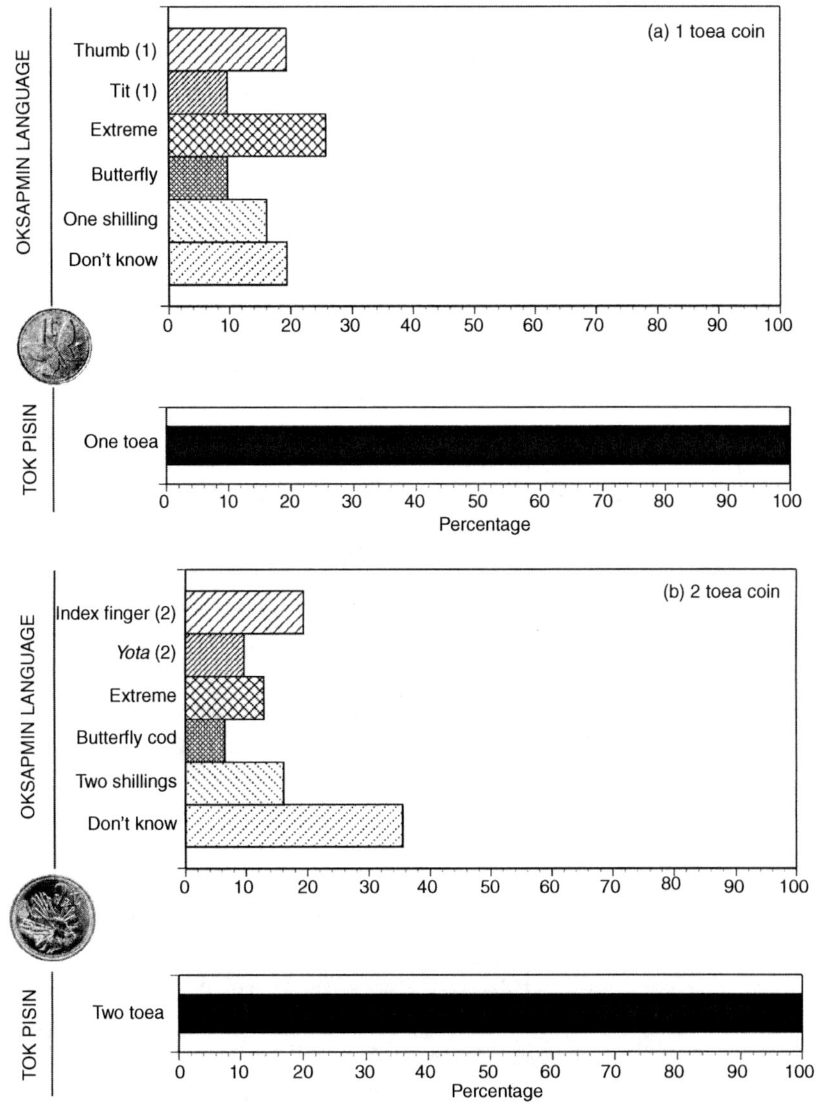

Figure 105. Percentage distributions of names used for the 1t coin and 2t coin in the Oksapmin language and Tok Pisin by third graders (2001).

were toea, not shillings (of course, the values of 1t, 2t, and 5t could only be expressed in fractions of shillings, words unavailable in the Oksapmin language). Sometimes the reference for number was using body parts, thumb (1) and index finger (2) for the 1t and 2t coins, respectively. Other times reference was made through the use of conversational numbers, *tit* (1) and *yota* (2). At still other times, children offered tropes. For the 1t and 2t coins, some children used the expression "extreme" (*sup*) for both coins, an expression that perhaps they had been instructed to use in their prior elementary school experience (see Chapter 11). Other tropes included the names of the animals printed on the coins, such as "butterfly" for the 1t coin (see Figure 105a) and "butterfly cod" for the 2t coin (see Figure 105b). A substantial minority of children said they did not know the names of the coins in the Oksapmin language (~20 percent for the 1t coin and ~35 percent for the 2t coin).

Of interest in children's variant names is their use of body parts or conversational numbers. In their abbreviated form (without the inclusion of the unit, like "toea" or "kina"), children's usage of the body part/conversational numbers clashed with the adults' usage. Indeed, when adults used the expression *tit* (1) or thumb (1) in reference to currency units, they generally meant the 10t coin (the shilling equivalent). In the cases here, children were using the same names to refer to the face value of the toea coins, *tit* (1) or thumb (1) to refer to the 1t coin. Note that some children made an exception to the use of toea as a numbered unit. Some children cited "one shilling" and "two shillings" for the 1t and 2t coins, respectively. But in this reference, the unit of shilling was not the normative use for the 1t or 2t coin in the adults' usage.[4] Of course, when children used tropes like "butterfly cod," they did not specify the numbered unit; it was merely a direct reference to the coin itself.

The lower distributions for the 1t and 2t coins in Figures 105a-b display students Tok Pisin names for the coins. The figures show that all students for each coin indicated "one toea" and "two toea" appropriately.

5t and 10t Coins. Figure 106a and Figure 106b provide the percentage distributions of names for the 5t and 10t coins, respectively, with the upper distributions for Oksapmin names and the lower distributions for Tok Pisin names. The patterns are similar to those for the 1t and 2t coins. For the Oksapmin names, some children used toea as the unit with either body part, pinky for the 5t coin and shoulder (10) for the 10t coin. As with the 1t and 2t names, this usage represented a clash with normative use in the community: Recall that many adults used shoulder (10) to refer to one kina

[4] The use of "shilling" in response to being asked the name for the coins in the Oksapmin language does suggest that the expression was no longer seen as a Tok Pisin or English expression, but an Oksapmin expression (though in these instances not used in normative ways).

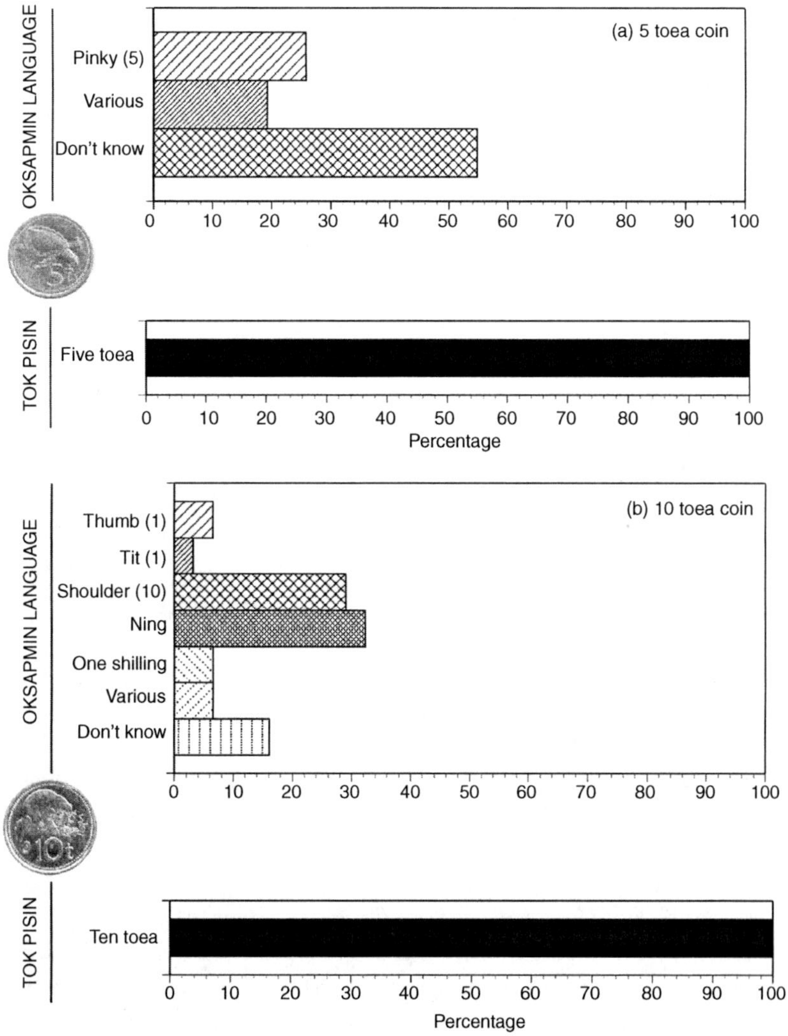

Figure 106. Percentage distributions of names used for the 5t coin and 10t coin in the Oksapmin language and Tok Pisin by third graders (2001).

(a tacit reference to 10 shilling equivalents (ten 10t coins)). A small number of responses did use shilling equivalents as units for the 10t coin, with the conversational word (*tit* (1) for the 10t coin) or the body part thumb (1). An additional small number of responses cited the Tok Pisin expression "one shilling" (*wan siling*) for the 10t coin. The modal response for the 5t coin was "don't know" for the name (55 percent), and for the 10t coin the modal response was the trope *ning*, the animal pictured on the coin.

20t and 50t Coins. Figure 107a and Figure 107b show the respective names used for the 20t and 50t coins, again in the Oksapmin language (upper distributions) and Tok Pisin (lower distributions). As in the case of the 1t, 2t, 5t, and 10t coins, some children cited body-part numbers to name the 20t coin, using toea as a unit, naming the coin as elbow (20). Others used the shilling as the unit (corresponding to normative use in the adult population). For example, some used body parts to name the 50t coin, calling it pinky (5); others used conversational numbers, using *yota* (2) for the 20t coin and *hanen* (5) for the 50t coin.

More frequent in students' names for the 20t and 50t coins was the use of tropes similar to those used normatively in the adult population, like "married" for the 20t coin and "pointed" for the 50t coin. Some used "cassowary" to refer to the 20t coin (the emblem on the coin), a variant not commonly reflected in the adult population. About 30 percent of the names cited for each of the expressions were coded as "various." The lower distributions in Figure 107a and Figure 107b show that in Tok Pisin, again, all children used the normative expressions "twenty toea" and "fifty toea."

K1 Coins and K2 Notes. Figure 108a and Figure 108b show the distribution of names cited for the K1 coin and the K2 note, again in the Oksapmin language (upper distributions) and in Tok Pisin (lower distributions). Some children used numbers for the K1 coin (but not the K2 note). For the K1 coin, the only numbers used were body parts: About 20 percent of the names cited were "shoulder (10)" using a shilling equivalent as the unit, which was similar to normative usage, and less than 5 percent cited "thumb (1)," using the kina as the unit, at odds with normative usage. As with the 20t and 50t coins, children more frequently cited tropes for both tokens, and these were identical to normative usage in the adult population. About 50 percent of the responses used "hole" (*temsi* or *temsi tana*) to refer to the K1 coin and about 50 percent used "leaf" to refer to the K2 note (for purposes of this analysis, dialect variants for leaf were combined, *haben* and *iratarin*). The lower distributions in Figure 108a and Figure 108b show that in Tok Pisin, again, all children used the normative expressions "one kina" and "two kina."

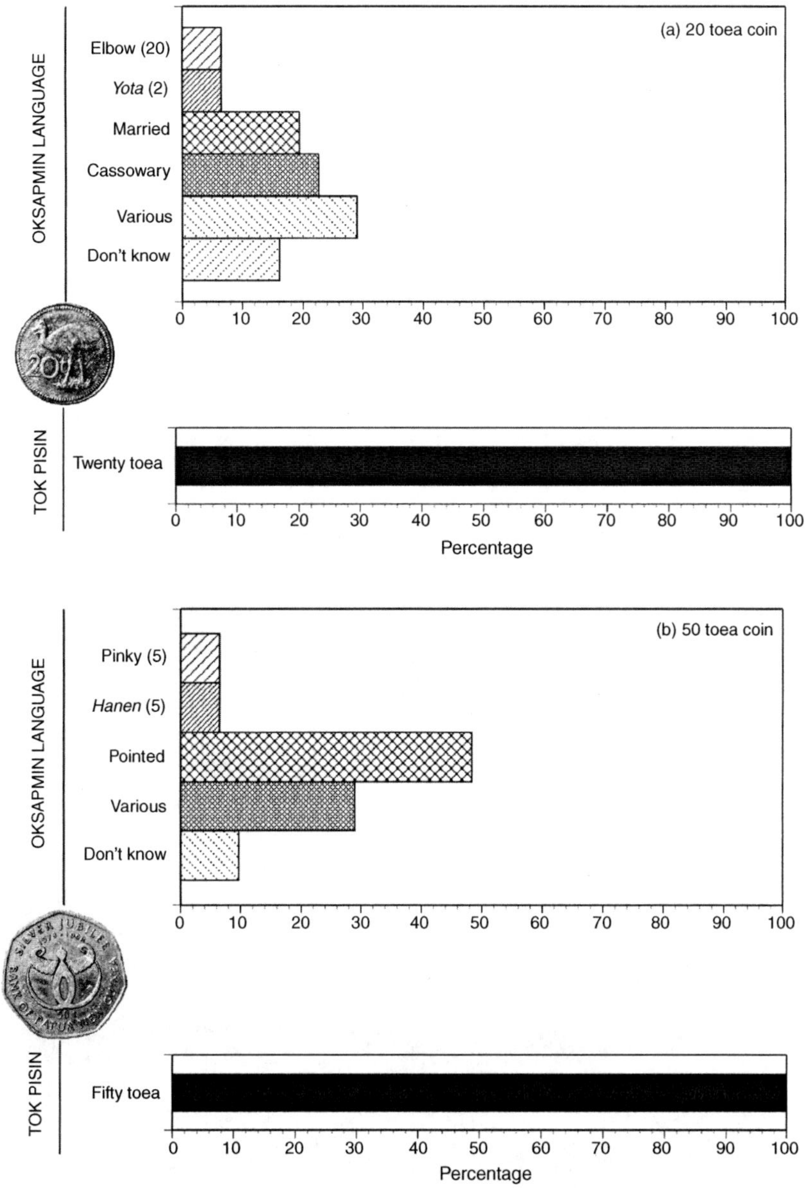

Figure 107. Percentage distributions of names used for the 20t coin and 50t coin in the Oksapmin language and Tok Pisin by third graders (2001).

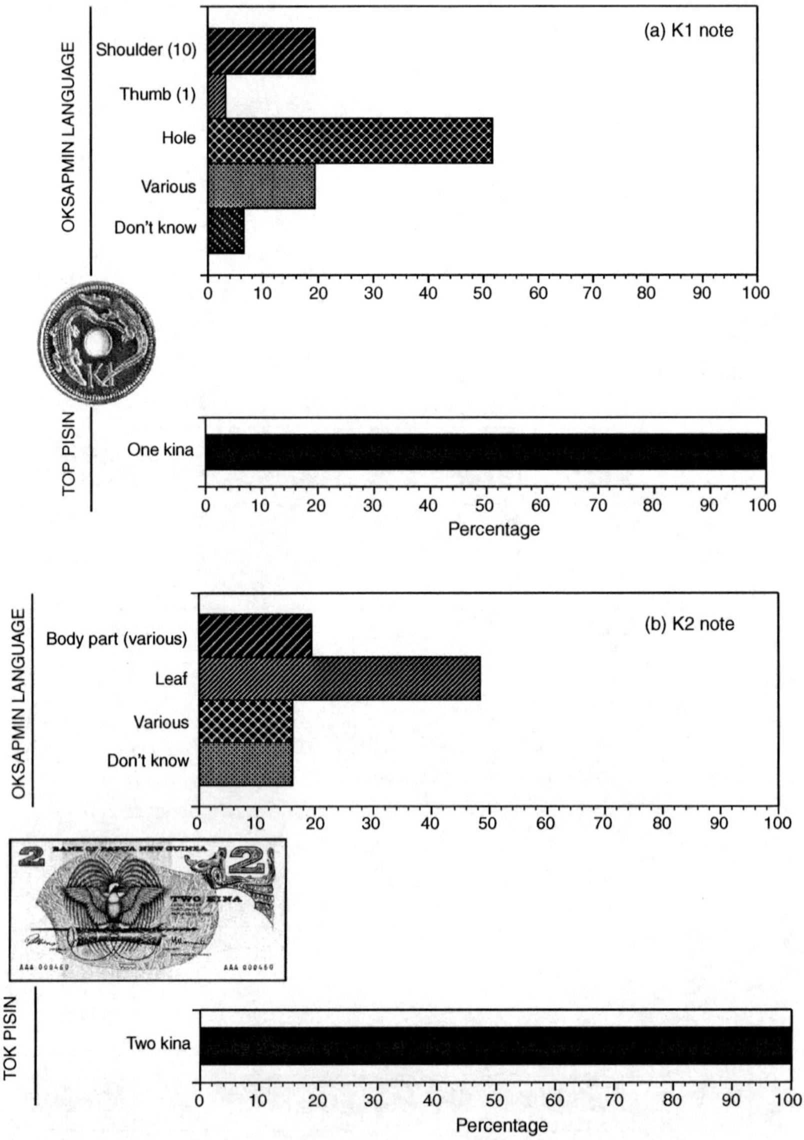

Figure 108. Percentage distributions of names used for the K1 note and K2 note in the Oksapmin language and Tok Pisin by third graders (2001).

Summary of Children's Names for Tokens

The distribution of variants in children's names for tokens in Oksapmin and Tok Pisin is comparable to that of the adult cohorts reported in Study 5-1 in several respects. First, like the adults, in Tok Pisin the children's use of names for tokens was uniform. The composition of Tok Pisin token names conforms to a composition rule: Use a Tok Pisin number word and follow it by the unit value of the token, in toea or kina as appropriate. Thus, the English/Tok Pisin names for the 1t, 2t, 5t, 10t, 20t, 50t, K1, and K2 tokens were always "one toea," "two toea," "five toea," "ten toea," "twenty toea," "fifty toea," "one kina," and "two kina," respectively.

Second, also like the adults, in the Oksapmin language the children gave multiple names for each token. Particularly when the "various" category was taken into account, we saw a striking heterogeneity of variants in naming in Oksapmin, in contrast with Tok Pisin.

A third similarity was children's use of tropes. Children often made reference to either physical features of tokens (e.g., a butterfly emblem on a token) or geometrical characteristics of the tokens (the corners on the 5t coin or the hole on the K1 coin). Many of the tropes reflected normative usage in the adult population. In using them, children could avoid the issue of quantifying a token in terms of numbers of particular unit values.

But a fourth feature of the children's names deviated from the adult cohort. Adults more frequently used the shilling (or pound) equivalent as units and children more frequently used units of single toea (or kina). When reference to the unit was tacit, as was normative practice (e.g., "shoulder (10)"), the resulting conversational worlds clashed. For some students, for example, shoulder (10) referred to the 10t coin, whereas for adults, shoulder (10) referred to the K1 coin.

Composing Currency Token Values and Reading Prices

To explore the children's use of token names in arithmetical compositions of token values and in reading written representations of prices, additional tasks were used. The tasks were similar to several used with adults in Study 5-2. Again, the focus was on solution of these tasks both in the Oksapmin language and Tok Pisin. Of particular interest was the extent to which children, as they shifted from one language to another, would preserve the normative use of units in each language.

Sixty Toea. Figure 109 displays the distribution of children's solutions to the quantification of K.60 represented as six 10t coins (Figure 109a) as well

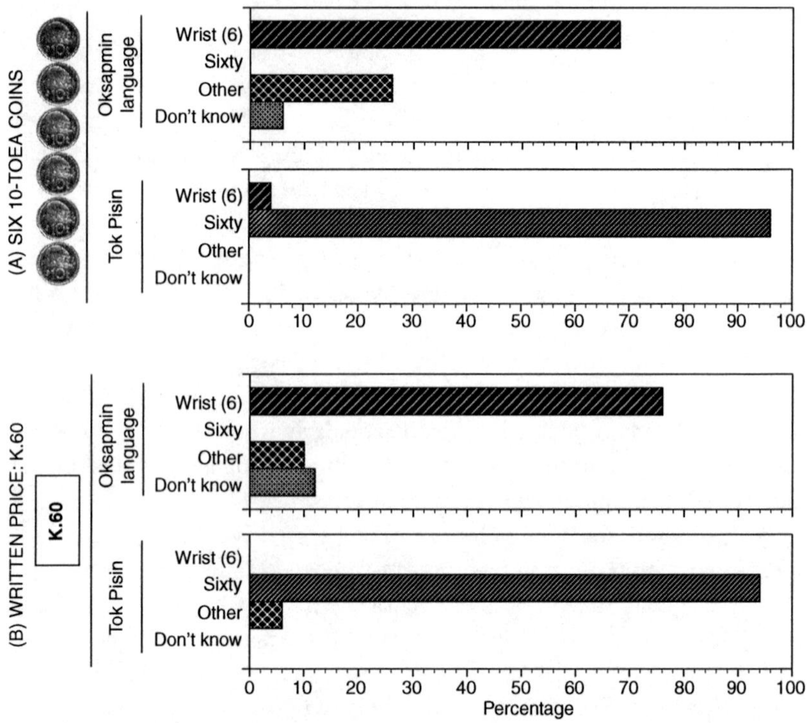

Figure 109. Variation in quantification of 60t as a function of form of presentation (coins or written price) and language of quantification (Oksapmin or Tok Pisin).

as their representation of the written price, K.60 (Figure 109b). Recall that for each presentation, the children were asked to tell the amount presented in the Oksapmin language and then in Tok Pisin. Children's solutions to the quantification and reading the written price are represented in the upper and lower distributions in each figure, respectively.

An analysis of children's quantifications of the six 10t coins in the Oksapmin language (Figure 109a, upper distribution) revealed that many children treated the value of 10 toea as equivalent to "one" with the body system, consistent with normative usage in the adult community; these students represented the sum of the values of coins as "wrist (6)" (about 70 percent of responses); most of the children who did not produce "wrist" (6) as an answer produced idiosyncratic representations that were taken to be errors (other). In contrast, students treated the same display in Tok Pisin as "sixty" (Figure 109a, lower distribution), using the toea as a unit (almost 100 percent of the responses).

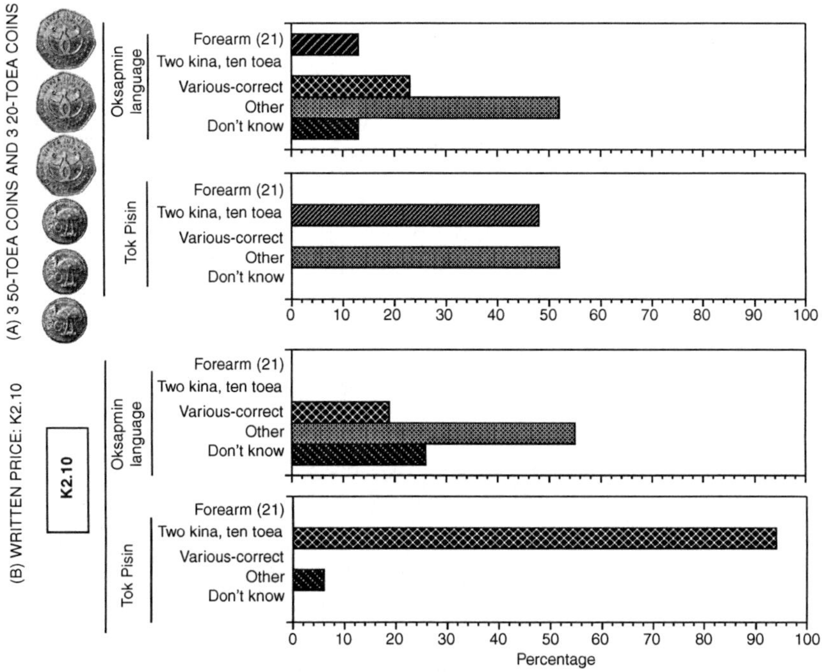

Figure 110. Variation in children's quantification of K2.10 as a function of form of presentation (coins or written price) and language of quantification (Oksapmin or Tok Pisin).

An analysis of the children's efforts to read the written representation of 60 toea in both languages revealed parallel findings. In the Oksapmin language, most children referred to the written price as a variant of "wrist" (6), using a shilling equivalent as the unit. In Tok Pisin, most children read the 60t price as "sixty" or "sixty toea," shifting to the toea as the unit.

Two Kina and Ten Toea. Figure 110 displays the findings for the children's efforts to quantify the value of K2.10. Once again, we presented this quantity both in the form of currency (Figure 110a) and as a written price (Figure 110b). As in the 60t case, we asked the children to quantify each presentation in both the Oksapmin language (upper distribution of each figure) and in Tok Pisin (lower distribution of each figure).

Children's quantifications varied by condition. Consider first the currency composition condition in the Oksapmin language in Figure 110a, upper distribution. A small minority of children (12 percent) used the body-part equivalent to quantify the currency, indicating that the amount was the forearm (21), the 21st position in the conventional sequence. Other children achieved an accurate representation in the Oksapmin language, though their expressions were idiosyncratic. For example, one child said the amount was "*temsi tan oha gangasi oha dop-hai*," which means "a K1 coin and a 50t coin and 60t more." Another child called the quantity "*tan-tipana*," meaning "thumb on the other side." This response is ambiguous, but interesting. It may reflect a familiarity with modified versions of the count system in which after the 20th body part, the count begins again at the thumb, indicating 21 (21-shilling equivalents or the equivalent of K2.10). Many children produced incorrect answers (other), and some simply indicated that they did not know.

An analysis of students' responses in Tok Pisin (Figure 110a, lower distribution) showed that almost half gave an appropriate representation of "two kina ten toea." The incorrect responses included "sixty toea" (where the child counted each coin as if it were 10t, not respecting the denominational value of the 50t coin), "one kina forty," and "three kina and ten toea," where the child's counting strategy was unclear. Unlike the case in the Oksapmin language, in Tok Pisin none of the children responded that they did not know.

For the written price problems in the Oksapmin language (Figure 110b, upper distribution), no student used the body-part convention of indicating the 21st body part, the forearm (21). Rather, many created idiosyncratic expressions such as "*yotana tita*" meaning "two flats and one," or "*tipnarip* (2) *tipana* (1)," meaning "index finger (2), thumb (1)." In these cases, the students appeared to create multiunit representations of one kina and ten toea. Nine of the children indicated that they did not know how to represent the value in Oksapmin, and many other children created inappropriate representations for this quantity. In contrast, when asked to read the inscription in Tok Pisin, almost all children indicated correctly that the value was "two kina ten toea," a more accurate performance than in any other condition.

Discussion

The two studies reported on in this chapter documented an interplay between (a) the changing collective practices of classroom life and (b) the reproduction and alteration of representational forms and functions for number in teachers' and children's activities.

Perhaps most remarkable about the findings is the way they contrast with those from classrooms at the same school in the postcolonial 1980 period (Chapter 9). In Ms. N's 2001 classroom (Study 10-1), teacher and students moved flexibly from Tok Pisin to English to Oksapmin representations, with the emphasis on English. There was very limited use of the body-counting system, perhaps due to its limited utility for classroom problems such as the multi-digit subtraction, time reckoning, or interpreting prices expressed in decimal notation like K139.97. The classroom strongly supported the use of Western computational strategies in ways that were made accessible to children, and children generally seemed to be more familiar with mathematical language and representations in English than in the Oksapmin language. Further, as a part of classroom activities, children were provided with resources for computation that supported their problem solving, such as the stones on the classroom floor.

In 1980, in contrast to 2001, children were being schooled in a language (English) that they were acquiring in the process of learning mathematics. Many children made use of the body system to represent and solve these problems, perhaps helping to make the foreign problems more sensible by using a known representational form. It was in this context that they developed body-part computational strategies to support keeping track of the addition and subtraction of terms in solving arithmetical problems and created hybrid body-part and Hindu-Arabic computational strategies (as documented in Chapter 9). In 2001, Indigo and I only rarely observed these body strategies in class. But there were other developments in students' thinking that marked the 2001 era of particular interest, like the use of stones in computations supported in Ms. N's classroom activities as well as a developing facility with Hindu-Arabic-based algorithms.

Study 10-2 provided insight into the students' uptake and understanding of representational forms used in class related both to number and currency. In the Tok Pisin conditions, all students displayed knowledge of Tok Pisin numeration when counting, identifying currency tokens, composing token values, and reading written prices. Further, most students performed competently, and generally their performance was more successful in Tok Pisin than in the Oksapmin language on tasks that involved counting, composing values, and reading written prices.

On the Oksapmin language versions of these tasks, students' performances were not as strong as they were in Tok Pisin, but the character of the solutions was fascinating. On some of the tasks using the Oksapmin language, children's performances reflected adults' normative practices as documented in Studies 5-1 and 5-2, with adult-like knowledge of the body-count sequence, units of quantification, and knowledge of language forms used to name tokens. But on some of the token naming and composition of token values tasks, their performance deviated markedly and clashed with normative practices. For example, sometimes students used the 10t coin as a unit, referring to the 10t coin as a shilling, or as thumb (1), or as *tit* (1). But other times they used the value of 1 toea as a unit, and referred to the 10t coin as shoulder (10), a naming approach that clashed with the extant practices in the community by a tenfold magnitude. The alteration in units was not restricted to the 10t coin but was evident with all of the tokens used in the interview protocol.[5]

Noteworthy is that children's performances revealed a clash between the history of economic exchange practices in Oksapmin and school-linked representations associated with the classroom life in 2001 that converged in Ms. N's classroom (and other classrooms in the 2001 era). On the one hand, there is the historical trajectory of economic exchange with normative uses of pound and shilling equivalent units. On the other hand, there is the trajectory of Western-style schooling in Oksapmin communities in which English, Tok Pisin, and Hindu-Arabic numerals have been privileged forms of representations that make use of decimal units. Many of Ms. N's students managed the clash in flexible ways; they switched from using decimal units when they used Tok Pisin, to using pound/shilling equivalent units expressed as body parts, or conversational numbers, or tropes when they used the Oksapmin language.

[5] Students used thumb (1), index finger (2), and pinky (5) to refer to the 1t, 2t, and 5t coins respectively. In normative practices of economic exchange, these body-part expressions are reserved for coins of 10t, 20t, and 50t.

11 Teachers and Students as (Unintentional) Agents of Change

My concerns with schooling in the prior chapters have principally focused on students and teachers in community schools. In this final chapter on schooling, I shift my concerns to recently established elementary schools, the schools that engage young children for their first three years. In this reform era (2001), teachers in elementary schools are expected to draw upon indigenous language and knowledge forms in instruction using these forms to serve as bridges to teaching Western knowledge, such as arithmetic using Hindu-Arabic numerals. I explore the idea that as teachers create links for children, teachers become levers of both the reproduction and alteration in the body-part and other pre-contact representational forms and functions.

Over the course of the chapter, I draw upon the distinction between the two complementary models of propagation in sociogenetic processes introduced previously, *local emergence* and *broadcast*. Though in any classroom activity there are elements of both, the broadcast model, which has been pronounced in teachers' reproduction and alteration of representational forms and functions for number, is the central concern here.

The Elementary School as a Research Site

By 2001, the 1998 school reforms were well under way in the Oksapmin region. The sites where these reforms are most visible are the new elementary schools that all children were to attend for their first three years. The grade levels included Preparatory (Prep), Elementary-1 (E1), and Elementary-2 (E2). The fourth year of schooling, Grade 3, begins at the community school (Ms. N's classroom, described in Chapter 10, was Grade 3). As noted, elementary schools were promoted to engage students with the language and practices of their own communities and their community histories, creating links between in and out of school knowledge. Consistent with this "bridging" function, teachers provided explicit instruction in the body system during all three years, framing Western problems in terms of the system.

I had an early clue that the elementary school might well be an important site of interest for a treatment of culture-cognition relations. It came with a chance 2001 encounter on an Oksapmin trail with the district coordinator for the eight Oksapmin elementary schools. My conversation with the coordinator covered a lot of ground. But one topic stood out. In the conversation, the coordinator mentioned that in the elementary school, children learn to use the indigenous body system initially to express small values and then move on to express very large values. For large values, he explained, Oksapmin children learn how to count by tens, showing with rapidity, "*tipan kat-hai* [thumb (1) shoulder (10)], *tipnarip kat-hai* [index finger (2) shoulder (10)], *bumrip kat-hai* [middle finger (3) shoulder (10)]..." while pointing to the thumb (1), index finger (2), middle finger (3), and so on, noting that this was a part of traditional practice. In other words, the coordinator explained that traditionally the number 10 is represented as thumb (1) shoulder (10), 20 is represented as index finger (2) shoulder (10), 30 is represented as middle finger (3) shoulder (10), and so forth.

The coordinator's remarks about traditional body counting were at odds with findings from my early work in 1978 and 1980 and interviews with elders in 2001, and this was striking to me. Recall that in Study 4-2, Indigo and I had asked participants, including elders, how they would represent 50 and 60. The questions were eventually dropped because the problem could only be adequately formulated in Tok Pisin or English, not the Oksapmin language; for elders who did not speak Tok Pisin, the questions were not easily communicated. But a number of elders and adults who understood the question in Tok Pisin said that in the Oksapmin language there was no way to count that high. Others said that the traditional way of counting could be used – once reaching the 27th body part (the pinky (27)) one would continue to the wrist (28) and back across the body to the other pinky (equivalent of the 49th position). If it were necessary to count higher, one would simply continue again to the wrist on the first side of the body (equivalent of the 50th position), et cetera. A small minority of schooled adults provided some corroboration of the coordinator's statement – to group numbers into tens and count how many groups of ten, and then how many units were left over. But we eventually dropped this question from our interviews and from our analysis due to difficulties both in expressing the question and in obtaining interpretable responses.

How might the coordinator have come to hold the view that he did about Oksapmin counting and traditional practices? He seemed so sincere in his convictions, and what he told me was so at odds with what I had learned over the course of my stays in Oksapmin. The interaction also gave me pause when I considered instruction in elementary schools. To what extent did the coordinator's portrayal reflect elementary instruction? If it did, this could introduce a new direction to sociogenetic processes in the reproduction and alteration of the Oksapmin body system.

I report on two studies in this chapter, each conducted in 2001 that were seeded by the interaction with the coordinator. In the first, Study 11-1, I observed elementary school mathematics lessons over the course of several weeks at a local school with a principal focus on one teacher's classroom. In the second, Study 11-2, I conducted interviews with two elementary teachers about their lessons, probing lesson structures, representational forms used, and mathematical content, while sampling more fully the year's curriculum. Across studies, my guiding concern was with the way teachers, in the course of instruction, reproduced and altered representational forms such as the body-count system and names for currency tokens to serve communicative and problem-solving functions.

Study 11-1. Observations of Classroom Instruction in 2001

Method

Participants

The two elementary teachers who were assigned to the target school were each raised in Oksapmin, and both were male. They both were engaged in implementing the new educational reforms, and both used the Oksapmin language and out-of-school practices in organizing and implementing classroom activities. Aligned with the reforms, their instructional concerns were to build on students' out-of-school knowledge as they taught mathematics in the Oksapmin language. Among the representations that teachers privileged was the body system. In addition to giving students explicit instruction in the system, they made connections between the use of the body system and Western mathematics (as was their charge). All students in their first year of elementary school were monolingual Oksapmin speakers.

The Setting

Made with indigenous materials (but not traditional architecture), the two-room schoolhouse (shown in Figure 111) is located off of a main trail

Figure 111. Two-room elementary school with dirt floor in 2001.

that runs through an Oksapmin valley. Inside the rooms, light was provided by the open door and small openings under the roof that served as windows. The rooms had shelving (also made from indigenous materials) and a teacher's desk at the front. There was no other furniture, and students sat on the dirt floor. The rooms each had a chalkboard at the front of the room and many posters covering many topics, such as mathematics and hygiene. Because there were three classes of students, Preparatory (Prep), Elementary–1 (E1), and Elementary–2 (E2), and only two rooms, the teachers combined all of the Prep and some of the E1 students into one class and E2 students and some of the E1 students into another class.

At the time the observations were conducted, teachers were using their own curricular materials, lessons that they had developed themselves.[1] Figure 112 shows one of the charts posted in each of the two classrooms in the schoolhouse. The chart is particularly noteworthy in that it makes visible the correspondence between three semiotic systems for number. The

[1] Just prior to our observations at the site, the teachers received a shipment of materials containing textbooks that students could use in their classes. These materials had been developed at the national level in such a way as to accommodate the myriad of local languages in Papua New Guinea: The activities in the texts involved no writing, and teachers were to adapt the materials using their local languages and traditional practices.

Figure 112. Chart posted on wall in each elementary classroom with corresponding representations of numbers to ten: Hindu-Arabic numerals (left column), body-part names in the Oksapmin language (middle column), and English number words (right column).

first column on the poster lists Hindu-Arabic numerals for numbers from 1 to 10. The second column lists the body-part names of each number in a written form of the Oksapmin language – for example, *tipana* means thumb (1), *tipnarip* means index finger (2), ... and *kata* means shoulder (10). The third column lists the English names for each number. A second chart in the upper-level classroom consisted of a multiplication table giving only Hindu-Arabic numerals.

Procedures

To document instruction in the elementary school classroom, I observed six lessons in the elementary grades. On a number of occasions, not having sufficient fluency in the Oksapmin language, I was accompanied by a translator and had postlesson discussions in which I asked for elaboration of lesson features that I had difficulty understanding.

Lessons: Their Organization and Content

Most lessons were organized around themes linked to out-of-school activities such as building a store or taking apart a pandanus nut, a broad flower-like nut whose large seeds are harvested and smoked. Other lessons had no thematic structure and were simply focused on operations. Regardless of the lesson structure, students often wrote lesson-related problems in notebooks and on pieces of lined paper.

In general, lessons were organized into prototypical phases: Lessons would begin with an opening whole-class activity that introduced a lesson's content. Next, students worked on problems linked to the content (during this phase, they were free to move around the room and to talk to one another as they worked), and during this work time, the teacher circulated around the room, noting and marking students' errors and then returning their papers so that students could correct them. For the last phase of the lesson, the teacher orchestrated a further whole-class discussion about how to solve the assigned problems.

Lesson structures were similar for students at all three grades, but problem difficulty and privileged representations varied by grade level. For the Prep-E1 students, most talk was in the Oksapmin language. For the E1-E2 class, mathematics problems were introduced and solved in both languages.

In the following sections, I describe classroom activities with concern for the forms of numerical representation used and the numerical functions that students and teachers used them to serve. I sketch two lessons, each well reflecting the lessons observed with regard to lesson structure and topics covered. Further, I selected these lessons also because they engaged recurring focuses of the volume – one on counting/arithmetic and the other on currency.

Lesson on Counting, Addition, and Subtraction

This lesson on counting, addition, and subtraction consisted of three sequenced activities. The lesson starts with a choral recitation of the body-count form to 27 (Activity 1), and then shifts to addition (Activity 2) and then subtraction (Activity 3). As depicted in Figure 113, the body-count system was a strand in each of the activities.

Activity 1: Practicing Body Counting. Children are sitting on the dirt floor inside one of the two rooms of the elementary school. Lighting is limited, due to the absence of windows. The teacher begins the math lesson by engaging students in the practice of recitation of the body-count system beginning with thumb (1); the teacher models and students count in chorus. As students count around their bodies, some show difficulty after

Problems and representational forms	Lesson Phase		
	Activity 1: Counting	Activity 2: Addition	Activity 3: Subtraction
Problems	Recitation of count to 27	5 + 3 = ? 6 + 4 = ?	5 – 3 = ? five similar problems
Privileged representational forms	Body counting to 27; variant	Body parts Graphical marks Written numerals English	Body parts Graphical marks Written numerals English

Figure 113. Coverage of the lesson on counting, addition, and subtraction that features various representational forms, featuring body counting.

reaching shoulder (10). What was particularly curious to me was that some students iterated the last five body parts of the sequence (23rd–27th) from little finger (23) to thumb (27), not from thumb (23) to little finger (27), the procedure that I had learned back in 1978. Figure 114a/b illustrates these two variants of the 23–27 spatial trajectory. My discussion with the teacher after the lesson revealed that, to the teacher, the proper way to count was the symmetrical ending.

Activity 2: Addition. The teacher had drawn on the blackboard the display depicted in Figure 115, which contained the English word "Maths" and the Oksapmin word, written in script, "*Tapdipti*."[2] To begin the activity, the teacher asked the children to count the circles in the addition problem in the figure using their bodies. They responded by counting in chorus: *tipana* [thumb (1)], *tipnarip* [index finger (2)], *bumrip* [middle finger (3)], *hadrip* [ring finger (4)], *hatatah* [pinky (5)]. The teacher repeated the request, this time directing them to count the lines. They again responded in chorus, counting the three lines using body-part words. Finally, the teacher instructed the children to count the circles and lines together, and they responded chorally by counting, *tipana* [thumb (1)] through *kir* [inner elbow (8)]. The teacher then emphasized that, altogether, there are *kir-hai* [elbow (8)], emphasizing the suffix that marks the position of elbow (8) as a cardinal number. Noteworthy in this exercise is that when children

[2] In the 1993 edition of Lawrence's Oksapmin dictionary (Lawrence & Snyder, 1993), he lists no entries for "math" or "add" (or "addition"). But there is an entry for a near match to the written expression "*Tapdipti*." Lawrence lists "*tap hapti*," with the definition "to come together."

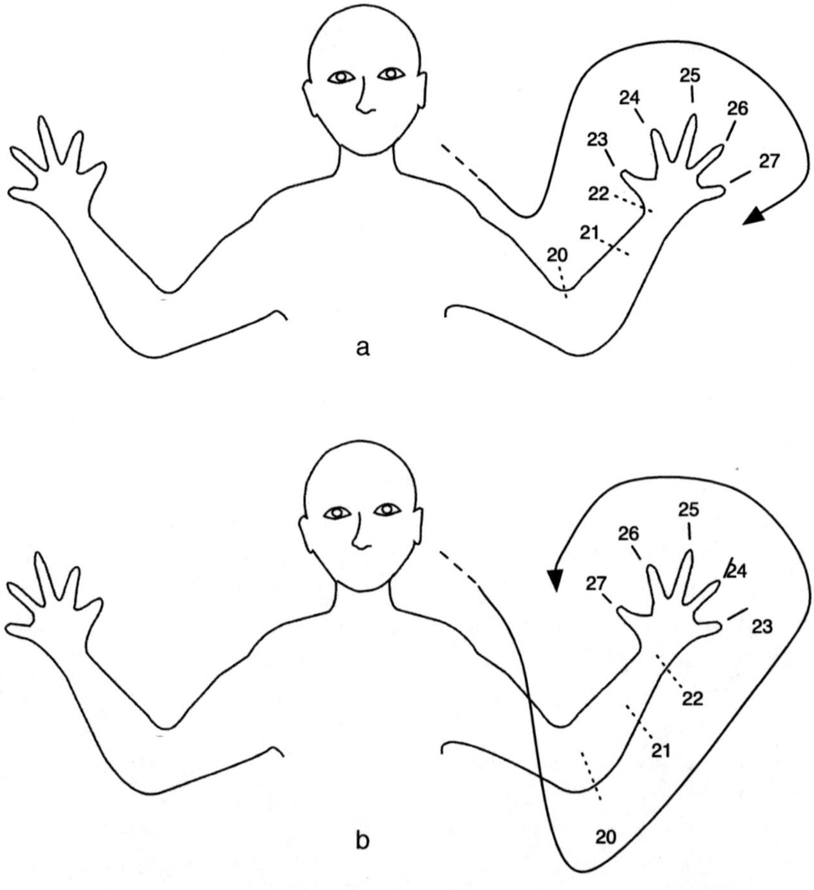

Figure 114. Two variants for enumerating 23rd to 27th positions on the body.

used body-part words in counting, only a few pointed to their bodies; most simply said body-part words.

Having completed the activity using body counting of the objects, the teacher then moved to the second line on the board – a Hindu-Arabic representation of the problem that he and the children just solved. The focus is on decoding the Hindu-Arabic representations in English, "five plus three equals eight." The next problem on the board was then taken up, and the teacher engaged the class in similar instructional moves, first working through the problem by counting with the body system and then decoding the Hindu-Arabic notation and reciting the sum in English.

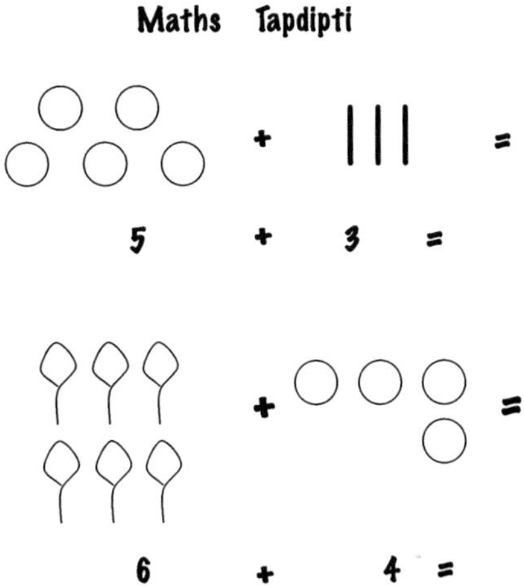

Figure 115. Teacher's blackboard representation for teaching addition.

Activity 3: Subtraction. In introducing the subtraction activity, the teacher drew on the blackboard the configuration of circles depicted in the subtraction problem in Figure 116. Guiding students through the blackboard representation, the teacher said (in the Oksapmin language), "There are five stones. Three are drawn to kill birds with slingshots. How many are left?" In chorus guided by the teacher, children counted all terms of the problem using the body system. As the teacher moved through the problem, he crossed out the last three of the five circles, along with the corresponding circles in the problem solution (see Figure 116b). Once the problem with circles was accomplished, the teacher then translated the problem into Hindu-Arabic numerals and solved this with the class (5 – 3 = 2). Finally, the teacher presented the second subtraction problem in the same format in Oksapmin: "There are three tomatoes and one is fed to a cassowary. How many are left?"

After completion of the introductory problems, the teacher moved to five previously prepared subtraction problems written on the blackboard (Figure 117). The teacher worked on the first problem with the whole class, asking the class to solve the problem by counting marks on the board using the body system. The teacher then asked the students to solve the remaining problems alone. Once the students had finished, the teacher solved the problems one by one, first in Oksapmin using the body system and then using Hindu-Arabic representation.

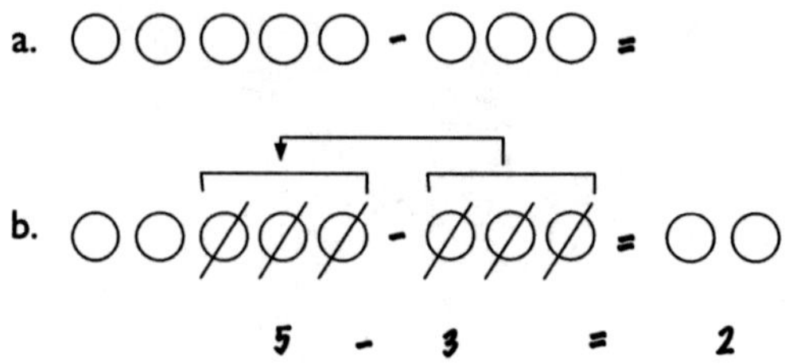

Figure 116. Subtraction problem: 5 – 3 = ?

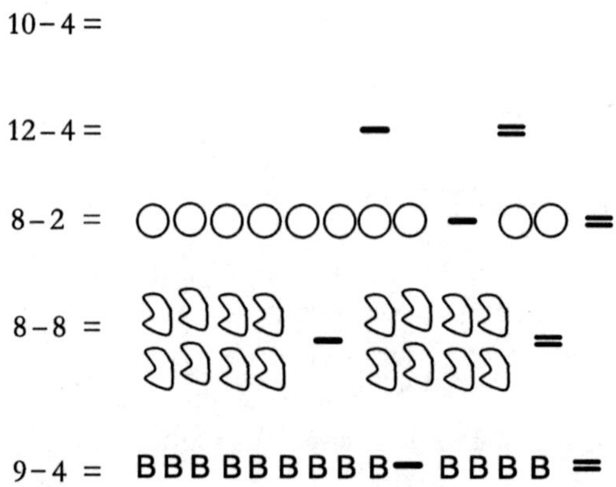

Figure 117. Teacher's blackboard representation for teaching subtraction.

Summary Remarks. The choral body-counting, addition, and subtraction activities of this lesson are particularly interesting because of the way the teacher guides students through the use of three semiotic systems. These include the body-counting system, English number words, and Hindu-Arabic inscriptions for number. Reciting the body system occurred at the lesson's opening, and it was used in the conduct of counting and representing geometric objects drawn on the blackboard. Hindu-Arabic numerals were used to create representations of the same objects, supporting a mapping between numerals and objects. English number words were used to represent the same objects in statements of operations and equations. In addition to the mapping across different representational forms for number in the context of arithmetic, the teacher's addition and subtraction activities were also marked by ways in which the teacher's talk reflected the reproduction and alteration of Oksapmin semiotic forms other than the body system. For instance, "*tapdipti*" and "*tabdis*" were used to signify the plus sign used to write addition sentences (trans. everyday use: come together), and "*madapti*" or "*madas*" were used to signify the written signs for subtraction.

Lesson on Currency and Numerical Naming Practices for Tokens

This lesson introduced currency token names in the Oksapmin language. It presents a second case in which the body system and other Oksapmin word forms appear to be inadvertently reworked and "standardized" in the process of being broadcast to children. As in the previous lesson, the teacher's concern here was to draw upon community knowledge in bridging to Western school subject matter. Figure 118 contains an overview of the three-phase lesson with reference to both the problem contexts used and the privileged representational forms.

Activity 1: Oksapmin Names for Tokens. The focus of the instructional activity is on names for currency tokens in the Oksapmin language. In the lesson, the teacher began with the smallest unit of currency – the 1t coin – writing its Oksapmin name (*suptan*) and proceeding up to the K2 note (*wanpon*).[3] The teacher discussed particular characteristics of some of the coins after displaying the written names as shown in the first column of Figure 119. For instance, he discussed the *gagasi* (meaning "points" in the Oksapmin language), which has points or corners, and the *temsi tan* (meaning "hole" and "flat," respectively). In this process, Tok Pisin borrowings and Oksapmin tropes become seamless parts of the same nomenclature; notably absent are body-part or conversational number words for tokens, which were common in local naming practices, as revealed in Study 5-1.

Following the teacher's review of names for tokens in the Oksapmin language, the teacher wrote adjacent to each expression (see Figure 119) the equivalent denominational value in Hindu-Arabic numerals, as can be seen in the third column of the table. During the presentation, the teacher engaged the class in some arithmetic – currency tokens were drawn with countable shapes on the blackboard, and emphasis was given to ratios between denominational units – how many tokens of one denomination were equivalent to tokens of another (e.g., two 50t coins being equivalent to one K1 coin)?

[3] In 2007, six years after the observations were made, the 1t and 2t coins ceased being legal tender.

	Lesson Phase		
Problems and representational forms	Activity 1: Oksapmin names for tokens	Activity 2: Arithmetical problems	Activity 3: Guided instruction on problem solutions
Problems	Presentation of token names in Oksapmin and Tok Pisin	Examples: 1 toea + 2 toea = 3 toea 20t + 50t =	Examples: 20t + 50t = 70t *tusring + gagasi = bes-hai*
Privileged representational forms	Oksapmin token names (tropes) Written numerals	Written numerals Operation signs Token names in: Oksapmin & Tok Pisin	Token names in: Tok Pisin, Oksapmin trops, body parts Written numerals

Figure 118. Coverage of lesson on token names and arithmetic with token names.

Activity 2: Arithmetical Problems. After constructing the chart and discussing correspondences between currency tokens, the teacher worked through a series of problems with the class. In each case, the teacher and class represented the problem first orally using the body-system and the Oksapmin names for the currency tokens. They then wrote Hindu-Arabic expressions, reading them in English with an occasional reference to body-part expressions: 1t + 2t = 3t, 5t + 4t = 9t, 6t + 4t = 10t, 7t + 5t = 12t, K2.00 + K1.00 = K3.00.

Once the whole-class problem-solving exchange was completed, the teacher directed students to practice what they had learned through individual work. In this activity, the teacher asked the students to copy the problems shown in Figure 120 into their notebooks and then to solve them. As students worked, the teacher circulated, checking students' answers and asking them to redo problems that were not correct.

Activity 3: Guided Instruction on Problem Solutions. Problem checking completed, the teacher turned to a whole-class presentation of solutions to addition problems. In his presentation, the teacher solved the problems using the naming system on the board shown in Figure 119. The problems and the teacher's solutions are shown in Figure 121. The note below the figure contains a translation of the terms written on the board. As in Figure 119, here the terms written were a mixture of Oksapmin and Tok Pisin borrowings. For example, problem 1 translates as "pointed plus pointed equals flat with hole," and problem 2 translates as "two shillings plus pointed equals forearm."

Summary Notes and Discussion. The present lesson provides a counterpoint to earlier chapters on processes of local emergence and the many variant

Oksapmin expression (written on blackboard)	English translation (not written on blackboard)	Denomination (written on blackboard after discussion)
Suptan	Extreme flat	1 toea coin
Gamintan suptan	Married extreme flat	2 toea coin
Gatep	Turtle	5 toea coin
Wan sring	One shilling	10 toea coin
Tusring	Two shillings	20 toea coin
Wan sring gamin tan	One shilling married flat	20 toea coin
Gagasi	Pointed	50 toea coin
Temsi tan	Hole flat	1 kina coin
Wanpon	One pound	2 kina note

Figure 119. A list of Oksapmin names for units of currency presented on the blackboard.

names for tokens. Chapter 5, for example, documented the many different Oksapmin names for currency tokens that were used by adults in 2001. I argued that the multiplicity of names could be well understood from the perspective of the local-emergence model. Interlocutors had created variant solutions to problems that arose in interaction at local sites, variant solutions that constituted the reproduction and alteration of Oksapmin word forms, with a gradual waning of many variants and waxing of others through time. The present lessons provide an interesting counterpoint to the processes of local emergence. In elementary classrooms, the teacher broadcasted a list in which each token type had one, and only one, name. Further, the broadcast was to a sizable population of children, supporting the standardization of names. The list had some interesting characteristics that I now discuss.

1. *Hybridity*. The names themselves are hybrids, drawing upon Tok Pisin forms, like *tusring*, and also Oksapmin forms, like *temsi tana*. The hybrid character draws on both prior names used in the Oksapmin community but is organized to better support a common ground of communication among teachers and students in instructional activities. Further, the lesson gives an orthographic form to the Oksapmin names, hybrid representations that afford greater durability than the spoken word.

Dahameng

1. 50t + 50t =

2. 20t + 50t =

3. 10t + 1t =

4. K1.00 + K1.00 =

Figure 120. Problems to solve (*dahameng*) listed on blackboard for independent work.

Dahameng

1. 50t + 50t =
gagasi + gagasi = temsi tana

2. 20t + 50t =
tusring + gagasi = bes-hai

3. 10t + 1t =
wansring + suptan

4. K1.00 + K1.00 =
temsi tan + temsi tan = wonpon

Figure 121. Problems written on the blackboard with inscribed solutions in a mixture of Oksapmin and Tok Pisin.[4]

2. *Exhaustive list of tokens up to K2.* The list of names is more exhaustive than everyday names. It begins with 1t and lists in sequence the tokens from 1t through K2. As noted in Chapter 5, 1t, 2t, and

[4] The translations for the Oksapmin and Tok Pisin words in the figure: *gagasi* (Oksapmin) means "pointed"; *temsi tana* (Oksapmin) means "hole flat"; *tusring* (borrowed from Tok Pisin) means "two shillings"; *bes-hai* (Oksapmin) means "forearm (7)"; *wansring* (borrowed from Tok Pisin) means "one shilling"; *suptan* (Oksapmin) means "extreme flat"; *wonpon* (borrowed from Tok Pisin) means "one pound."

5t coins are rarely seen in Oksapmin, and many adults do not know Oksapmin names for them.

3. *Clashes avoided.* Recall that Study 10-2 reported that many Grade 3 children called the 1t, 2t, and 5t coins by the body parts thumb (1), index finger (2), or pinky (5), or their conversational number word equivalents. These names made numerical sense in the translation of token values into Oksapmin representations. But the names clashed with normative usage in the world of Oksapmin economic exchange. In the elementary school classroom, the teacher appeared to take on this potential for ambiguity in his systemization. He avoided the use of body-part names for the 1t, 2t, and 5t coins, instead drawing on the tropes used in everyday talk to refer to tokens. Thus, he referred to 1t as a unit – *suptan*. Using *suptan* as a root, 2t became *gamintan suptan*, or a married *sup* (again drawing upon the use of the marriage trope in the 20t coin, a "marriage" of two shillings, or 10t coins); the 5t coin was referred to by the emblem on the coin. The naming was motivated by prior naming conventions, but regularized in new ways. Other currency names listed above typically reference either a feature of the coin (e.g., "*temsi*," meaning "hole," referring to the hole in the center of the K1 coin). "*Gamintan*" was used twice – for the 2t coin, as in the phrase "*gamintan suptan*" (meaning roughly "married worthless coin"), and also for the 20t coin as in the phrase "*wansring gamintan*" (meaning a marriage of two 1-shilling coins). The currency names appeared to have been selected in ways that made clear the relationships between the different denominations.[5]
4. *Units.* The list of tokens presents currency units that privilege the unit structure of 100 toea = 1 kina, rather than 20 shillings = 1 pound.

This study on elementary school lessons shed light on the interplay between ongoing shifts in collective practices of schooling at the elementary level and valued forms and functions related to number representation. But due to the limitations in my own proficiency in the Oksapmin language as well as the restricted set of lessons that I observed, I sought additional data on elementary instructional practices to corroborate the observations I made in this study.

[5] An exception to the teacher's practice can be found in Figure 121, where "*bes-hai* (7)" (forearm (7)) is used to refer to 70t in the context of an addition problem.

Study 11-2. Elementary Lesson Plans

The purpose of the study was to corroborate and extend what was documented in the classroom observations of Study 11-1, further exploring the structure of lessons and the representational forms used. The interview technique supplemented my direct observations of the prior study in two ways. The interviews circumvented the need for a translator, since the teachers were fluent in English, and they also supported a greater understanding of the scope of the mathematics lessons across grades, allowing a review of lessons over the course of a broader span of time.

Most of the lessons that were the teachers' focus in the interviews had a structure similar to observed lessons. In the interviews, it became clear that the teachers' general practice was to organize lessons by narrative themes, engaging children in a series of activities linked to the theme.

Methods

Participants

The same two male Oksapmin teachers who were observed in Study 11-1 participated in the present study.

Procedures

Teachers completed a written protocol that prompted them to write lesson plans that they had previously taught. To support their efforts, I created a model protocol based upon an observed lesson, "Building a Trade Store." As noted earlier, the thematic organization of lessons was a common practice for the teachers. I used teachers' completed protocols to organize the interviews, probing teachers' written lesson descriptions. Each teacher completed four lesson plans.

Results

The teachers' written lesson plans were analyzed to determine whether and how Oksapmin body counting and Oksapmin words were adapted for use in the mathematics classroom. I selected lessons that cut across the three years of elementary instruction. Of the lessons teachers described, I selected four that cut across years. They included lessons with the following themes: a medicinal herb (Prep), a bird (E1), planting a garden (E2), and rubbish (E2).

Overview of Lesson Themes

Brief overviews of the lesson themes are presented below. The overview section is followed by an analysis of representational forms used in the lesson plans and the functions that the forms were intended to serve.

Lesson 1 (Prep): Planting Medicinal Herbs (Wan/Sarat). The teacher partitioned the lesson into three phases. In the first phase, the teacher accompanied the students to a nearby garden and they considered the garden area. The teacher asked students to count medicinal herbs; after they produced counts, the teacher asked each of them to collect ten pieces of *pit pit* (a type of thin bamboo); upon return to the classroom, the pieces of *pit pit* would serve in class as tokens for herbs. In the second phase, the teacher moved back to the classroom and engaged in a counting exercise. The bamboo pieces were put into two groups, "good herbs" and "bad herbs," and students counted the number in each group and then summed up the number across groups (the activity was repeated again and again, with the teacher dividing up the groups in different ways). In the third phase, the children engaged in a similar activity but without direct guidance.

Lesson 2 (E1, Used Later in Year): Bird. Papua New Guinea is well known for its myriad species of indigenous birds, among them the bird of paradise and the cassowary. This lesson had a bird theme, and the teacher described a lesson consisting of several phases. In the first, the teacher showed the class a picture of a bird, noting and counting its principal parts: legs, wings, beak, head. Each student also drew a whole bird and counted its parts. In the second phase, the children assembled in groups of four with their pictures, and these groupings became the context for doing problems of addition and multiplication. How many parts does each bird have? How many parts does the group of birds have altogether? In the third phase, the lesson shifted to sharing or division: With four children sharing one bird, how many parts would each person receive? In the fourth and final phase, the teacher assigned to the class exercises organized around the bird theme involving addition and multiplication.

Lesson 3 (E2, Used Early in Year): Garden. Taro and sweet potato are the traditional staples in Oksapmin gardens, but after the coming of government patrol officers and missionaries, and with an expanding export economy, today people plant a range of vegetables. The theme of the lesson was making a garden, and the focus was on the materials required

Figure 122. A piece of rubbish on a muddy Oksapmin trail near a trade store.

for making a big garden. As an ingredient of the lesson, the children were provided with materials – stones, sticks, and seeds – that represented vegetables such as sweet potatoes. The lesson consisted of four sequenced activities. In the first, students counted sweet potatoes (bamboo sticks) to 20, guided by the teacher. In the second and third activities, the teacher guided students through subtraction problems, first with sticks (Phase 2) and then with stones (Phase 3). In the fourth phase, students solved addition and subtraction problems written on the board, using sticks (representing sweet potatoes) as an aid.

Lesson 4 (E2, Used Later in Year): Rubbish. It was common to find on trails an occasional discarded snack bag bought at a trade store, perhaps dropped by children (see Figure 122). In this lesson, consisting of four activities, the teacher engaged students with arithmetic using the theme of picking up rubbish, which was represented by pictures of sticks, stones, and seeds drawn on the blackboard. In the first, students used bamboo sticks as rubbish, and were asked to use the sticks to show their solutions to addition and subtraction problems that the teacher presented on the board using Hindu-Arabic representations. In the second, the teacher asked each student to retrieve 50 pieces of rubbish in groups of 10 sticks of bamboo. In the third, students solved subtraction problems by holding up pieces of rubbish, keeping a specified number in their hand, and dropping the rest. They then had to determine the number on the ground, summing those in their hand and those on the ground. In the fourth, they solved addition and subtraction problems involving "rubbish" independently, continuing to use the manipulatives.

Oksapmin Body System and Word Forms Extended to Western Mathematics

To understand the bridges teachers were building between representational forms of the community and Western arithmetic, I coded the teachers' descriptions of their lessons. I targeted for coding the ways in which the teacher (a) employed the body-counting system in the lesson and (b) used expressions in the Oksapmin language to refer to Western arithmetical representations and ideas. In coding teachers' lesson plans, I focused on activity phases within lessons as units, and as appropriate I applied multiple codes for each phase. (Examples of lesson phases are contained in Figure 113 and Figure 118.) For any given phase, I did not apply a single code more than once; the reason for this was that it was not clear what the boundaries were between the first occurrence of a particular code and when a new instance of the same event recurred within the same phase.

Codes for Body System and Adaptation of Language Forms. Uses of the body-counting system included the oral recitation of the body parts as well as applications of the system to classroom problems. The applications included: counting physical objects; marking a body part with a suffix that turned it into a cardinal number; and applying body parts to arithmetical problems.

Codes for Adaptations of Oksapmin Words to School Mathematics. The uses of Oksapmin expressions included the extension of Oksapmin words to refer to representations and ideas used in Western arithmetic, such as operation signs for addition or multiplication. Figure 123 shows the frequency counts of teachers' extensions of the body-counting system or the use of indigenous language expressions by each of the four targeted lesson plans. An inspection of the figure reveals that the only code with no instances was that for oral recitation of the body system. In retrospect, this is surprising, since oral recitation of the body-counting system was common in the

Context of use	Lesson				Total
	Medicinal herb	Bird	Garden	Rubbish	
BODY COUNTING					
A. Applied to objects	1	2	1	...	4
B. Cardinality	1	...	...	...	1
C. Applied to addition	2	1	3	2	
D. Applied to subtraction	2	...	2	2	6
E. Applied to multiplication (repeated addition)	...	2	1	...	3
F. Applied to division (fair share)	...	1	...	...	1
G. Base-10 or 20 interpretations	1	...	...	1	2
H. Body sequence variants	...	1	...	...	1
I. Recitation	...	...	...	...	
SEMIOTIC ADAPTATIONS					
J. Oksapmin names for Western expressions	...	1	2	2	5

Figure 123. Oksapmin body and language extended to classroom mathematics.

lessons that I observed (two of which were described in Study 11-1). I expect that oral recitation was a common warm-up activity, and teachers therefore did not note its use in their descriptions. All other uses of the body-system categories noted in the left column of the figure did occur, though with varying degrees of frequency across lessons.

As indicated in Figure 123, in the lessons sampled, the most frequent instructional application of the body system was in addition (eight occurrences) and subtraction (six occurrences). Indeed, addition with body parts occurred in every lesson, and in most lessons it occurred multiple times across lesson phases. Subtraction was similarly frequent, though it was absent in the bird-themed lesson, which focused on multiplication and division. Other common foci for the use of body parts across lessons included counting of objects (four occurrences) and multiplication (three occurrences). Of interest was the treatment of the body system as base- or cycle structure organized by cycles of 10 (two occurrences). Also of note in instruction was the support that teachers gave for a variant of the use of all 27 body parts, one that ended symmetrically with the thumb as the 27th position (as opposed to the pinky as the 27th position).

The adaptation of Oksapmin words to refer to Western mathematical representations and operations was also noteworthy in the lessons and lesson phases. Oksapmin word use reveals a shift in function of these word forms as they were recruited to serve instructional purposes. These expressions included adapting Oksapmin word forms to refer to arithmetical signs, like the signs for addition and subtraction, as well as adapting word forms to refer to arithmetic operations, like multiplication.

In my descriptions below, I illustrate each of the codes used in Figure 123. The illustrations were extracted from the teachers' reports on the four lesson plans, along with the interviews with teachers about those reports.

A, B. Body Counting: Applications to Objects and Cardinality. Planting Medicinal Herbs, Activity 1. In Phase 1 of the lesson on planting medicinal herbs, the teacher emphasized the cardinal representation of the count as it was applied to pieces of bamboo. The activity received two codes, the body counting applied to objects code and the cardinality suffix code.

> I begin the lesson by taking students outside to a village and tell them to count the plants with their eyes. After counting [using the body], I ask them to collect pit-pit [thin bamboo] to count how many plants are there. The number of pit-pit students were ask[ed] to get was *kat-hai* (shoulder (10)).

Planting Medicinal Herbs, Activity 2. Back in the classroom after the visit to the garden, the teacher asked students the number of the sticks they collected. Like the prior example, this activity involved body counting.

> In the classroom, I ask how many plants do you have, expecting them to count [using body]. They look at their sticks, counting them, beginning with *tipana* (thumb (1)).

In the lesson plan, the teacher drew the sticks on the board and modeled the solution to the problem using body counting.

C, D. Body Counting: Applications to Addition and Subtraction. Building a Garden, Activity 4. During the last phase of the garden lesson, the teacher

	Madas	**Tabdis**
(a)	○○○ - ○○	○○○ + ○○
(b)	○○○○ - ○○	○○○○ + ○○
(c)	○○○○○ - ○○	○○○○○ + ○○
(d)	○○○○○○ - ○○	○○○○○○ + ○○

Figure 124. Addition and subtraction problem on board as represented in the teacher's written lesson description.

engaged his students in small groups with addition and subtraction problems that were written on the board, with circles representing sweet potatoes. Students in turn were to use short pieces of bamboo to support their solutions – they were to count the bamboo using the body system to solve the problems. The teacher reported that the subtraction problems were completed with the sticks, and then students worked on the addition problems just using the figures on the board.

> Towards the end of the lesson, I draw four other subtractions and four other additions on the board [he illustrates this as shown in Figure 124]. Students in each group work out or solve the problems with their sticks [that stand for *kau kau* (sweet potato)]. They do subtraction first and then do addition from the board. Groups do the exercise with counting in Tok Ples (the Oksapmin language, body counting). The groups that finish raise their hands and I check if their answer is incorrect. If it is incorrect, I ask them to do it again until they finish.

Garden, Activity 3. In the addition problems that follow, the teacher framed them in a way that resembled the arithmetical problem solving practiced by the 1980 trade store owners (described in Study 4-1). The teacher asked the students to represent one term of the problem on one side of the body and the second term on the other side of the body. But the teacher did not elaborate the solution processes supported. Nonetheless, the arithmetical problem was framed on the body, and children were instructed to solve the problem with the body.

> The student group leader picks up seven sticks in one hand and three in other hand [the sticks are represented as circles in the teacher's sketch in

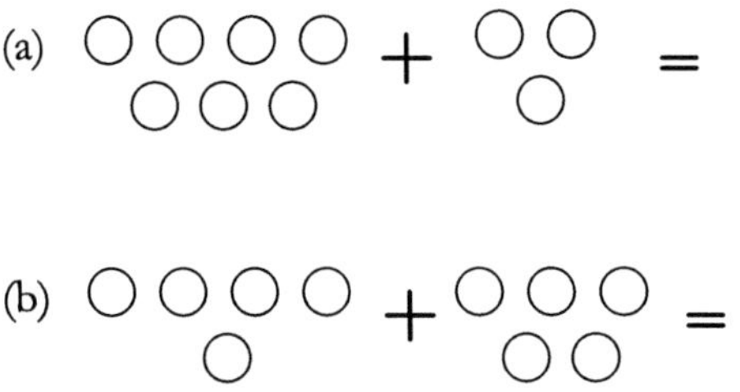

Figure 125. Addition of 7 + 3 = ? and 5 + 5 = ? represented in the teacher's written lesson description for a sticks problem.

> Figure 125a]. Leader asks how many? Students count in using their bodies and say the answer in the Oksapmin language. Same for no. 2 [the teacher's sketch is in Figure 125b].

E. Body-Part Counting: Applications to Multiplication (or Repeated Addition). Bird, Activity 2. In the lesson on birds, the teacher described his introduction of multiplication, again in the context of the body system. The presentation appeared to support the use of the body in a repeated addition approach to multiplication.

> I now count through, with the student groups, some addition and multiplication problems. For each, I express pictures in Tok Ples [the Oksapmin language]. I begin with pictures about the parts of birds each have [the teacher refers to Figure 126a in his written description]. I describe that I had one bird with six parts. I then ask, how many parts do four birds have? [The teacher sketches as in Figure 126b to support the question.] [The teacher states that he explains to the class the following] Four children working in groups should have 24 parts of birds. It means "*tan hatrip*" [ring finger on the other side].[6] I teach signs of plus [*tapdis*] "+" and times [*ammuk*] or groups "×."

F. Body-Part Counting: Applications to Division (Fair Share). Bird, Activity 3. The teacher described an instructional activity that involved the construction of fair shares of parts of a bird. The activity led to the representation of fractions. The teacher pointed out that in each student group there were four members, and six parts of the bird.

6 Note that the teacher's use of *tan hatrip* (ring finger on the other side) was to represent the 24th body part, using the symmetrical variant of the full count of 27 body parts. In what I take to be the more traditional system in the Oksapmin area, *tan hatrip* would be the 26th body part.

(a)

(b)

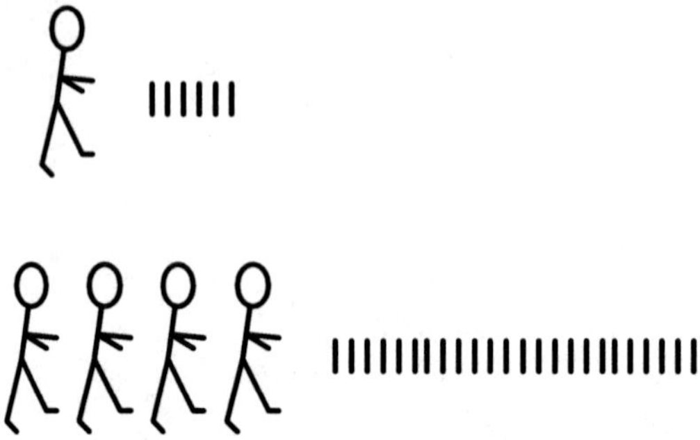

Figure 126. (a) The teacher and parts of a bird as represented on the blackboard; (b) four children and parts of their birds as represented on the blackboard.

> "With four children, each child will have how many parts?" [The teacher indicates that the answer is one and a half in the Oksapmin language [*dapsi ohe kat*]].

In the Prep year, teachers do not use written notation for number, and the teacher explains this. He states that the point is to get across the idea that with four children, each child will have one and a half. He writes in the Oksapmin language that this is "*dapsi ohe kat*."[7]

G. Body-Part Counting: Base-10 or -20 Interpretations. Rubbish, Activity 2. During this phase of the lesson, the teacher had students collect 50 pieces of rubbish using body counting with a specified strategy, assembling it in groups of 10. In the interview, the teacher explained the drawing (Figure 127). He shows 50 pieces of rubbish on the blackboard. He calls each group of 10 "*ammuk*" (group), writing "50" in Hindu-Arabic numbers, as he explains that this is *hathat kat-hai* ("pinky (5) shoulder (10)" or "five tens").

H. Body Sequence Variants. Bird, Activity 2, Ending. At the end of the teacher's description of the lesson on birds, he refers to the count of 24 bird parts as "*tan hatrip*" (ring finger other side (24)).

> Four children working in groups should have 24 parts of birds. It means "*tan hatrip*." I teach signs of plus (*tapdis*) "+" and times (*ammuk*) "×."

[7] Noteworthy is that one meaning of *kat* or *kut* is "part of something" (Lawrence & Snyder, 1993), which here is used to refer to a fraction.

{lines represent pieces of rubbish}

Figure 127. Representations of 50 pieces of rubbish.

As noted earlier, to use *tan hatrip* (ring finger) to refer to 24 is a variant of the Oksapmin counting system that I had learned, though a variant that the teacher takes as traditional Oksapmin counting. This variant terminates with a trajectory through the digits of the second hand that is symmetrical with the first hand.

J. Semiotic Adaptations: Oksapmin Names for Western Expressions. Bird, Activity 2, Ending. In the prior example, the teacher noted that he introduced the Oksapmin words *tapdis* and *ammuk* to refer to the written signs for addition and multiplication. As I noted previously, to my knowledge there are no specialized words in the traditional Oksapmin language for either addition or multiplication, and certainly not for the written operation signs for "plus" or "multiply." In Marshall Lawrence's 1993 Oksapmin dictionary, he lists "*tap hapti*" as meaning "to come together." In the use of *tap hapti* here, the word's meaning appears to have been extended to the idea of addition and plus sign. Lawrence lists two words with perhaps some qualitative sense of multiply: *ardeipat*, a transitive verb that Lawrence defines as "to cause to increase; multiply," and *artahapat*, an intransitive verb that Lawrence lists as "to become fruitful; increase; multiply." In the lesson, the teacher extended a different Oksapmin word – the term "*ammuk*," or group. Lawrence defines *muk*, a noun, as "collection of things; group; bunch."

Garden, Activity 2. In the lesson on gardens, the teacher wrote on the board subtraction problems for the class to review and solve (see Figure 128). The problems were written in the Oksapmin language, accompanied by supporting drawings of circles used to represent stones. The written expressions are mixtures of body-part numbers and conversational numbers, along with a word form "*madas*" that is taken to mean "subtract." The first expression, "*hatrip-hei ohe yot madas ohe*," translates as "pinky (5) *yot* (2) take." The second, "*hathat-hei ohe yetir madas ohe*," translates as "pinky (5) *yetir* (3) take."

Figure 128. Subtraction problems in the teacher's written lesson description as drawn on the blackboard.

> I tell students to get ready to do two examples of subtraction from the board. I draw the pictures of stones on the board. Students get their sticks and work out the subtraction problems with their sticks. They also count in Tok Ples and solve the problems.
>
> [The teacher continues with reference to Figure 128.]
>
> They get four sticks and drop two sticks. The group leader asks group members to tell how many sticks left. All count and say together the answer to the same question.

The literal translations of the problems are: (a) "ring finger (4) and then two *madas*"; (b) "ring finger (4) and then three *madas*." For the teacher and his students, the expression "*madas*" is used to mean subtract. Note that the closest expression to *madas* that I could find in Lawrence's Oksapmin dictionary was "draws attention to something being offered" (Lawrence & Snyder, 1993).

The Unintended Consequences of Teachers' Instructional Efforts

In teachers' efforts to accomplish mandated instructional reform objectives, they created bridges for students between the community's traditional use of representational forms like body counting, names for currency tokens, or everyday Oksapmin nonmathematical language, and the use of Westernized Hindu-Arabic representations for number and arithmetical functions for these Westernized representations. In their bridging efforts, teachers appeared to alter, unwittingly, the form and function of Oksapmin representations. These alterations emerged in classrooms as the target of instruction.

These unintended consequences of national reforms in teachers' reproduction and alteration of indigenous knowledge take on particular interest as evidence of sociogenetic processes. As Oksapmin students take up more uniform uses of Oksapmin representational forms in their classroom activities, both teachers and students become agents in the propagation of altered representational forms and functions in Oksapmin communities. My research extended only to 2001, and it is unclear how far particular variants would propagate through the community and through time. I expect that the durability of the altered representational forms in classrooms depends upon the occasions in which the hybrid forms become useful in supporting the solution of coordination problems between interlocutors in and out of the collective practices of classroom life.

Part IV

Toward an Integrated Treatment of Sociohistorical and Cognitive Developmental Processes

I have provided considerable nuance in documenting the representational forms for quantity, the functions they serve, and the collective practices in which people are engaged in Oksapmin communities. The documentation is unusual in that it draws upon a wide range of data sources over a stretch of marked social change. In the closing chapters, I summarize key findings and revisit the conceptual framework that organized the inquiry. In doing so, I point to ways that the framework builds upon, departs from, and opens up new strands of research in the cognitive sciences vis-à-vis the reproduction and alteration of representations and ideas.

I have organized the discussion around three focal questions. In Chapter 12, I review perspectives on *what* develops in mathematical thought, briefly situating the treatment of form and function in relation to other programs of work. In Chapter 13, I consider *how* development occurs as I discuss trajectories of form-function relations in collective practices within and across historical periods. And in Chapter 14, I consider the ways in which the analyses reported in this volume inform our understanding of *why* development occurs, with a focus on regulative processes in the micro-, onto-, and sociogenesis of form-function relations in collective practices.

12 What Develops?

A Focus on Form-Function Relations

Early in the volume, I compared the structural-developmental, information-processing, and cognitive-neuroscience approaches to the study of cognitive development, and I commented that their units of analysis did not adequately capture emergent relationships between the cognitive processes of individuals and the collective activities in which they participate. Hutchins (2006) captures well the problem with analytic units associated with these approaches when he states that investigators' units of analysis should be structured in a way that "do[es] not leave important things unexplained or unexplainable" (p. 376). In these prior treatments, the focus was on units that bifurcate the individual and the collective. What is left invisible and thus unexplainable are the dynamic processes in the reproduction and alteration of ideas over historical time. It is these dynamics that are at the crux of understanding culture-cognition relations.

The units of analysis that I used to organize inquiries presented in this volume are cultural forms of representation and the functions they are used to serve in collective practices. I have not rejected psychological and biological constructs in my focus on forms, functions and collective practices. Indeed, forms and functions, as they become embodied in activity, are cognitive constructions, enabled and constrained by anatomical structures and psychological information processing. But the investigation of culture-cognition relations requires a frame that captures emergent relations between the individual and the collective.

Key Properties of Form-Function Relations in Quantification

In synthesizing the analyses of forms and functions of the prior chapters in relation to the question "what develops?" I would point to three principal features of forms and the functions that they afford. These include their conventional, hybrid, and instrumental properties.

1. *Conventionality: Presupposed Meanings*

Cultural forms used in communication are conventions with no inherent or fixed communicative functions. To be useful in communications, forms – whether they be body parts, currency tokens, or quantifiers – must be produced with an audience in mind and interpreted with attention to the speaker's communicative intent. The indeterminate meaning of forms used in communicative acts allows for flexibility in communicating intended meanings but also contributes to tensions in producing and interpreting conventions. In Chapter 10, for example, I reported contemporary third graders' presupposition that the shoulder (10) was used to refer to the 10t coin, while Oksapmin adults, in contrast, assumed that the shoulder (10) referred to the value of 1 kina. For schoolchildren, the presupposed meanings of shoulder (10) are linked to the decimal interpretation of kina and toea, but for adults, the presupposed meaning of shoulder (10) was rooted in the obsolete Australian shilling and pound, and their kina and toea equivalents. Children's and adults' presuppositions would clash in communications in the Oksapmin language.

2. *Hybridity: Forms Modified or Drawn into Activity as Problems Shift in Collective Practices*

Representational forms can be usefully understood as hybrids (Gutierrez, Baquedano-Lopez, & Tejeda, 1999), emerging as interlocutors blend forms to support communication and problem solving as they engage with new communicative and problem-solving challenges. Consider three illustrations discussed in prior chapters related to the body form for number.

a. *The body as a representational form for number.* As I discussed in Chapter 2, Laycock surmised that in pre-contact times in the Mountain-Ok area, people recruited the use of their bodies to measure strings of shells (*bonang*) in trade relations; he argued that *bonang* would have been stretched across the arms and, over historical time, some body positions were eventually taken as indicative of lengths. Based on Laycock's argument, the body-counting system emerged as a hybrid form from two converging strands – (a) people's use of their own bodies (bodies that have many different nonmeasurement functions) and (b) the recruitment of the body to measure an artifact of trade.

b. *Modified body system from 27 to 20 body parts*. In Chapter 5, I reported the shift from the use of the full 27 body parts in the Oksapmin system to the use of 20 body parts and its varieties. The modification can be understood as people's efforts to use the body system to accomplish new problems linked to the introduction of pounds and shillings. The modified body system has a form hybridized from two converging strands: (a) the prior history of the 27-position body system and (b) the history of Australian shillings and pounds in the Oksapmin area in relation to emerging collective practices of exchange.

c. *Fu*. Chapter 6 provided evidence that the shifting functions of *fu* were linked to processes of hybridization. For instance, I dated *fu* back to its everyday meaning in pre-contact Oksapmin communities to meaning "a complete group of plenty of things." Subsequent hybrid meanings of *fu* occurred as (i) *fu* became an expression (accompanied by raised fists) linked to the completion of all 27 body parts (a complete group of all of the plenty of body-part positions), (ii) to the first 20 body parts with the link to pounds and shillings, (iii) to "double" with the introduction of kina and toea and the specialized use of *fu* to refer to kina notes.

The illustrations are a few instances of ways that the crossing of historical trajectories leads to the emergence of new collective problems in communications and can lead to the production of hybrid forms. An explication of this process will be the focus of subsequent chapters – the how and why of form-function shifts.

3. *Instrumentality: Means and Goals*

The prior chapters are replete with illustrations of the uses of forms as "instruments" of communication and problem-solving activities that emerge in collective practices. I reported instrumental uses of the body form in comparing discontinuous quantities (Study 8-1), measurement (Study 8-3), composing and communicating about currency token values (Studies 5-2 and 10-2), and arithmetical compositions of tokens (Studies 4-1 and 9-2), to name a few. In these activities, forms serve quantitative functions as they are tailored into means to accomplish quantitative goals. It is through this process that communication is achieved and problems are solved. An important entailment of the process is that as forms are used, variants are generated – a process that seeds the possibility of new lines in the sociogenesis of form-function relations.

I find it noteworthy that investigation of the instrumental functions of cultural forms for quantification has often been overlooked in prior scholarship on culture-cognition relations. In comparative ethnographies and historical analyses, the properties of material and nonmaterial artifacts are often catalogued (Gordon, 2004; Lancy, 1983; Laycock, 1975; Lean, 1992; Schmandt-Besserat, 1992), but how these artifacts are used in activity is often treated as unproblematic (or circumstances do not permit the study of their uses). In Chapter 1, I reviewed Schmandt-Besserat's conjectures about uses of ancient tokens in Mesopotamia, Lancy's studies contrasting the cognitive development of various Papua New Guinea groups that differed in numeration systems, Lean's cataloguing of more than 800 number systems in the South Seas, and Laycock's description of the many number systems in the New Guinea highland areas. These treatments catalogued varied and fascinating representational forms for number. But without the analysis of how these forms emerge and are used in activity, conjectures about relations between culture and quantitative thought or about the genesis of form-function relations remain problematic.

Form-Function Relations in Quantification Practices as Units of Analysis

Form-function relations are at the crux of mathematical activity. To investigate how body parts can refer to shillings or be used to keep track of arithmetical compositions, for example, we must examine relationships between forms and functions through an analysis of activity. As units of analysis, form-function relations afford the exploration of intrinsic relations between cognizing activities of individuals, on the one hand, and collective life, on the other.

The next two chapters build on the focus on conventional, hybrid, and instructional properties discussed in this chapter. In Chapter 13, I take up historical trajectories of form-function relations, discussing the interplay between continuity and discontinuity in historical time. In Chapter 14, my focus shifts to regulative mechanisms that constrain and enable historical shifts.

13 How Do Form-Function Relations Develop?

How questions seek to investigate the trajectory of change in relations between representational forms and their functions. These questions share affinities with those asked in evolutionary biology – for example, how do anatomical forms and their functions evolve in the emergence of new species? Of course, the issues of conventionality, hybridity, and instrumentality that mark form-function relations in collective practices are not issues in the case of biological evolution. Nonetheless, the questions regarding form-function relations in the case of biological evolution present a useful parallel, because they bring issues of historical trajectory and the analysis of progenitor forms and functions into clear relief.

In his account of the origins of a bird's wing, Stephen J. Gould asked, "How could the wing have evolved?" (Gould, 1996). The question leads to a conundrum that often emerges in accounts of evolution: The precursory forms of a wing – those that did not afford loft – could not offer a survival advantage for flight. The discontinuity of function – from no flight to flight – is an evolutionary puzzle. How did proto-wings themselves emerge, and how did they endure? Clearly the evolution of the wing could not be a story of continuous perfection of the flight function. As Gould and others make clear, a small fraction of a wing is not adaptive if adaptation is measured by the function of flight, for a fraction of a wing does not allow a creature to fly. At the same time, a working wing is a complex anatomical structure and could not have appeared suddenly as a fully formed organ.

Like other evolutionary biologists, Gould argues that, to understand the evolution of the wing, we must appreciate that evolutionary change is a complex of quirky shifts between anatomical forms and the functions that they serve. The wing's progenitor may well have offered organisms a selective advantage by serving a function other than flight. On this argument, the seeds of flight would have been born from the structural properties of these proto-wings that served other functions. Like other evolutionists, Gould points to thermal regulation as a likely candidate for the transitional function. Proto-wings that regulated temperature may have supported a degree of loft by coincidence, and the nature of the adaptive advantages that they afforded shifted, giving their possessors a novel selective advantage. The selective advantage of thermal regulation that at the same time happened to afford loft, in turn, led to new lines of evolutionary change and new ecological niches.

Though there is a world of difference between adaptations in the origins of species and quantification practices that have emerged in the social history of number, the parallelism is instructive. From my perspective, the evolutionary conundrum poses three challenges for argumentation and evidence with regard to form-function relations in the case of number. These involve (a) the genetic continuity and discontinuity of representational *forms* in numerical activity, (b) the genetic continuity and discontinuity of *functions* in numerical activity, and (c) changing *niches* (the particular ecosystems to which a species is adapted) and the implications of changing niches for genetic shifts in *form-function relations*.

Genetic Continuity and Discontinuity of Forms in Quantifications

The first challenge is to establish genetic continuity of forms. In the analogical case of the wing, how do we identify the progenitor form? The answer is not straightforward and is well represented in problems of cladistics – the identification of ancestral species. Consider, for example, the bat (a mammal) and the sparrow (a bird). The wings of both these creatures serve the function of flight, but current evolutionary wisdom has it that their genetic roots are quite different. In other words, though we find similarity of form and function in the wings of these creatures today, the progenitors of these forms and functions differed. Similarly, we can ask, are the early forms and functions associated with Oksapmin mathematics the progenitors of current forms used in practices of quantification? For instance, is the traditional expression "*tit fu*" and its accompanying gesture, which most likely date back to Oksapmin prehistory, the progenitor of the contemporary use of *fu* as a suffix serving to double the value of a body-part name? Similarly, is "*temsi-tana*," the Oksapmin expression meaning "flat with hole," the progenitor of the current expression for the value of 1 kina?

Problems about the genesis of representational forms are not trivial. It may well be that the archaic and modern forms are like bats and birds – whether archaic and modern uses of *fu* (a complete group of plenty; double the value of a body part), *temsi-tana* (flat, hole; 1 kina), or body-part names (used as names of anatomical structures; used for keeping track in order to add or subtract quantities). Indeed, older and more recent uses may have been spawned from quite different (semiotic) evolutionary lines. On the other hand, they may indeed be related by a common genetic lineage, and an understanding of that lineage would help us understand the origins and development of mathematical thought in Oksapmin communities.[1] Of course *fu*, *temsi-tana*, or body parts used in body-part substitution strategies for arithmetic are only instances of many contemporary forms. One could ask the same historical questions about the progenitor relations of any contemporary forms. The genetic analysis of representational forms is one of several important keys to understanding the social history of mathematical cognition in human communities.

Genetic Continuity and Discontinuity of Functions in Quantification Practices

A second challenge is to establish the genetic continuity of functions. The complexity of this analytic task is hard to overestimate. In the case of the wing, I noted that, for Gould, the evolutionary answer to the quirky shift problem is that the proto-wing could have evolved only if it afforded an adaptive advantage. Further, this adaptive advantage could not have been the function of flight itself, because these relatively small proto-wing appendages could not support flight in the ancestors of modern birds. The same argument applies to *fu*, *temsi-tana*, body-part names, or any of the many other lexical forms used in quantification practices. A genetic perspective implies that the new functions that appear discontinuous should exhibit transitional functions in activities that would be progenitor functions of quantification. The difficult challenge here, of course, is to discover how novel functions, such as the multiplicative meaning of *fu*, might emerge from other candidate progenitor functions, like the meaning "complete group of plenty." The analyst's task, then, is to develop conjectures about, as well as corroborative evidence for, how discontinuity emerges through continuous change. In Chapter 6, I used this analytic approach in a treatment of the social history of *fu*. As further illustrations of shifts in functions of progenitor forms, let's consider the case of *temsi*

[1] In the case of *fu*, the empirical challenge became a systematic search for evidence that would indicate whether the doubling *fu* had roots in the counting *fu*.

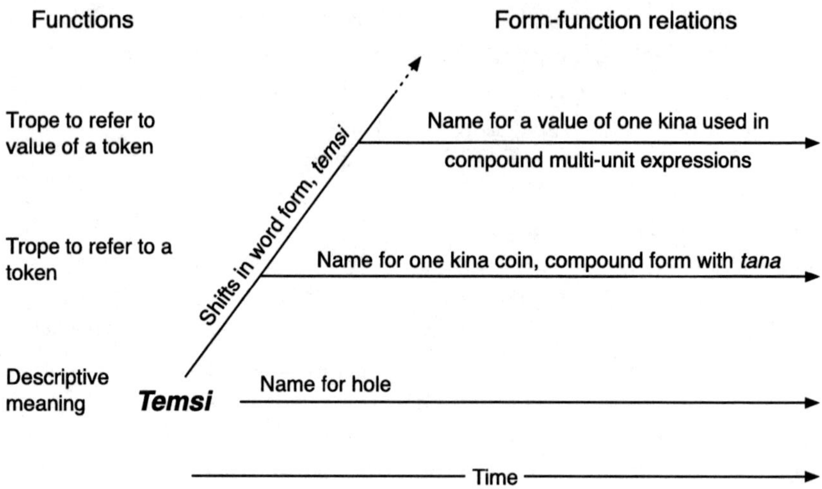

Figure 129. The word form *temsi* (hole) and its shifting function in quantification practices.

("hole") to signify the value of 1 kina and the case of *haben* ("leaf") to signify the value of 2 kina.

Figure 129 is a depiction of a set of conjectures about the form-function shifts in the use of *temsi* over historical time. The vertical axis displays shifting functions for *temsi*, and the horizontal axis represents time (Studies 5-1 and 5-2) with horizontal lines that represent onset time and duration of use of new functions of *temsi*. As shown in the figure, the earliest function of *temsi* was as a name for hole, and as shown in the figure this function persisted through 2001. Upon the issuance of the K1 coin – a coin with a hole in the middle – *temsi* was recruited as a trope to support reference to the K1 coin, and this later emerging function for *temsi* persisted. Still later, an emerging function was *temsi*'s use to mean the value of 1 kina. This use is found in the many expressions that involve odd-numbered values like the K5 note or the composition of currency values that sum to an odd number.

Conjectures for the case of *haben* are represented in Figure 130. Some corroborative support for these conjectures is contained in Studies 5-1 and 5-2, which documented the use of *haben* outside of quantification and its usage in the quantification of currency. As indicated in the figure, *haben*'s earliest known function was as a lexical term to refer to the leaf of a specific kind of tree (a function that persisted through 2001). *Haben* was recruited as a trope to support reference to the paper note, a leaf-like object, perhaps

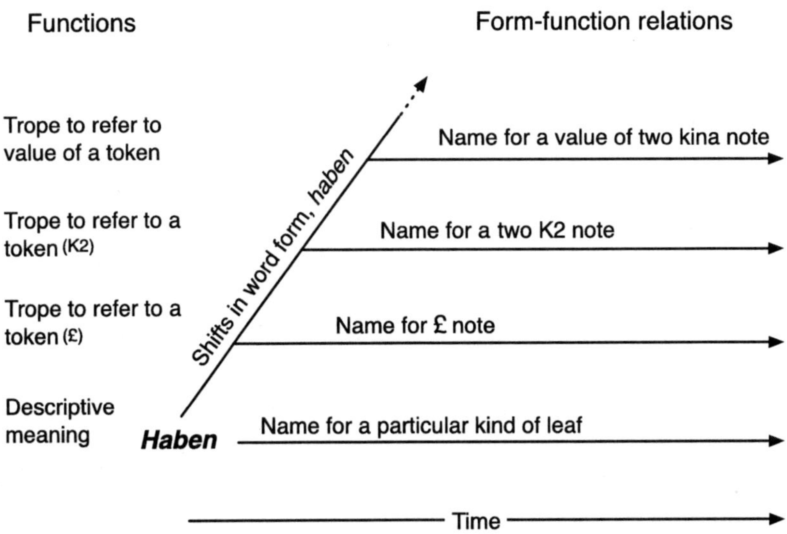

Figure 130. The word form *haben* (leaf) and its shifting function in quantification practices.

not long after the initial penetration of paper currency into the Oksapmin area. This use of *haben* persisted as an expression for any pound equivalent. It was used for the Australian 2-dollar bill (a pound equivalent) during its short life in the area, and then the word came to be applied to the K2 note (also a pound equivalent).

Changing Niches and Implications for Genetic Shifts in Form-Function Relations

A third challenge for the evolutionary analysis that bears on *how* questions concerns the idea of a niche (or larger ecosystems, of which niches are constitutive subsystems). There is a twofold analytic challenge here: (a) to understand the conditions in which progenitor forms were viable and (b) to trace the shifts in properties of niches that support diverging evolutionary trajectories between parent and child forms and functions. Consider again the progenitor of the wing. The ancient ancestors of modern birds were adapted to particular niches. Through environmental changes such as shifts in climate, some mutations afforded a survival advantage. If the progenitor of the wing provided a survival advantage linked to thermal regulation in a shifting climate, it may have supported a divergent evolutionary line, creating closer terrestrial ancestors to modern birds, though the proto-wings of these ancestors would have served thermal regulation rather than flight functions. Understanding the complex interplay between shifts in properties of organisms and properties of their niches is key to an evolutionary treatment.

Collective practices, like those that are reproduced in activities such as economic exchange or schooling, are usefully conceived as analogous to ecological niches. But rather than being semistable biological worlds of organism-environment relations, collective practices are semidurable, socially organized activities, constituted by individuals engaged in joint actions through the use of semiotic and instrumental forms. In the semiotic world of collective practices, emergent problems are analogous to the selection pressures of ecological niches (Hawkins & Pea, 1987; Toulmin, 1972). As in the biological case, processes external (as well as internal) to the niche may lead to alterations of conditions. In the case of the biological niche, changes may be in climate, geology, or the migration of new life forms. In the case of collective practices, external changes may be in the form of introduction of new units of currency, increases in the availability of commodities and cash, or the introduction of national school reforms – external processes that lead to new collective problems internal to practices. For example, shifts in the character of arithmetical problems in exchanges may emerge with the penetration of new token denominations into Oksapmin communities. The result is shifting "selection pressures" for variants in the semiotics of form-function shifts.

Some have conceptualized form-function shifts (or their equivalents) over historical time as inevitably proceeding toward greater differentiation and integration akin to Werner's orthogenetic principle (Werner, 1948; Werner & Kaplan, 1963) or Tomasello's ratchet effect (see Boesch & Tomasello, 1998; Cole, 1996; Tomasello, 1999). But this is not necessarily the case. Indeed, with shifts in collective practices, some achievements may become no longer useful, and over generations in a population they may be lost to the past. There is some evidence that such may have occurred in the pre-contact history of the Oksapmin body system – a shift from what may have been previously a system with a repeating cycle of 27 to a system with no cyclic structure at all. Let's consider this possibility as a thought experiment that sheds insight into relations between form-function shifts and collective practices.

Before reviewing evidence that addresses this issue, let's first consider a New Guinea highlands community, the Kapauku, that had developed a rich and vibrant cowry shell money economy in traditional life. The Kapauku occupy an area to the west of the Mountain-Ok. The character of the group's numerical activities and traditional number system contrasts with that of the Oksapmin.

In fieldwork conducted in the 1950s on the economy of the Kapauku, Pospisil (1963) noted the Kapauku people's "obsession with quantification" and was impressed with their number system:

> These people are so fascinated by numbers that they indulge in counting, in recalling precise sums paid in specific exchange transactions, and in discussing these data at any opportunity that presents itself. As I stated in my monograph on law: "They place value upon higher numbers and larger value. The emphasis upon quantity assumes forms which come as a shock to a Western observer. My informants when confronted with a magazine picture of a smiling girl failed completely to react to her beauty. Instead they started to count her teeth." (Pospisil, 1963, p. 14)

On the Kapauku number system used in collective practices of economic exchange, Pospisil wrote, "The paramount emphasis on sales and money in general is facilitated by a quite adequate numerical system. The Kapauku possess a decimal system that stops at 60 and starts over again, having as higher units 600 and 3,600."

In 1978, I saw nothing in the Oksapmin world comparable to the elaborate system of trade that Pospisil detailed in his 1963 monograph, and certainly nothing comparable to the Kapauku number system. The Oksapmin body system did not have a cyclic structure; to count beyond the pinky (27) one looped back to the 22nd position (wrist), which became the 28th position, and proceeded back toward the origin (see Figure 12).

Might there have been some time in the Oksapmin prehistoric past a similar world involving more complex collective practices of exchange? If so, perhaps the 27 body-part system had had a cyclic structure and had been used in more powerful ways. Consider the following evidence.

1. *Telefol body counting: A Mountain-Ok 27-body-part system with a cyclic structure.* In Chapter 2, I noted that the Oksapmin are one of many Mountain-Ok groups that use a 27-body-part counting system (see Figure 10), though languages vary markedly across Mountain-Ok groups. Phyllis Healey's work with the Telefomin, a Mountain-Ok neighbor about 80 kilometers west of Oksapmin, suggests that the Oksapmin body system may have had a more complex structure during earlier times. In her 1965 monograph, Healey described the Telefol 27-body-part counting system as follows:

> Numerals [sic] are based on a BODY SYSTEM of counting, commencing at the little finger of the left hand, working up the left arm, across the face,

> and down the right arm to the right hand little finger. A few people go the other way, beginning with the right hand little finger, but this variant is non-significant. Once round the body totals 27...[sic]
>
> One unit of 27 is known as deéng, and multiples of 27 are formed thus: deéng maakub "one unit of 27" (= 27), deéng alob "two units of 27" (= 54)....

Thus, in Telefol, Healey described a 27-cycle system in *deéng* as a unit of 27, a unit that can be modified by successive body parts, e.g., *deéng maakub* (*maakub* meaning left hand little finger (1)) and *deéng alob* (*alob* meaning left hand ring finger (2) and so on). In Lean's terminology, the Telefol system had a 27-cycle structure, one unlike the Oksapmin system as used in 1978.

2. *The composite Oksapmin expression, tit fu or tit fu gona*. In Chapters 5 and 6, I documented uses of *fu* or *tit fu!* as the completion of the 27 body parts in the Oksapmin system. This body of evidence provides some support for the argument that in prehistory the Oksapmin used the body in ways that were more powerful with a 27-cycle structure. Recall that *tit* is the conversational word for "one" in Oksapmin, and the conjunction with *fu* suggests a meaning of "one round is complete" (1 × 27), with *yot fu* two rounds (2 × 27), et cetera. A small handful of interviews provide some support for this idea (consider the statement of Elder no. 13 reflecting on the use of *fu!* in Chapter 6).

3. *Brumbaugh's records of Telefol numeration*. Robert Brumbaugh collected data related to his doctoral dissertation with Telefol speakers in 1976, and he recently reviewed his archive of 37 traditional Telefol stories (Brumbaugh, 2010). He found references to number, but no numbers greater than 27, and he noted that his observations during that time were that the body system was not exploited to its full extent in number representation.

What can we make of these observations? It may well be that in Oksapmin communities, as in Kapauku, there was a time when collective practices associated with trade were more complex and number representations more powerful. The body system may have indeed been used in a complex shell-money economy, with counts that entered into the equivalent of many hundreds. The use of *tit* (1) with the expression *fu*, the cyclic structure of the Telefol 27-body-count system, and the lack of use of the cyclic structure in Telefol speakers in 1975 (like the Oksapmin of the same time period), suggests that in a prior time, Oksapmin number may have been a more elaborated domain of understanding. But this is highly speculative. Whatever the truth, it is buried in a long ago prehistory.

Continuities and Discontinuities in Forms, Functions, and Collective Practices over Historical Time

People reproduce and alter numerical forms and functions through their participation in collective practices as they engage with problems of quantification. In this process, we find the evolutionary engines of variation and selection. But, unlike the biological case, in the semiotic ecosystem of collective practices, the survival of actors is not at stake. The variants that are produced are semiotic means of accomplishing emerging communicative and problem-solving goals in activities. These means become recognizable cultural forms as they are taken up by others when communities adapt to shifting collective problems. There are complex regulative processes involved in the production and reproduction of collective practices and the semiotic forms and functions constitutive of them. These regulative processes, which lead to continuities and discontinuities over time in form-function relations, are the focus of the final chapter.

14 Why Do Form-Function Relations Develop?

The answer to the question "Why do forms and functions develop?" may seem to be addressed in the prior chapters on the "what" and "how" of development. Forms and functions develop because children and adults participate in collective practices, like economic exchange or schooling, in which new needs for quantification arise with distal events such as the issuance of currencies, changes in government policies about education, or shifts in demographics in a community. Though such events can lead to people's needs for developmental change, *why* questions require an analysis that addresses regulative mechanisms. In his treatment of the emergence of conscious awareness, Vygotsky expresses this well, arguing that an explanation of the *why* of development cannot be based solely on the *need* for development:

> to say that awareness appears as a result of a child's need to become aware of something amounts to a claim that wings originate in a bird's need to fly. In what concerns scientific epistemology, such an explanation seems antiquated. It also endows the need with the ability to create apparatus necessary for its satisfaction. (Vygotsky, 1986, p. 164)

Answers to *why* questions must articulate the processes that constrain and enable the emergence, reproduction, and alteration of form-function relations in collective activities. I take up the challenge in this chapter, and consider regulative mechanisms in micro-, socio-, and ontogenesis as these processes are situated in collective practices.

Regulative Processes in Microgenesis

Numerical problems and their solutions take form in microgenetic activity. In microgenetic activity, people bootstrap shifting intentions, intuitions,

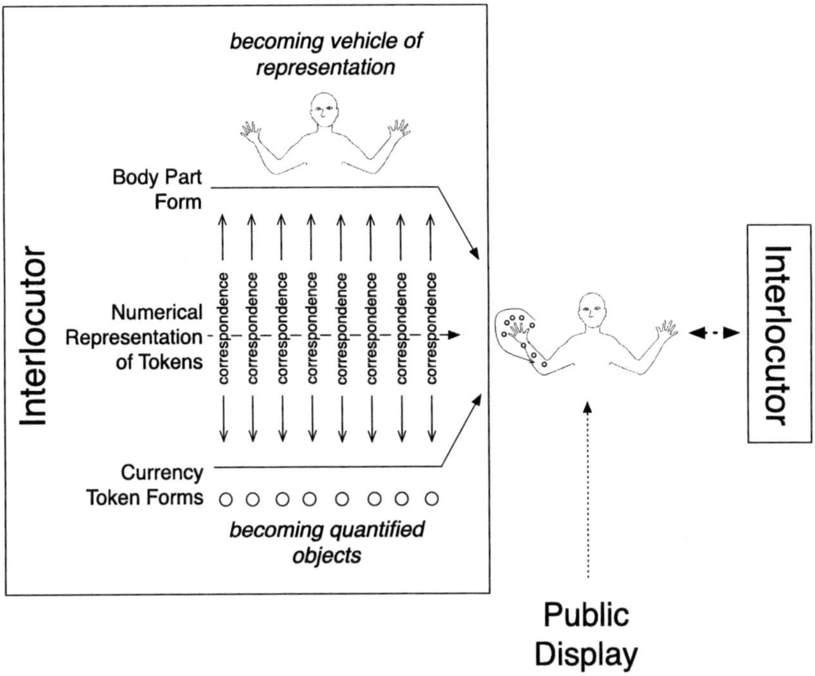

Figure 131. The microgenesis of a numerical meaning for the elbow.

and physical actions as they structure and accomplish numerical problems. In the context of microgenetic constructions, people adapt and tailor numeric forms and functions to the exigencies of situations as they work to render their own numerical activity coherent for themselves and to others. To illustrate, consider an Oksapmin man's microgenesis of a counting activity as he engages in a purchase of goods in a trade store.

The man identifies a commodity that he wants and its cost; he retrieves coins from a can in his string bag to determine the value of the tokens that he has in relation to his purchase. He produces a count of the coins, coordinating denominational value with body parts in his enumeration in units of shillings, then translating them into units of Papua New Guinea currency, K.80, or inner elbow (8) in the body-part form. The microgenesis of his determination of coins in his string bag is represented in Figure 131. In producing a count, the man quantifies coins as he puts them in one-to-one correspondence with body parts, and reciprocally, he quantifies body parts as he puts them in correspondence with the coins. In this process, the numerical functions of word forms and counted objects are constructed in relation to one another, such that the vehicle – in this case body parts – come to index objects, in this case a summation of the coins. The representation is not strictly for the man himself, but is a public display of currency value that the man believes is interpretable by the clerk.

In the illustration, the man's construction of the representation of K.80 is part of a microgenetic process. Such a focus on the microgenetic aspects of representational phenomena is not new in psychological treatments; indeed, it was a central concern of prior developmentalists, like Werner and Kaplan (1963), Piaget (1962), and Vygotsky (1978; 1986). Vygotsky, for example, in his focus on relations between speaking and thinking, pointed to the idea of microgenesis many times in statements like the following:

> The relation of thought to word is not a thing but a process, a continual movement back and forth from thought to word and from word to thought. In that process, the relation of thought to word undergoes changes that themselves may be regarded as development in the functional sense. Thought is not merely expressed in words; it comes into existence through them.... [T]his flow of thought occurs as an inner movement through a series of planes. An analysis of the interaction of thought and word must begin with an investigation of the different phases and planes a thought traverses before it is embodied in words. (Vygotsky, 1987, p. 218)

In numerical activities like the Oksapmin man's counting, we can identify two regulative processes in the back-and-forth as individuals turn forms into means to accomplish quantification goals. One process is the construction of coherent relations between body parts and the quantified objects (numerical coherence), and the other process is the construction of communications that adequately convey one's intent to another (communicative coherence). Let's consider each kind of regulation in microgenetic activity.

Constructing Numerical Coherence

The concern for building numerical coherence regulates the microgenesis of form-function relations in quantification practices. In the production of a count, individuals create numerical coherence by monitoring one-to-one correspondence relations between body parts and target objects. When they skip or double-count objects, their knowledge of one-to-one correspondence serves as a basis for corrections. Concerns for numerical coherence also guide anticipations, as when one arranges a haphazard group into a linear array to avoid errors in the production of a one-to-one correspondence constitutive of counting. I am speaking here of older children and adults who have knowledge of one-to-one correspondence, because, of course, young children may have limited understanding of the numerical significance of one-to-one correspondences and thus may use body parts to serve nonnumerical functions in enumerative activities (Fuson, 1988; Saxe et al., 1987b). The central idea of the generation of coherence as a regulative process in microgenetic constructions is a general one. It can be extended to any of the meaning-making activities covered in the prior chapters, whether arithmetic, measurement, or token naming.

Constructing Communicative Coherence

Building communicative coherence is a second regulative process in the microgenesis of numerical representations. Consider that an Oksapmin person could use any haphazard series of body parts to represent number in a count like nose, elbow, pinky, et cetera, even though the convention is to use thumb (1), index finger (2), middle finger (3), and so on. Noteworthy is that whether an individual uses an idiosyncratic order or the conventional order of body parts, the set of body parts could be considered accurate if the count were treated in terms of the principles of one-to-one correspondence. But using the idiosyncratic order in communications with others in the Oksapmin community would lack communicative coherence. Indeed, Oksapmin interlocutors presuppose that there is a standard sequence of body parts in counting and that the use of the last body part in a count in a standard order represents the number in a collection.[1] For interlocutors, precedent – the canonical use of the body part order in prior interactions – is useful in the production and interpretation of representational forms in communication. Indeed, precedent provides a useful way to get across and read one another's intended meanings in communications (Lewis, 1969).

Grice's "Cooperative Principle" and his treatment of maxims in communication is at the crux of regulative processes in building communicative coherence in interaction (Grice, 1989; Keller, 1994). Keller's (1994) extensions of Grice's maxims are particularly apt for my concerns

[1] Grice (1989) contended that the Cooperative Principle and its associated maxims were universal. Others have argued otherwise (e.g., Spencer-Oatey, 2003). For my purposes here I simply contend that some conversational constraints are constructed in ontogeny in collective practices of everyday communication. These constraints operate in ways akin to Gricean maxims that lead to regulation of forms and functions in collective practices related to quantification.

here – regulations like "Talk in such a way that the other can recognize your intentions" and "Indicate your intentions in such a way that the other can recognize them" (p. 98). Consider the way that such maxims may be useful in understanding the use of tropes in the microgenesis of a communicative act, with a focus on *fu*. In a 1970s exchange between a store owner and an elder, the owner may try to explain the equivalence between 20 shillings (or 20 10t coins) and a pound (or a K2 note). To this end, he refers to the 20th shilling in a count as the elbow on the other side (20), or a *fu*, indicating a K2 note, calling the 20th body part both a pound and a *fu* (a complete group of 20 shillings). In this case, the trade store owner's efforts may be to draw on what he presupposes is the elder's meaning of *fu*, a complete group of plenty. In this illustration, to make communications coherent, the trade store owner makes an effort to accommodate his audience, anticipating sources of miscommunication and thereby addresses potential problems of communicative coherence.

The productions of mathematical and communicative coherence are interwoven with one another in microgenetic constructions, and together they shape how individuals tailor forms to serve numerical and communicative functions in the microgenesis of representational activity.

Regulative Processes in Sociogenesis

The sociogenesis of form-function relations includes regulative processes constitutive of microgenesis (the production of numerical and communicative coherence). It also includes population-level processes distributed across many sites where individuals are engaged in producing numerical representations in interactions.

In this section, I consider again the matters of local emergence and broadcasts, the two broad population-level processes involved in sociogenesis discussed in prior chapters. Local emergence is more clearly manifested in the propagation of form-function relations as Oksapmin people seek to establish common ground in local interactions. In local emergence, social positions in collective practices often emerge as more or less symmetrical, as interlocutors engage in the back-and-forth talk (Davies & Harré, 1990; Harré et al., 2009; van Langernhove & Harré, 1999). In this process, forms are used to serve communicative and mathematical functions as a common ground of talk is negotiated. Broadcasts, in contrast, are not negotiated; they often are produced by a point source and are inherently asymmetrical – as when a teacher lectures to a class, or a missionary or a patrol officer uses a particular way of representing quantity in exchanges. Missionaries, patrol officers, and teachers position themselves and are positioned by others as authoritative brokers of knowledge. Of course, issues of social position and power relations emerge in processes of local emergence as well as broadcasts, but a dominant social position is particularly apparent in broadcasts.[2]

Local Emergence – Sociogenetic Processes from the Bottom Up

Let's first revisit the population-wide process of local emergence. In prior chapters, I documented diverse Oksapmin names used for tokens, names that were relatively recent innovations in Oksapmin communities. How did these names come to be collective representations in the community, and why the marked variation in some names and marked uniformity in others? The local-emergence model provides a plausible account, one that is rooted in the idea of social networks constituted by local interactions in which people are working through problems of coordinated action (or interpretation).

Imagine the time in which trade stores were beginning to populate Oksapmin valleys. At multiple sites in local communities, owners and customers were engaged with coordination problems related to exchanges. To work through the new coordination problems, interlocutors at different locations had to generate ad hoc solutions to create a common ground and recreate that common ground in subsequent interactions with shifting partners in communications. What might be their communicative solutions? How might these solutions change through time as interlocutors crisscross sites? To address these questions, let's return to the data from the study of naming practices for currency tokens. Of special concern here will be variation and uniformity in names for common tokens.

In Study 5-1 on names for tokens, I reported variant expressions for the K1 coin. Two common variants appeared to have roots in pragmatic efforts to get across communicative intents. One variant was *kat-hai* (10), the Oksapmin equivalent of "shoulder (10)" or tenth body part in the

[2] The treatment of sociogenetic processes presented here is different from Dawkins's (1976) portrayal of "memes" replicated through a passive process of imitation. In the form-function account, people reproduce and alter variant forms (names for currency tokens, body-part arithmetic strategies, and ways of using the body to measure), tailoring them to serve emergent purposes, and in this process, the products become subject to others' interpretations. In contrast to Dawkins's treatment, sociogenesis is a constructive process of active interpretation and production, and this perspective is the root of a quite different analytic treatment.

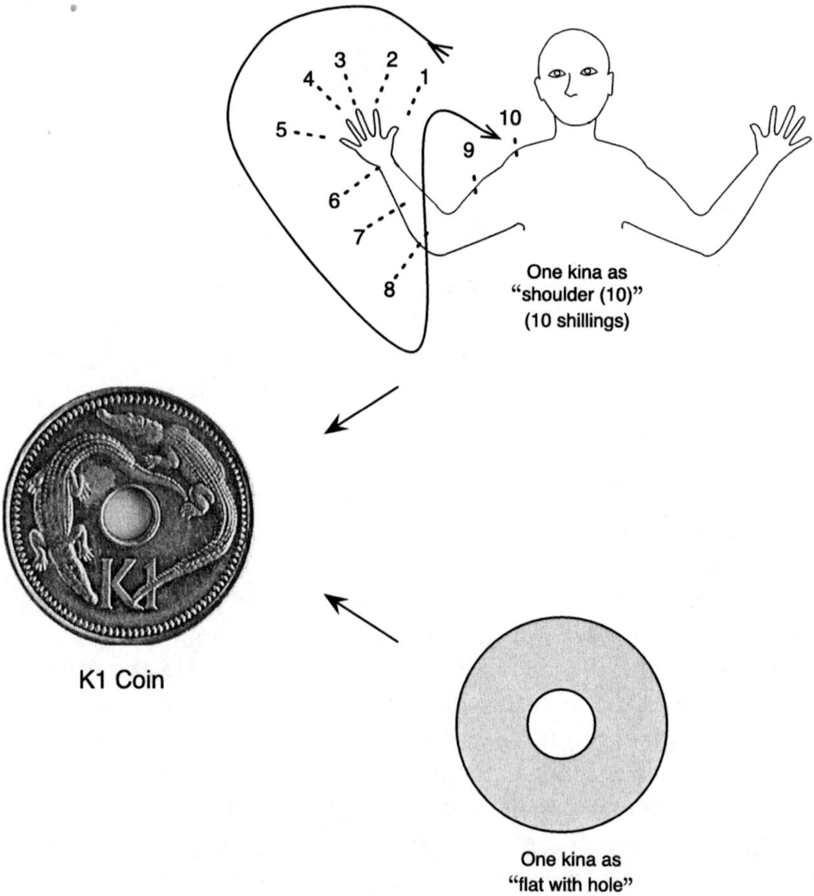

Figure 132. Two conventions used to refer to the K1 coin.

Oksapmin counting system; the other was *temsi*, or the Oksapmin equivalent of "hole." These variants for the same coin may well have been initially used to orient other interlocutors. As depicted in Figure 132, the reference to *kat-hai* signals that a count of 10 shillings (or ten 10t coins, the shilling equivalent) is equivalent in value to the K1 coin. The reference to *temsi* signals the hole in the K1 coin. Although we may not have direct access to the historical processes that elevated these two forms to serve their modern functions, the descriptive and orienting properties of "hole" and "shoulder (10)" to the K1 coin should be clear. Of course, there may have been many

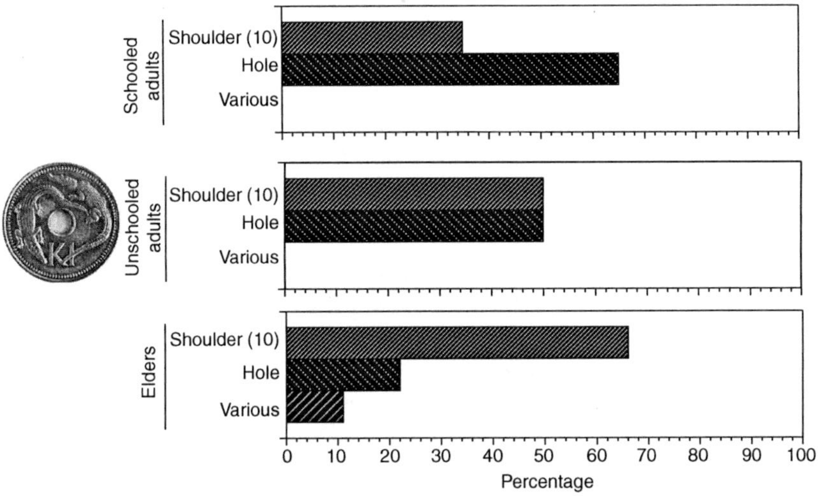

Figure 133. Percentage distribution of use of "shoulder (10)" or "hole" to refer to the K1 coin as a function of cohort.

other communicative solutions aimed at getting across intended meanings in many different early sites, with "hole" and "shoulder (10)" surviving.

Figure 133 shows the percentage of usage of *temsi* and *kat-hai* (10) for the three cohorts of elders, unschooled adults, and schooled adults (adapted from Figure 58, Chapter 5). The figure shows that elders used *kat-hai* (shoulder (10)) three times more frequently than they did *temsi* (hole), whereas schooled adults showed the reverse profile. The unschooled adults used the two expressions with equal frequency.[3]

Drawing upon the idea of local emergence, I find a plausible explanation for the relationship between cohort and representational form (hole vs. shoulder (10)) in the semi-isolation of communicative networks (Agha, 2003; Croft, 2000; Milroy, 1987) and in the regulative microgenetic processes of numerical and communicative coherence. Consider that at multiple trade stores, interlocutors sought to get across their communicative intents and repair miscommunications as they solved problems of coordinated action about currency tokens, prices, computations, payments, and change. Over time, interlocutors also interact with others who have different interactional histories. As interlocutors produce communicative acts regulated by communicative and mathematical coherence, over time, semistable naming conventions emerged for tokens.

[3] A few people produced more than one expression to refer to the K1 coin. In the bar chart, only the first expression offered by an individual was counted.

Individuals' developmental histories have implications for variant preference within cohorts. Consider that a plausible reason for why elders (but not schooled populations) frequently make use of *kat-hai* (shoulder (10)) for the K1 coin is that schooled individuals have access to the Tok Pisin language, with its semiotic system of Tok Pisin numerical expressions; the system references tokens in units of kina (or toea), like 1 kina, 2 kina, 5 kina, 10 kina, and 20 kina. In contrast, elders, who do not have access to this Tok Pisin representational form for number, may prefer the body-part form more than the figurative features (e.g., *temsi* (hole)) of the tokens to support numerical meaning-making in transactions; their preference for body parts is a choice that supports their reasoning about numerical relations between tokens. Of course this is a conjecture. The point is that preferences of interlocutors in a community may be rooted in their efforts to create mathematical and communicative coherence in communicative acts; this is a process that would lead to semistable precedents in the emergence and stabilization of collective systems of representation.

The schematic in Figure 134 illustrates a general case in the local emergence of representational forms. It shows a community of 12 interlocutors represented by uppercase letters *A* through *L*. While the letters indicate hypothesized interlocutors, the numbers (1 through 4) index time periods – that is, moments of synchronic time. Thus, each column identifies pairs of interlocutors interacting in the same period but at different locations. For example, at Time Period 1, *A* and *B* interact at one location, *C* and *D* at another, and so forth. In their local interactions, interlocutors draw upon their prior knowledge to solve coordination problems that emerge in communicative interactions. As their productions are made public (denoted in Figure 134 as the variant forms *u*, *v*, *w*, *x*, *y*, *z*), they are decoupled from the communicative intents of the speaker and may be taken up and interpreted in varied ways, some intended and others not. Over diachronic time (moving from left to right in Figure 134), the process of reproduction and alteration takes variant paths as people communicate with different interlocutors. At Time 1, *A* interacts with *B*, but at Time 2, *A* and *L* move to a common venue and interact with one another. Other people pursue different geographical trajectories over time, interacting in new pairings (e.g., at Time 2, *B* and *C* interact).

The figure is based on assumptions about the emergent coordination problems that have implications for the reproduction and alteration of variants. I make the assumption that, at Time 1, the coordination problem that emerges in an interaction is relatively novel. To denote the token, individuals bring forward forms that are all idiosyncratic, linked to ad hoc

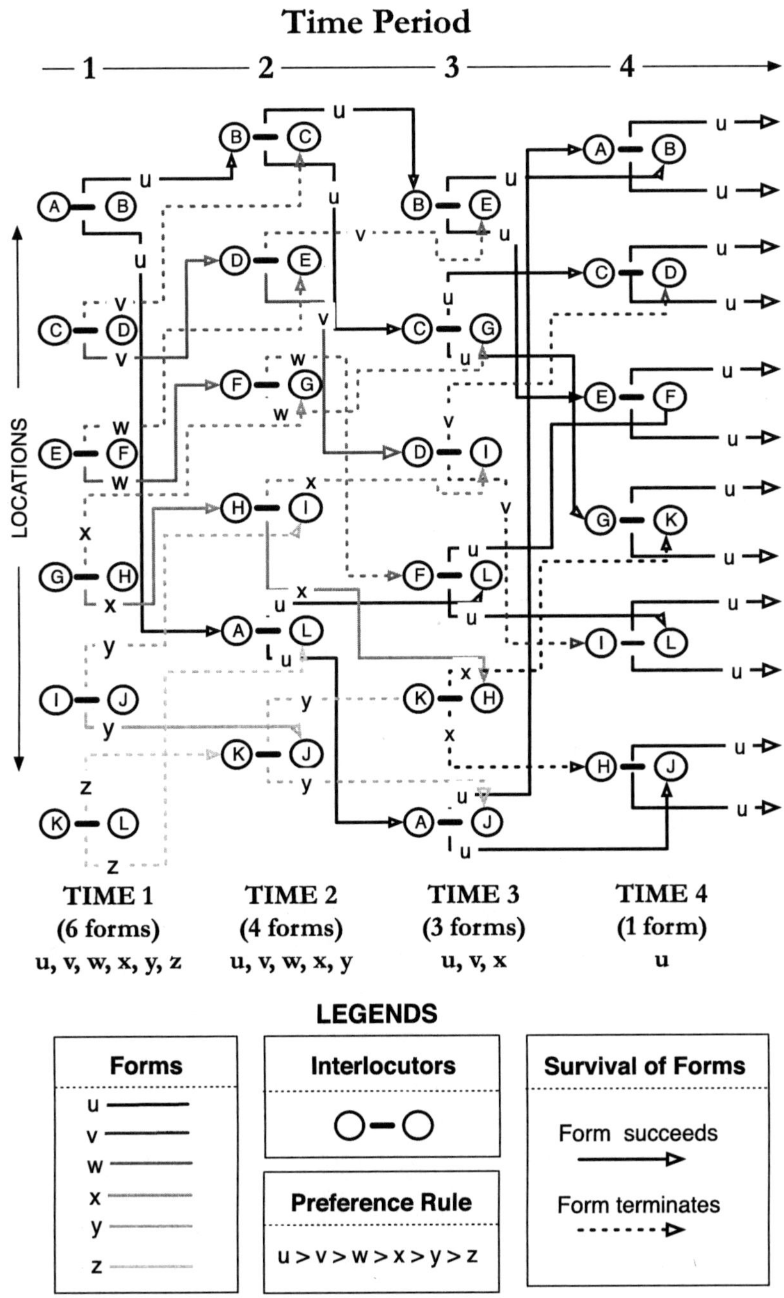

Figure 134. Schematic of dyadic interactions in a community of interlocutors in synchronic and diachronic time.

constructions made up by individuals trying to recruit one another's attention to a common object like a token. The word form may be an effort to orient the other person to the token by reference to a particular dimension, whether it be the token's value, its physical appearance, or a thematic property. For each dyad, I assume that only one form is brought forward. For instance, at Time 1, *u* is brought forward by each member of the *A-B* dyad; *v* is brought forward by each member of the *C-D* pair, and so forth.

At Time 2, the coordination problem recurs at the various sites, but there is a shift in pairings of interlocutors. Because each member of the pairing has a different interactional history, speakers bring forward different forms that have been previously used. Further, I assume that the forms generated at Time 1 have different degrees of attractiveness to the community, as represented by alphabetical order; that is, *u* is the most compelling representation to all individuals in relation to other forms, and *z* is the least compelling. Thus, in the interaction between *C* and *D* at Time 1, both bring forward *v*, but when *C* interacts with *A* at Time 2, *u* is taken up and *v* is not; whereas when *D* interacts with *E*, *v* is taken up and *w* is not. I indicate forms that are taken up in interaction at Time_{n+1} by a solid line that connects Time_n and Time_{n+1} and forms that are *not* taken up by a broken line. The effect is that from Time 1 to Time 4, we observe a propagation of forms, such that some forms are not taken up, while others endure. Thus, over time, one can observe a shift from heterogeneity to uniformity of forms: At Time 1, we observe six different forms; at Time 2, we observe only four; at Time 3, only three; and at Time 4, only one is used across interlocutors.

The schematic model grossly oversimplifies, and this is very clear when we consider the complexity of the process of reference in the Oksapmin world. A more complete model would require assumptions that are more nuanced. Consider the following:

1. *Preferences*. In the model, all individuals were assumed to have uniform preferences, but of course this cannot be the case. Elders, for example, preferred the body-part representation of the K1 coin (shoulder (10)) and schooled individuals preferred the motivated expression "hole." Preferences may be systematically linked to subnetworks of interlocutors, leading to different patterns of propagation within a community as common networks cross and crisscross.
2. *Context/audience sensitivity*. Individuals' preferences are probably context sensitive. For example, when interacting with an elder, a

store clerk may prefer one form, but in interacting with a peer, she may prefer another. Consider the use of *kat-hai* (10) and *temsi* (hole) to refer to the K1 coin as cases in point. The clerk may well use the expression *kat-hai* (shoulder (10)) with an elder but *temsi* or *wan kina* with a schooled individual. Such context sensitivity would certainly add nuance to the modeling and lead to different forms being favored to the extent that the communities are isolated and/or "dialect" differences exist in subnetworks in communities.

3. *Network size.* Not only are there many more than 12 participants in the Oksapmin community, but also interactions involve more than two people, further complicating any plausible model.
4. *New variants.* The model assumes that no new variants are produced, when of course the introduction of new variants is an intrinsic part of the sociogenetic process.

Of course, shifts in assumptions would alter the model of sociogenetic trajectories of form-function relations.

Broadcasts, Hegemony, and the Propagation of Quantification Practices from the Top Down

A class of cases in Oksapmin cannot be well explained by models of local emergence. The uniformity in Tok Pisin names for tokens presents an illustration. A strong case can be made that the lack of variation of token names is linked to the role of missionaries, teachers, and trade store owners in the collective practices in which they participate. These people serve as brokers between worlds outside and inside Oksapmin communities, and their roles appear to stabilize and lead to more uniform names for tokens.

Missionaries, teachers, and trade store owners use Tok Pisin and/or English in communications with those outside of the Oksapmin communities. Indeed, Tok Pisin and English names for tokens and quantitative expressions are entrenched in leading institutions like education, banking, and commerce, and names assume durable form in written language and Hindu-Arabic numerals.

The other world in which missionaries, teachers, and trade store owners participate is within Oksapmin communities. As holders of valued goods as well as knowledge about Western practices, they occupy positions of power. When customers and students participate with these brokers in collective practices, they must interpret and accommodate new representational forms and functions associated with these local institutions.

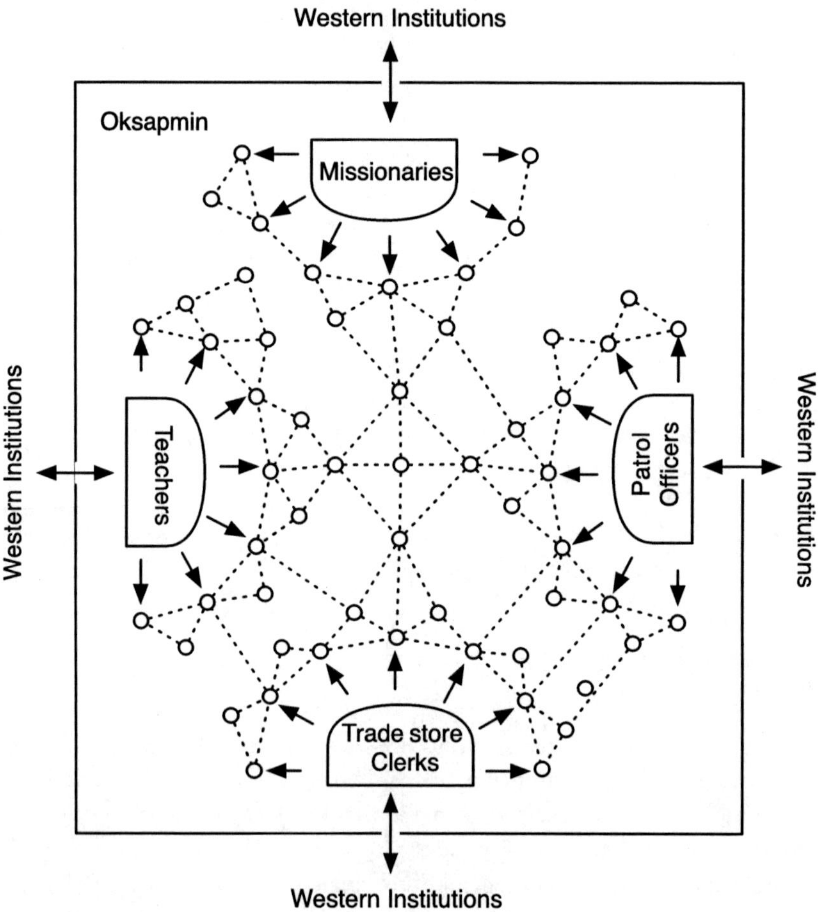

Figure 135. Broadcast of token naming in synchronic processes over diachronic time.

The diagram in Figure 135 depicts brokers at multiple sites communicating with the Western world external to Oksapmin and with local people in the Oksapmin community. In local interactions within the Oksapmin community, the brokers' production of variants is constrained by the forms of representation that they use (or have used) in their interactions with Western institutions and populations. They use these forms in interaction with local Oksapmin people in the context of collective practices, whether economic exchange or schooling. Over diachronic time, brokers often broadcast these Tok Pisin forms as a "point source" (e.g., a lecture to a class by a teacher). The brokers' recurrent production and reproduction of precedents, constrained by the brokers' links outside the Oksapmin community, support the uptake and maintenance of these forms by local Oksapmin people.

Local Emergence and Broadcasts: A Synthesis

Population-wide processes that regulate the reproduction and alteration of form-function relations in communities are spread across sites in synchronic time, and the reproduction and alteration of form-function relations recur through diachronic time. In synchronic and diachronic time, precedent, social position, and power relations constrain the waxing and waning of representational forms and the functions they serve. Shifting collective problems produced by any number of external factors (the introduction of a new currency token, for example) can give rise to new communicative and mathematical conundrums internal to collective practices. In these synchronic and diachronic processes, interlocutors are also historical actors, unwittingly reproducing and altering collective representational forms, functions, and collective practices.

Regulative Processes in Ontogenesis

Individuals construct their own developmental trajectories in ontogeny as they use forms to serve communicative and problem-solving functions in microgenetic productions. Over time, these repeated microgenetic constructions adapted to local situations lead to both continuities and discontinuities in form-function relations in ontogenetic developments.[4]

To illustrate, let's return to the form-function shifts documented in 1980 in the study of the emergence of arithmetic using the body in collective practices of economic exchange (Study 4-1). Though the study made use of a cross-cohort (not a longitudinal) design, it provides some insight into possible ontogenetic trajectories of individuals engaged with collective practices of economic exchange who had different levels of participation in the money economy. For instance, I documented that, when solving arithmetical problems, some people with limited participation in

[4] My discussions of regulative processes in ontogenesis build on constructivist formulations that treat development as a dialectical process of individuals' efforts to conserve the integrity of developing knowledge systems while at the same time adapting to novelty. These treatments include those articulated by Langer (Langer, 1969a, 1969b, 1974), Piaget (Piaget, 1952, 1963, 1970, 1977), Vygotsky (Vygotsky, 1978, 1986), and Werner (Werner, 1948, Werner & Kaplan, 1963).

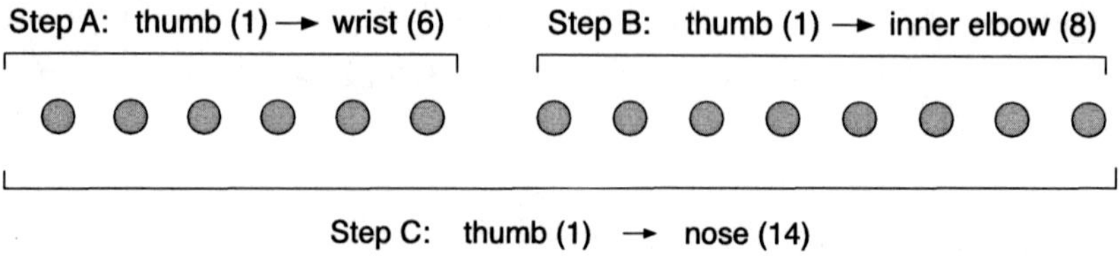

Figure 136. Solving 6 + 8 = ? with coins present: count each term and then count all together.

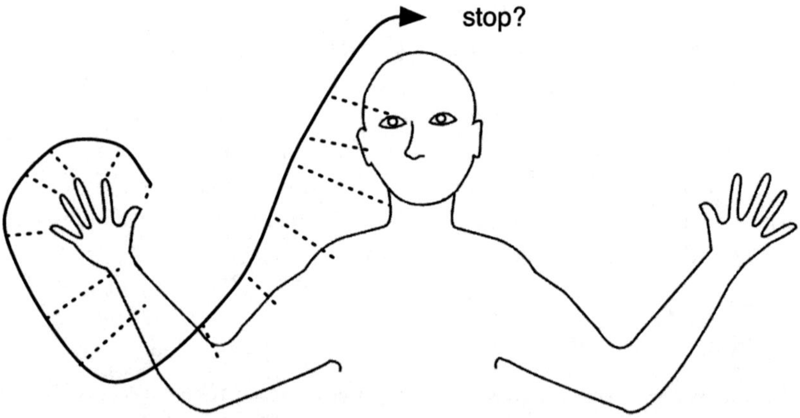

Figure 137. Solving 6 + 8 = ? without coins present: the global-enumeration strategy and the difficulty of keeping track.

the money economy used the body-part counting system only for external one-to-one correspondences (correspondences between the body system and objects). Figure 136 illustrates a prototypical approach to adding six and eight coins with the coins present, first counting both terms separately (Figure 136a and Figure 136b) and then both terms together to achieve a total (Figure 136c). In this approach, the one-to-one correspondences produced were between the Oksapmin series of body parts and the coins, and when this method itself is used, keeping track of the addition of a non-present set does not precipitate a cognitive problem. When coins were not present, however, the same individuals did not create ways of keeping track of the addition or subtraction of terms. Rather, they made use of global enumerations, and that method resulted in the problem of where on the body to stop the count. Figure 137 illustrates the approach to solution of the problem 6 + 8 = ? that was used by people with limited participation in the money economy. To add 6 + 8, for example, an individual would count thumb (1) to wrist (6) and would then continue counting to add eight, forearm (7), biceps (8) etc., but did not appear to conceptualize the need to keep track of where to stop in the addition as a problem, or if he did so, to construct a means to accomplish this. Most elders in Study 4-1 used this approach in the absence of coins.

How would keeping track emerge in ontogenesis? A plausible conjecture is that, with repeated use and in interaction with others in exchanges, individuals would understand the global-enumeration approach as problematic; this would be likely to occur if individuals sought a precise representation for an arithmetic solution, and indeed a concern for precision took hold in collective practices of economic exchange with the cash economy. The turning of externally oriented correspondences inward to create internal body-part to body-part correspondences would be a plausible resolution to the keeping-track predicament (and mark a shift to the use of body parts for a keeping-track function). Such a transformation in solution approach is not straightforward. It requires people to make new coordinations in which they reconstruct external correspondences in a new way and subject to new constraints. Nonetheless, an understanding of external correspondences offer "tools" for people's construction of internal correspondences.

In Study 4-1, I documented varied forms of internal correspondences, and these variants may well reflect ontogenetic trajectories of shifting relations between the body form and the functions that it serves. The double-enumeration approach shows an inward turning of external correspondences, as depicted in Figure 138a. To solve 6 + 8 with a double-enumeration strategy, individuals created correspondences between one term of the problem, wrist (6), and its addition to the other term, inner elbow (8). To compose these two numbers, a prototypical solution was to first enumerate the first term, thumb (1) through wrist (6), and then use the enumeration of the second term, thumb (1) through inner elbow (8), to keep track of the addition. In this process, thumb (1) and forearm (7) were paired, index finger (2) and inner elbow (8) were paired, middle finger (3) and biceps (9) were paired, and this process continued until finally inner elbow (8) and nose (14) were paired, yielding the sum, nose (14).

I documented other internal correspondence approaches to serve keeping-track functions that may well have developed out of the double-enumeration approach. For example, in Figure 138b, body substitution provides a more efficient means of solving the problem – users violate a linguistic convention by calling one body part by the name of another in

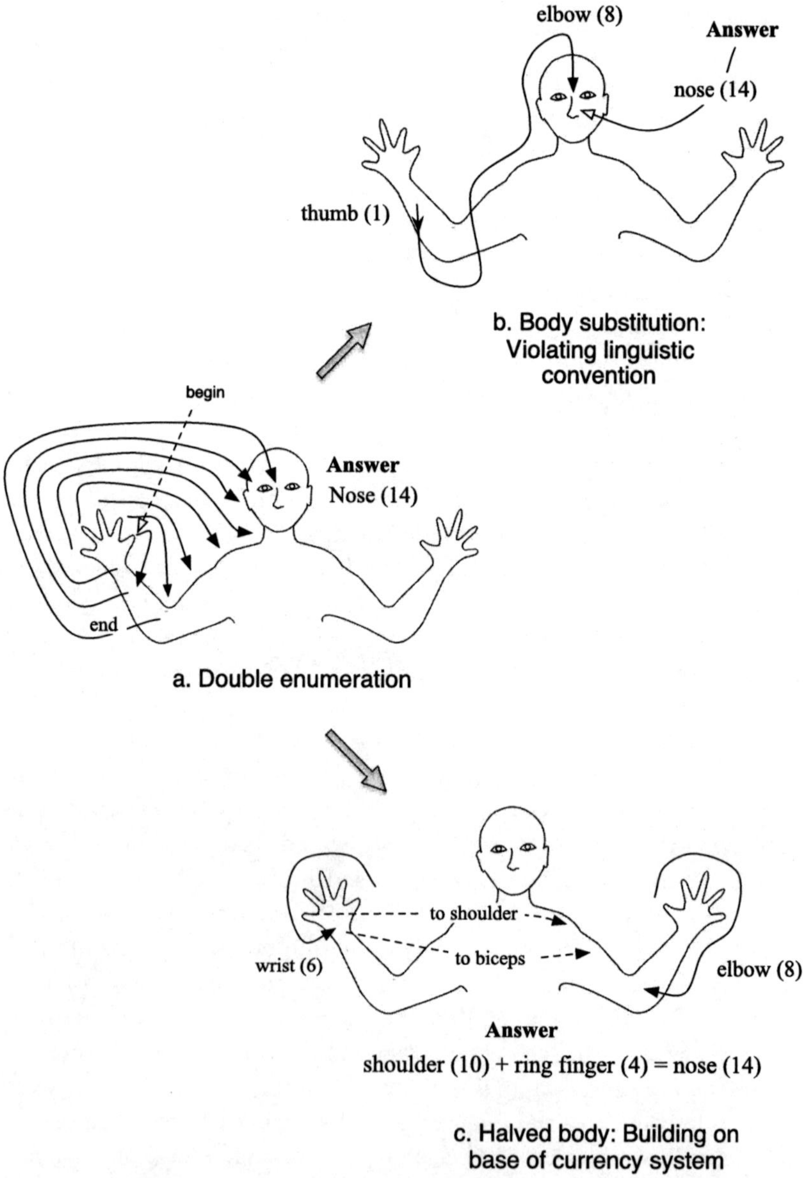

Figure 138. Solving 6 + 8 = ? without coins present: internal correspondences and the specializations in the use of the body form to serve the function of keeping track.

order to coordinate the two series in a strategy analogous to the double-enumeration procedure. A second kind of specialization was the halved-body procedure (Figure 138c). The solver distributes the solution across the two sides of the body. Eight is represented on one side and six on the other, "translating" two of the body parts on the "6" register (the wrist and the pinky) to the "8" register to the positions of the biceps (9) and shoulder (10), creating the representation of shoulder (10) and ring finger (4), or nose (14). Whether the latter two approaches (body substitution and halved body) have a necessary order in ontogenetic development is unclear. Both may be rooted in the correspondences used in the double-enumeration approach.

People's construction of arithmetical operations as linked to participation in collective practices of exchange is an example of form-function shifts in ontogeny, and there are other cases reported throughout the chapters. For example, schooled children also demonstrated a turning inward of external correspondences that served keeping-track functions in arithmetical problem solving (Study 9-2). Ontogenetic shifts were also documented in children's developing use of body parts as a counting system to represent external correspondences in comparison and reproduction tasks in Studies 8-1 and 8-2, as well as in the development of measurement operations in Study 8-3.

The Interplay between Micro-, Socio-, and Ontogenetic Developments

The three genetic strands have some independence, in that they lead to quite separate developments. Microgenesis is the source of meaningful representations in individual activity. Sociogenesis is the source of the reproduction and alteration of collective representations in the social history of communities. Ontogenesis is the source of the construction of new knowledge organizations in the developmental trajectories of individuals. Over the course of the volume, I have argued that these genetic strands are rooted in the same representational activities that are constitutive of collective practices. But their distinct trajectories take shape through a complex interplay with one another, each one taking on direction in relation to one another's products. In this closing section, I discuss the interplay in light of some of the key findings.

Figure 139 contains a schematic of microgenetic, sociogenetic, and ontogenetic developments similar to one presented in Chapter 1 (Figure 4). It depicts the interrelations between the activities of three individuals (I_1,

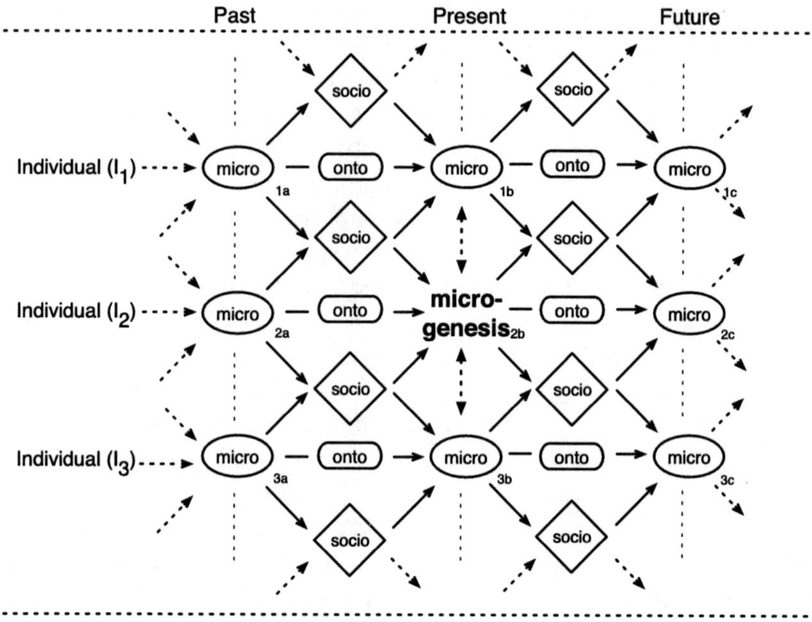

Figure 139. Three individuals interacting in a collective practice: the interplay between micro-, socio-, and ontogenetic strands over time.

I_2, and I_3) over time (past, present, future), and the microgenetic, sociogenetic, and ontogenetic processes through which the construction of form-function relations is produced and altered. To illustrate, I use the production of a reference to the K1 coin, with a focus on the use of *kat-hai* (shoulder (10)).

The bolded microgenetic construction of individual no. 2 that appears in the center of the figure involves the production of *kat-hai* (10) to refer to the K1 coin (microgenesis$_{2b}$). The use of *kat-hai* (10) occurs in a back-and-forth with the other two interlocutors, individual no. 1 (microgenesis$_{1b}$) and individual no. 3 (microgenesis$_{3b}$). The use of *kat-hai* (10) to refer to the K1 coin and the coin's value is rooted in prior sociogenetic reproductions and alterations in the community. But in the present, the actors' microgenetic constructions of *kat-hai* (10) to refer to a K1 coin constitute a reproduction and alteration of the expression – a reproduction of the same expression as well as its alteration in how it is used in a new local context. Thus the microgenetic acts in the present (microgenesis$_{1b,\ 2b,\ 3b}$) are constitutive of sociogenetic reproductions and alterations that can be drawn upon in subsequent microgenetic acts of the future, just as these microgenetic acts are moments in the ontogenetic developments of all three individuals.

The gap between historical treatments of numerical forms and functions in human communities, on the one hand, and developmental treatments of number in human ontogeny, on the other, is remarkable. This gap has left fundamental problems of the origins of numerical representations and ideas unapproachable. I have endeavored to show that filling the gap cannot be achieved by simply putting treatments of culture and cognition side by side, correlating one with the other. Productive approaches in the cognitive sciences will lie in generating units of analysis that reveal intrinsic mutually constitutive relations between cultural and cognitive processes in activity. My efforts have been to explore the origins of representations and ideas with a focus on the reproduction and alteration of collective practices of daily life. To be sure, the analytic perspective on the Oksapmin case elaborated over these pages is skeletal. But I do believe that the body of work presented here shows the utility of the approach as a means of exploring core questions. My hope is that the approach will contribute to better framing of problems of culture-cognition relations, and in so doing, further our understanding of each.

Epilogue

Over a period of 23 years, I conducted studies in the Oksapmin world using a wide range of data sources and empirical techniques; these included archival documents produced by early patrol officers and government records as well as a range of empirical techniques drawn from the social and behavioral sciences. Some of the empirical studies were observational, others were interview studies, and some of the latter involved the elicitation of numerical, arithmetical, and measurement problem solving. The studies made use of varied designs, some correlational, others quasi-experimental, and still others with elements of true experiments (for example, counterbalancing of tasks, using participants as their own controls). The resulting collection of studies and analyses of historical records is extensive, and yet this body of research constitutes a single case of the origins of numerical representations and ideas.

I acknowledge that the scientific merit of case studies is a matter of some dispute. For the critics, the problems relate to generalizability, the sine qua non of scientific inquiry: case studies, regardless of how rich, systematic, and extensive, present findings of a single case sampled from a larger universe of potential cases. Nevertheless, as Robert Yin (2008) argues in *Case Study Research – Design and Methods*, cases have the potential to advance generalizations. In his argument, Yin distinguishes between two types of generalizations – statistical and analytical. Statistical generalization is a revered canon, with its positivistic roots in the social and behavioral sciences. Using experimental, quasi-experimental, and correlational designs, an investigator samples individuals from a defined population (samples of people, classrooms, cultural groups) in order to generalize to that population to test theoretically motivated hypotheses. In contrast, the target of analytical generalization is theory construction, and here the endeavor is to explore the utility of theoretical constructs and, through the analysis of varied data sources associated with a case, to corroborate and refine constructs. In this process, analytical generalization proceeds through the differentiation and elaboration of theoretical constructs. Case studies, guided by the process of analytical generalization, are the method of choice when phenomena and context have fuzzy boundaries, as with my Oksapmin research on culture-cognition relations, in which cultural and cognitive processes are understood to be jointly rooted in human activity.

In my Oksapmin research, I coordinated methods in an effort to support both statistical and analytical generalization in the construction of the case. I began with descriptive ethnography – observing local interactions, talking with people in informal conversations, consulting archival records. From this phase of work came conjectures about the Oksapmin community and then the design of experimental studies and empirical techniques that are standard in the tradition of research seeking statistical generalizations. For instance, I sampled from the population of trade store owners to generalize to that population. With statistical generalization in mind, I designed the studies to mitigate threats to the validity of inferences regarding the larger population of Oksapmin cohorts in order to test and explore theory-motivated hypotheses. At the same time, I was building a case study of culture-cognition relations in a process of analytical generalization. In constructing the case, my objective was to corroborate the utility of the theoretical framework in the Oksapmin world. Further, my expectation that the case would lead to further differentiation and elaboration of theoretical constructs related to collective practices and the microgenesis, sociogenesis, and ontogenesis of form-function relations over historical time.

In the end, the methods used in the Oksapmin research have led to three principal contributions. The first is the descriptive contribution of the work. This volume produced rich description related to a period in the social history of number in Oksapmin communities. In his book *Dying Words*, Nicholas Evans (2010) expresses well the importance of capturing aspects of communities in which representational forms are rapidly vanishing: "Never before in history have languages and the knowledge they hold been disappearing at a faster rate. But equally, never before have we been aware of the dimensions of what is being lost, or had the curiosity, appreciation and technology to document what is still hanging on" (p. 231). This volume on Oksapmin number is responsive to Evans's call. Using the varied resources to which Evans alludes (solar powered laptops, digital media, film, and an enduring interest in the Oksapmin world), I expended a great effort to get facts right and perspectives straight, though I have no doubt there have been wrong turns along the way.

The second contribution is to what might be best characterized as "little theory." The pages of this volume are filled with conjectures and findings about the social history of number in the Oksapmin world. I reported analyses of the social history of word forms like *fu*, *temsi*, or body counting. I posited theoretical models for the interplay between genetic processes that led to continuities and discontinuities in form-function relations, and I evaluated these models in relation to the data I collected. I used similar approaches in studies of form-function relations in arithmetical problem solving in collective practices of schooling and economic exchange. By drawing upon a wide range of investigative techniques, I pursued ideas that led to assertions about processes of reproduction and alteration of representational forms and functions in collective practices specific to Oksapmin life.

The third contribution is to "big theory," a conceptual treatment of culture-cognition relations with particular regard for the origins of representations and ideas. I took the Oksapmin world as a case in which shifting collective practices involving number provided a remarkable window into the interplay between microgenetic, sociogenetic, and ontogenetic processes and the shifting collective practices of everyday life. Though my focus has been on number in Oksapmin communities, I believe that the broad contours of this approach – one in which intrinsic relations between cognitive developmental and sociohistorical processes are the nexus of productive analyses – can be useful to investigators whose concerns lie in other communities and with other knowledge domains.

Appendix A

The Five Oksapmin Conversational Number Words

The body system is not the only way of expressing number in the Oksapmin language. Oksapmin use a number word system that is limited to five values. These number words have a cyclical structure organized by a modified two-cycle structure (Lean, 1992) shown in Figure 140. In the figure, dialect differences and suffix variants are shown in braces.

English	Oksapmin word form (and dialect variants)
One	*tit {fitil, fitir, fitle}*
Two	*yot (-a), {ot (-a), otlip (-a)}*
Three	*yetir, (orup (-a), olup (-a)}*
Four	*yota yota, (ota ota)*
Five	*hanen, {hatlen}*

Figure 140. The five "conversational" numbers (not used for counting).

Appendix B

Currency Tokens, Images, and Physical Descriptions

Figure 141 and Figure 142 contain images of coins and notes that were in circulation during the three waves of research. The 1t, 2t, and 5t coins were rarely found in the Oksapmin communities during any period of fieldwork, though in 2001 the coins and their values were discussed in community schools.

Release date: *1975*; Color: *Brown*;
Dimensions: *17.65 mm (diameter)*

1 toea

Birdwing Butterfly — Crest, PNG

Release date: *1975*; Color: *Brown*;
Dimensions: *21.72 mm (diameter)*

2 toea

Butterfly Cod — Crest, PNG

Release date: *1975*; Color: *Silver*;
Dimensions: *19.53 mm (diameter)*

5 toea

Plateless turtle — Crest, PNG

Release date: *1975*; Color: *Silver*;
Dimensions: *23.72 mm (diameter)*

10 toea

CusCus — Crest, PNG

Release date: *1975*; Color: *Silver*;
Dimensions: *28.65 mm (diameter)*

20 toea

Cassowary — Crest, PNG

Release date: *1980*; Color: *Silver*;
Dimensions: *30mm (diameter)*

50 toea

Logo, SP Festival of Arts — Crest, PNG

Release date: *1975*; Color: *Silver*;
Dimensions: *33.30 mm (diameter)*

1 kina

Sea & river crocodiles — Logo, SP Festival of Arts

Figure 141. Papua New Guinea coins used in the 1978, 1980, and/or 2001 study.

Release date: *1975*; Color: *Dark green and black design elements on light green*

2 Kina

Release date: *1975*; Color: *Dark violet design elements on light multicolored underprints*

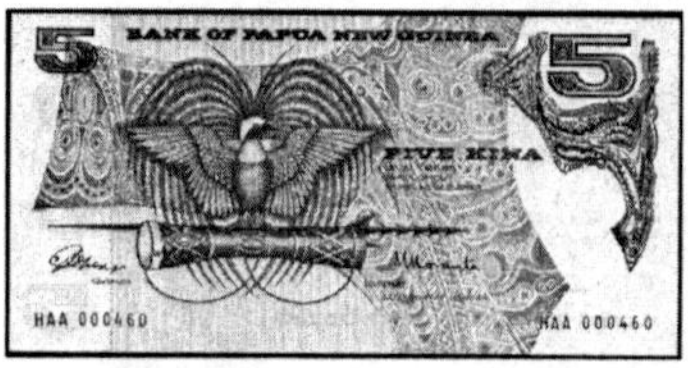

5 Kina

Release date: *1975*; Color: *Blue and multicolor*

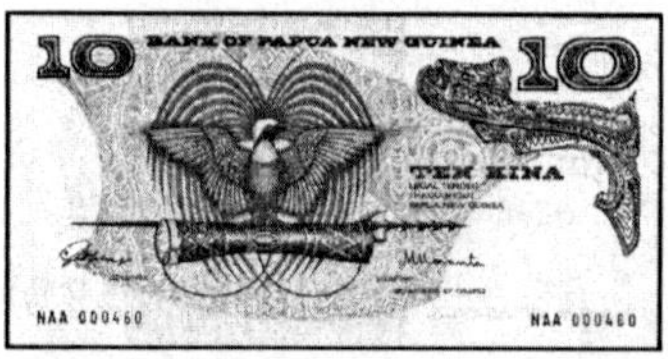

10 Kina

Release date: *1977*; Color: *Red and multcolor*

20 Kina

Figure 142. Papua New Guinea notes used in the 1978, 1980, and/or 2001 study. *Note*: Images of notes were obtained from http://numismondo.com/pm/png/index100.htm. Information on the reverse and obverse sides of tokens was obtained from the Bank of Papua New Guinea: http://www.bankpng.gov.pg/index.php?option=com_content&task=blogcategory&id=0&Itemid=113.

Appendix C

Notes on Age Determination, Back-Translation, and Interview Settings

Age Group Classifications

The following procedure was used to categorize individuals into one of several cohorts for the 2001 studies. Indigo and I identified two well-known individuals in the community, one whom we regarded as elderly and the other as middle-aged. We then asked participants whether they were closer in age to one or the other.

Translation of Interviews

All interviews were conducted in Tok Pisin and Oksapmin. Translation of interview materials from English into Tok Pisin and Oksapmin was an iterative process that took place during and after pilot interviews. I first wrote the formal interview procedures in English. Next, Indigo and I worked with local teachers and our research assistants to translate these materials into Tok Pisin and Oksapmin. Typically, one of us worked with a research assistant (who spoke English, Tok Pisin, and Oksapmin) to translate a segment of the interview from English into one of the other languages. After the initial translation was complete, other research assistants and local teachers were asked to evaluate the accuracy of the translation. This process involved discussion of the aims of the English-language questions and discussion of the goals and principles of Western research in general.

In the interviews, the written translations in Oksapmin were used as a rough guide, but translators didn't adhere strictly to the translated material. Our translators and local informants felt that reading exact translations in Oksapmin was awkward for two reasons. First, Oksapmin is largely an oral language. Community members, including our translators, have extremely limited experience in writing and reading in Oksapmin. Second, speech events in which an interviewee reads a set question for an interviewer to answer are not practiced within the traditional Oksapmin community (though such knowledge-display questions have become a practice of schooling).

Location of and Privacy of Interviews

When interviews were conducted in public places (sometimes we used a private house on the grounds of Bak or Divanap schools), on occasion Oksapmin people came close to observe. We asked them to step back so that they could not overhear the interview, with varying degrees of effectiveness. Another related problem was that sometimes participants appeared nervous.

Appendix D

Record Form Used in Trade Store Diary Study (Sample Page)

Figure 143 contains an example of a record sheet that trade store clerks made use of in the records of transaction study. For every transaction, the clerk would fill out or circle the appropriate cell of the table, with each row representing a single transaction.

Many of the columns contain Tok Pisin headers. English translations are as follows. *Man or meri*: man or woman; *Krismas*: age match of customer with individuals of known ages; *Skul*: amount of schooling; *Tok usim*: language used; *Senis note*: whether the customer asked for change; *Amas samting*: number of items requested in a transaction; *Kostama paid*: amount of money the customer produced for the transaction; *Em i kostam hamas?*: the cost of the item.

Time	Man or Meri	Kostama	Krismas	Skul	Tok usim kaudim	Senis note?	Amas samting	Kostama paid	Em i kostim hamas?
AM PM	Man Meri	________	Sel Dikitet Henry Sul age Pikinini	No skul Grade 1-5 Grade 6+	Tok Ples Tok Pisin Both	Yes No	0 1 2 3 4 5 6 moa		
AM PM	Man Meri	________	Sel Dikitet Henry Sul age Pikinini	No skul Grade 1-5 Grade 6+	Tok Ples Tok Pisin Both	Yes No	0 1 2 3 4 5 6 moa		
AM PM	Man Meri	________	Sel Dikitet Henry Sul age Pikinini	No skul Grade 1-5 Grade 6+	Tok Ples Tok Pisin Both	Yes No	0 1 2 3 4 5 6 moa		
AM PM	Man Meri	________	Sel Dikitet Henry Sul age Pikinini	No skul Grade 1-5 Grade 6+	Tok Ples Tok Pisin Both	Yes No	0 1 2 3 4 5 6 moa		
AM PM	Man Meri	________	Sel Dikitet Henry Sul age Pikinini	No skul Grade 1-5 Grade 6+	Tok Ples Tok Pisin Both	Yes No	0 1 2 3 4 5 6 moa		
AM PM	Man Meri	________	Sel Dikitet Henry Sul age Pikinini	No skul Grade 1-5 Grade 6+	Tok Ples Tok Pisin Both	Yes No	0 1 2 3 4 5 6 moa		
AM PM	Man Meri	________	Sel Dikitet Henry Sul age Pikinini	No skul Grade 1-5 Grade 6+	Tok Ples Tok Pisin Both	Yes No	0 1 2 3 4 5 6 moa		
AM PM	Man Meri	________	Sel Dikitet Henry Sul age Pikinini	No skul Grade 1-5 Grade 6+	Tok Ples Tok Pisin Both	Yes No	0 1 2 3 4 5 6 moa		

Figure 143. Diary record form used by trade store clerks.

Appendix E

Auxiliary Words Used in Quantification

A wide range of lexical forms are used in quantitative expressions in people's daily talk and in the context of interviews conducted over the course of fieldwork. Figure 144 contains some of the expressions used, along with their English translations.

Word form	English translation
fok (wi), pok (wi)	only {not an independent word – only used with number}
ham (-a)	and {and this one] or and this one, after that
ham	and
hatforhan	and or and then or after that
iré	adding on, theirs
kut, kat	half of an object (when applied to money); reports by elementary school teachers that *kut* refers to fourths and that *tan* refers to "half"
moh (-a)	this one, close that is object of pointing
muk {lower Bak}, *moca* {upper Bak}	a group or a heap
mukmuk	two or three heaps
oh (-a) {used in Takap, Divanap} *yoh (-a)* {used in Bak}	that one, more distant that may not be object of pointing
ohepe {*ohohep(-e)*}	adding on {adding to the end of that}
(-si)	and or add, as in *humrip-si* {and a third one} *yotit-si* {two more}
tan (-a)	elementary school teachers report that *tan(a)* means half, as in other half of the body when counting beyond the nose (e.g., *tan-besa*) – body split into two halves; other meanings reported include flat, like a valley floor (or a coin)
tit fok wi	only one
timbas	no, nothing, without
wasa	going down (sometimes used with trajectory of body count)
wuris, rusa	going up (sometimes used with trajectory of body count)
yeet, eet	and or again

Figure 144. Auxiliary words used in quantitative expressions in the Oksapmin language.

Appendix F

Tit Fu Interview Used with Adults

Figure 145 contains the complete interview protocol for probing varied aspects of *fu* or *tit fu* among different Oksapmin cohorts. The figure contains English, Tok Pisin, and Oksapmin translations.

ENGLISH	TOK PISIN	OKSAPMIN
Where is "tit fu" if you count in Oksapmin?	*Supos you kaundim long bodi bilong yu. Inap yu soim mi "tit fu" i stap we?*	*Tit fu ob diradire mengti saya hen dati pat?*
What does "fu" mean?	*Mining bilong fu em wanem?*	*Fu kana?*
Show me how much money is "tit fu."	*Soim mi hamas mani "tit fu" i stap.*	*Nimgitina kinhe mani tit fu oh.*
	{MONEY PRESENTED}	
Is there another coin or note that can be called "tit fu"?	*I gat narapela we long kolim "tit fu" long coins o notes?*	*Tit fu ma porishe oh kwikor o iratapin ohe ti kisdis henda?*
Can "tit fu" refer to a number of sweet potatoes?	*Inap "tit fu" makim namba bilong kau kau?*	*Tit fu oh tuhana nong he kismatin ina da.*
If yes: How many?	If yes: *Hamas pela?*	If yes: *Kinhae?*
Have you ever heard of the expression, "tit fu gona"?	*Yu bin harim mining bilong "tit fu gona"?*	*Go tit fu gona poriptin he ti amratip da.*
Is "tit fu gona" the same as "tit fu"?	*Mining bilong kaundim "tit fu gona" em wankain o em i narapela kain we long kaundim "tit fu"?*	*Tit fu gona sapti o topta yota da dokwah tit fu ohsi?*
IF SO FAR KIR-HAI (20) IS THE ONLY EXPRESSION FOR "TIT FU"		
Show me the number that is one more than "tan kir (20)".	*Soim mi dispela namba wan moa long "tan kir (20)'.*	*Go nimgitin sat prusa tan-kir oh mandama hotdohnung ma dip tioh.*
IF COUNTS TO TAN-HATATAH (27) OR USED TAN-HATATAH (27) FOR "TIT FU"		
Can you count further than that? (beyond 27) (If yes:) Show me.	*Inap yu kaundim i go moa long dispela?* (if yes:) *Soim mi.*	*Go sat hondoh poson hetin da?* (If yes:) *Nomgitin.*
IF STILL HAVE NOT USED EITHER TAN KIR (20) OR TAN HATATAH (27) FOR "TIT FU"		
Do you know any other places on the body where people stop and say "tit fu" or "tit fu gona."	*Yu save long narapela ples long bodi we planti manmeri save tok "tit fu" o "tit fu gona"?*	*Go amda kuhan ire dira dire bok nung sat posos kimiripti ti fu gona o tit fu ma porishe oh.*
Someone told me that "tit fu" means__ {which ever was not cited previously: tan-hatahta (27) or "tan-kir (20)}."	*Wanpela meri tokim mi olsem "tit fu" em*___. {which ever was not cited previously: *tan-hatahta* (27) or "*tan-kir* (20)}.	*Tit han tit nohnung nopor oh tit fu ma poriptioh* ***tan-kir*** *mur.* OR *Tit han tit nohnung nopor oh tit fu ma poriptioh* ***tan-hatatah*** *mur.*
Do you think that's right or wrong? (what do you think? True or false?)	*Yu ting wanem, giaman o tru?*	*Go dahah oh arpin da do dasup.*
Do you think Oksapmin counting has changed or stayed the same since times of ancestors.	*Yu save olsem wanem tok ples kaundim em i senis o wankain long taim bilong tambuna?*	*Go kanhan dahoh dira dire meng ti ma sashe oh awher ihir dupurgetpa da do poptesbe.*
How has it changed?	*Em senis olsem wanem?*	*Kinhat meng dupurgetpa.*
Is there any way to count 50 in Oksapmin?	*I gat we long kaundim fiftipela long tok ples?*	*Dira dire mengti sat prus "fiftipela" he porishe da.*
Is there any way to count 60 in Oksapmin?	*sixtipela?*	*Dira dire mengti sat prus "sixtipela" he porishe da.*
Is there any way to count 70 in Oksapmin?	*seventipela?*	*Dira dire mengti sat prus "seventipela" he porishe da.*
Did our ancestors say 50, 60, or 70 in the same way?	*Bipo tru, ol tambuna i save kolim fiftipela, sixtipela, o seventipela long wankain we?*	*Rih ma awher patesbe ihir "fiftipela o sixtipela o seventipela" por he sashe da dira dire meng ti oh.*

Figure 145. *Tit fu* interview written in English, Tok Pisin, and Oksapmin translations.

References

Agha, A. (2003). The social life of cultural value. *Language & Communication*, *23*, 231–273.

Antell, S. E., & Keating, D. P. (1983). Perception of numerical invariance in neonates. *Child Development*, *54*(3), 695–701.

Ascher, M., & Ascher, R. (1981). *Code of the Quipu: A study in media, mathematics, and culture*. Ann Arbor: University of Michigan Press.

Barth, F. (1987). *Cosmologies in the making*. Cambridge: Cambridge University Press.

Bateson, G. (1972). *Steps to an ecology of mind*. New York: Balantine Books.

Berger, P. L., & Luckmann, T. (1966). *The social construction of reality: A treatise in the sociology of knowledge*. Garden City, NY: Anchor Books.

Bever, T. G., Mehler, J., & Epstein, J. (1968). What children do in spite of what they know. *Science*, *162*(3856), 921–925.

Boesch, C., & Tomasello, M. (1998). Chimpanzee and human culture. *Current Anthropology*, *38*, 591–604.

Boram, C. (1980). *Oksapmin children*. New Haven, CT: Human Relations Area Files.

Bourdieu, P. (1977). *Outline of a theory of practice*. London: Cambridge University Press.

Bourke, R. M., Allen, B. J., Hide, R. L., Fritsch, D., Grau, R., Lowes, E., Nen, T., Nirsie, E., Risimeri, J., & Woruba, M. (2002). *West Sepik province: Text summaries, maps, code lists and village identification*. Canberra: Australian National University, Research School of Pacific and Asian Studies.

Brenner, M. E. (1998). Meaning and money. *Educational Studies in Mathematics*, *36*(2), 123–155.

Brongersma, L. D., & Venema, G. F. (1962). *To the mountain of the stars* (A. G. Readett, Trans.). London: Hodder & Stoughton.

Brumbaugh, R. (1990). Afek sang: The 'old woman' myth of the Mountain-Ok. In B. Craig & D. Hyndman (Eds.), *Children of Afek: Tradition and Change among the Mountain Ok of Central New Guinea*. (Vol. 40, pp. 54–87). Sydney: University of Sydney.

Brumbaugh, R. (2010). E-mail exchange on use of body counting in Telefol, 1976.

Bruner, J. S. (1966). On cognitive growth. In J. S. Bruner, R. P. Olver, & P. M. Greenfield (Eds.), *Studies in cognitive growth* (pp. 30–67). New York: Wiley.
Bruner, J. S., Olver, R. R., & Greenfield, P. M. (1966). *Studies in cognitive growth*. New York: Wiley.
Brutti, L. (2000). Afek's last son: Integrating change in a Papua New Guinea cosmology. *Ethnohistory*, *47*(1), 101–111.
Butterworth, B. (1999). *The mathematical brain*. London: Macmillan.
Carpenter, T. P., & Moser, J. M. (1982). The development of addition and subtraction problem-solving skills. In T. P. Carpenter, J. M. Moser, & T. A. Romberg (Eds.), *Addition and subtraction: A cognitive perspective* (pp. 9–24). Hillsdale, NJ: Erlbaum.
Carpenter, T. P., Moser, J. M., & Bebout, H. C. (1988). Representation of addition and subtraction word problems. *Journal for Research in Mathematics Education*, *19*(4), 345–357.
Chochon, F., Cohen, L., van de Moortele, P. F., & Dehaene, S. (1999). Differential contributions of the left and right inferior parietal lobules to number processing. *Journal of Cognitive Neuroscience*, *11*(6), 617–630.
Chomsky, N. (1967). Recent contributions to the theory of innate ideas. *Synthese*, *17*, 2–11.
Chomsky, N. (1980). *Rules and representations*. Oxford: Basil Blackwell.
Clark, H. H. (1996). *Using language*. Cambridge: Cambridge University Press.
Clark, H. H. (2006). Social actions, social commitments. In N. J. Enfield & S. C. Levinson (Eds.), *Roots of human sociality: Culture, cognition and interaction* (pp. 126–150). New York: Berg.
Closs, M. P. (1986a). *Native American mathematics*. Austin: University of Texas Press.
Closs, M. P. (1986b). The mathematical notation of the ancient Maya. In M. P. Closs (Ed.), *Native American mathematics* (pp. 291–369). Austin: University of Texas Press.
Cohen, L., Dehaene, S., Chochon, F., Lehéricy, S., & Naccache, L. (2000). Language and calculation within the parietal lobe: A combined cognitive, anatomical and fMRI study. *Neuropsychologia*, *38*(10), 1426–1440.
Cole, M. (1996). *Cultural psychology: A once and future discipline?* Cambridge, MA: Harvard University Press.
Cole, M., Gay, J., Glick, J. A., & Sharp, D. W. (1971). *The cultural context of learning and thinking*. New York: Basic Books.
Cole, M., & Griffin, P. (1980). Cultural amplifiers reconsidered. In D. R. Olson (Ed.), *The social foundations of language and thought: Essays in honor of Jerome S. Bruner* (pp. 343–364). New York: Norton.
Cole, M., & Hatano, G. (2007). Cultural-historical activity theory: Integrating phylogeny, cultural history, and ontogenesis in cultural psychology. In S. Kitayama & D. Cohen (Eds.), *Handbook of Cultural Psychology* (pp. 109–135). New York: Guilford Press.
Cole, M., Levitin, K., & Luria, A. R. (2006). *The autobiography of Alexander Luria: A dialogue with the making of mind*. Mahwah, NJ: Erlbaum.
Cole, M., & Scribner, S. (1974). *Culture and thought: A psychological introduction*. New York: John Wiley & Sons, Inc.

Craig, B. (1988). *Art and decoration of Central New Guinea*. Haverfordwest: Shire Publications Ltd.

Craig, B., & Hyndman, D. (Eds.). (1990). *Children of Afek: Tradition and change among the Mountain Ok of Central New Guinea* (Vol. 40). Sydney: University of Sydney.

Cranstone, B. (1990). The Tifalmin and their environment: Material culture of a Mountain Ok people. In B. Craig & D. Hyndman (Eds.), *Children of Afek: Tradition and change among the Mountain Ok of Central New Guinea*. (Vol. 40, pp. 35–53). Sydney: University of Sydney.

Croft, W. (2000). *Explaining language change: an evolutionary approach*. Harlow, England; New York: Longman.

Croft, W. (2008). Evolutionary linguistics. *Annual Review of Anthropology*, *37*, 219–234.

Crump, T. (1990). *The anthropology of numbers*. New York: Cambridge University Press.

D'Andrade, R. (1990). Some propositions about the relationship between culture and human cognition. In J. W. Stigler, R. A. Sweder, & G. Herdt (Eds.), *Cultural psychology: Essays on comparative human development*. New York: Cambridge University Press.

Dasen, P. R. (1972). Cross-cultural Piagetian research: A summary. *Journal of Cross-Cultural Psychology*, *3*(1), 23–40.

Dasen, P. R. (1975). Concrete operational development in three cultures. *Journal of Cross-Cultural Psychology*, *6*(2), 156–172.

Dasen, P. R. (1977). *Piagetian psychology: Cross-cultural contributions*. New York: Gardner Press, 379.

Davies, B., & Harré, R. (1990). Positioning: The discursive production. *Journal for the Theory of Social Behaviour*, *20*(1), 43–63.

Dawkins, R. (1976). *The selfish gene* (2nd ed.). New York: Oxford University Press.

Dehaene, S. (2003). The neural basis of the Weber-Fechner law: A logarithmic mental number line. *Trends in Cognitive Sciences*, 7(4), 145–147.

Dehaene, S., Piazza, M., Pinel, P., & Cohen, L. (2003). Three parietal circuits for number processing. *Cognitive Neuropsychology*. Special Issue: *The organisation of conceptual knowledge in the brain: Neuropsychological and neuroimaging perspectives*, *20*(3–6), 487–506.

Dong, T., Anderson, R. C., Kim, I., & Li, Y. (2008). Collaborative reasoning in China and Korea. *Reading Research Quarterly*, *43*(4), 400–424.

Education, PNG Dept. of. (2003). *National curriculum statement for Papua New Guinea*. Port Moresby: Department of Education.

Engestrom, Y. (1987). *Learning by expanding*. Helsinki: Orienta-Konsultit Oy.

Esmonde, I. (2006). *"How are we supposed to, like, learn it, if none of us know?": Opportunities to learn and equity in mathematics cooperative learning structures*. Unpublished dissertation, University of California, Berkeley.

Evans, N. (2010). *Dying words: Endangered languages and what they have to tell us*. West Sussex: Wiley-Blackwell.

Everett, D. L. (2005). Cultural constraints on grammar and cognition in Pirahã: Another look at the design features of human language. *Current Anthropology*, *46*(4), 621–646.

Everett, D. L. (2007). Challenging Chomskyan linguistics: The case of Pirahã. *Human Development*, *50*(6), 297–299.

Frank, M. C., Everett, D. L., Fedorenko, E., & Gibson, E. (2008). Number as a cognitive technology: Evidence from Pirahã language and cognition. *Cognition, 108*(3), 819–824.

Fried, M. (1975). *The notion of tribe*. Menlo Park, CA: Cummings.

Fuson, K. (1982). An analysis of the counting on solution procedure in addition. In T. P. Carpenter, J. M. Moser, & T. A. Romberg (Eds.), *Addition and subtraction: A cognitive perspective* (pp. 67–82). Hillsdale, NJ: Erlbaum.

Fuson, K. C. (1988). *Children's counting and concepts of number*. New York: Springer-Verlag.

Gallistel, C. R. (1992). *Animal cognition* (1st MIT Press ed.). Cambridge, MA: MIT Press.

Gallistel, C. R., & Gelman, R. (2005). Mathematical cognition. In K. Holyoak & S. Morrison (Eds.), *The Cambridge handbook of thinking and reasoning* (pp. 559–588). Cambridge: Cambridge University Press.

Gammage, B. (1998). *The sky travellers: Journeys in New Guinea 1938–1939*. Victoria: Melbourne University Press.

Garfinkel, H. (1967). *Studies in ethnomethodology: Some essential features of common understandings*. Englewood Cliffs, NJ: Prentice-Hall.

Gearhart, M. (1980). *Divanap Community School field report*.

Gelman, R., & Cordes, S. (Eds.). (2001). *Counting in animals and humans*. Cambridge, MA: MIT Press.

Gelman, R., & Gallistel, C. R. (1978). *The child's understanding of number*. Cambridge, MA: Harvard University Press.

Giddens, A. (1986). *Constitution of society: Outline of the theory of structuration*. Berkeley: University of California Press.

Goodwin, C., & Duranti, A. (1992). Rethinking context: An introduction. In A. Duranti & C. Goodwin (Eds.), *Rethinking context: Language as an interactive phenomenon* (pp. 1–42). New York: Cambridge University Press.

Gordon, P. (2004). Numerical cognition without words: Evidence from Amazonia. *Science Express*. Special Issue, *Cognition and Behavior, 306*(5695), 496–499.

Gould, S. J. (1996). Creating the creators. *Discover, 17*(10), 42–54.

Greenfield, P., Maynard, A., & Childs, C. (1997). History, culture, learning, and development. *Biennial Meeting of the Society for Research in Child Development*.

Greenfield, P. M. (1993). Historical change and cognitive change: A two-decade follow-up study in Zinacantan, a Mayan community of southern Mexico. *Biennial Meeting of the Society for Research in Child Development*.

Greenfield, P. M. (2004). *Weaving generations together: Evolving creativity in the Maya of Chiapas* (1st ed.). Santa Fe, NM: School of American Research Press.

Grice, H. P. (1989). *Studies in the way of words*. Cambridge, MA: Harvard University Press.

Guberman, S. R. (1996). The development of everyday mathematics in Brazilian children with limited formal education. *Child Development, 67*(4), 1609–1623.

Guberman, S. R., & Saxe, G. B. (2000). Mathematical problems and goals in children's play of an educational game. *Mind, Culture, & Activity*, 7(3), 201–216.

Guilford, V. (1994). *Oksapmin trade stores*. Unpublished manuscript, La Jolla, CA.

Gutierrez, K. D., Baquedano-Lopez, P., & Tejeda, C. (1999). Rethinking diversity: Hybridity and hybrid language practices in the third space. *Mind, Culture, and Activity, 6*(4), 286–303.

Hallpike, C. R. (1979). *The foundations of primitive thought*. New York: Oxford University Press.

Harper, D. A. (2010). Numbers as a cognitive and social technology: On the nature of conventional number sequences used in economic systems. *Journal of Institutional Economics*, *6*(2), 167–190.

Harré, R., Moghaddam, F. M., Cairnie, T. P., Rothbart, D., & Sabat, S. R. (2009). Recent advances in positioning theory. *Theory & Psychology*, *19*(5), 5–31.

Hauser, M. D., & Carey, S. (2003). Spontaneous representations of small numbers of objects by rhesus macaques: Examinations of content and format. *Cognitive Psychology*, *47*(4), 367–401.

Hawkins, J., & Pea, R. D. (1987). Tools for bridging the cultures of everyday and scientific thinking. *Journal of Research in Science Teaching*, *24*(4), 291–307.

Healey, A. (1964). *A survey of the Ok family of languages*. Unpublished manuscript, Canberra.

Healey, P. M. (1965). *Telefol noun phrases* (Vol. 4). Canberra: Australian National University.

Hicks, J. S. (1962a). Patrol report no. 2 of 1962/1963–Oksapmin. *National archives & public records services of Papua New Guinea*.

Hicks, J. S. (1962b). Patrol report no. 3 of 1962/1963–Oksapmin. *National archives & public records services of Papua New Guinea*.

Hurford, J. R. (1987). *Language and number: The emergence of a cognitive system*. Oxford: Blackwell.

Hutchins, E. (1995). *Cognition in the wild*. Cambridge, MA: MIT Press.

Hutchins, E. (2006). The distributed cognition perspective on human interaction. In N. J. Enfield (Ed.), *Stephen C. Levinson* (pp. 375–398). New York: Berg.

Hyndman, D. (1990). The Ok Tedi mining project: Problems with the pot of gold. In B. Craig & D. Hyndman (Eds.), *Children of Afek: Tradition and change among the Mountain-Ok of central New Guinea*. (Vol. 40, pp. 161–168). Sydney: University of Sydney.

Hyndman, D., & Morren, G. (1990). The human ecology of the Mountain-Ok of central New Guinea: A regional and inter-regional approach. In B. Craig & D. Hyndman (Eds.), *Children of Afek: Tradition and change among the Mountain-Ok of central New Guinea*. (Vol. 40, pp. 9–26). Sydney: University of Sydney.

Keller, R. (1994). *On language change: The invisible hand in language*. London; New York: Routledge.

Klein, A., & Starkey, P. (1988). Universals in the development of early arithmetic cognition. *New Directions for Child Development No. 41*, 5–26.

Kroeber, A. L., & Kluckhohn, C. (1952). *Culture: A critical review of concepts and definitions*. New York: Vintage Books.

Lagrange, J. B., & Erdogan, E. O. (2009). Teachers' emergent goals in spreadsheet-based lessons: Analyzing the complexity of technology integration. *Educational Studies in Mathematics*, *71*, 65–84.

Lakoff, G., & Nuñez, R. E. (2000). *Where mathematics comes from*. New York: Basic Books.

Lancy, D. (1983). *Cross-cultural studies in cognition and mathematics*. New York: Academic Press.

Langer, J. (1969a). Disequilibrium as a source of development. In P. H. Mussen, J. Langer, & M. Covington (Eds.), *Trends and issues in developmental psychology*. New York: Holt, Rinehart, & Winston.

Langer, J. (1969b). *Theories of development*. New York: Holt, Rinehart, & Winston.
Langer, J. (1974). Interactional aspects of cognitive organization. *Cognition*, *3*(1), 9–26.
Langer, J. (1986). *The origins of logic: One to two years*. Orlando, FL: Academic Press.
Langer-Osuna, J. (2010). *The co-construction of identity and engagement in collaborative tasks*. Paper presented at the Research Presession of the National Council of Teachers of Mathematics.
Lave, J., & Wenger, E. (1991). *Situated learning*. New York: Cambridge University Press.
Lawrence, M. (2003). Personal communication. Statement from Mission Aviation Aircraft, which runs vegetable exports out of Oksapmin area.
Lawrence, M. (2010). *The world at my door*. Belleville, Ontario: Guardian Books.
Lawrence, M., & Snyder, D. M. (1993). *Oksapmin dictionary*. Ukarumapa, Papua New Guinea: Summer Institute of Linguistics.
Laycock, D. (1975). Observations on number systems and semantics. *New Guinea Area Languages and Language Study*, *1* (Pacific Linguistics, series C, no. 38), 219–233.
Leach, E. (1954). *Political systems of highland Burma: A study of Kachin social structure*. Cambridge, MA: Harvard University Press.
Lean, G. A. (1992). *Counting systems of Papua New Guinea and Oceania*. Unpublished dissertation, Papua New Guinea University of Technology, Lae.
Leontiev, A. N. (1981). The problem of activity in psychology. In J. V. Wertsch (Ed.), *The concept of activity in Soviet psychology* (pp. 37–71). New York: Sharpe.
Levy-Bruhl, L. (1979). *How natives think*. New York: Arno Press.
Lewis, D. K. (1969). *Convention: A philosophical study*. Cambridge, MA: Harvard University Press.
Lin, T., Anderson, R. C., Hummel, J. E., Jadallah, M., Miller, B. W., Nguyen-Jahiel, K., et al. (2010). *Children's use of analogy during collaborative reasoning*. Unpublished manuscript, Champaign, IL.
Litteral, R. (2000). Four decades of language policy in Papua New Guinea: The move towards the vernacular. *Radical Pedagogy*, *2*(2).
Lohmann, R. (2004). Personal communication about fieldnotes with the Asabano: Duranmin, April 8, 1994.
Loughnane, R. (2009). *A grammar of Oksapmin*. Unpublished manuscript, Melbourne.
Loughnane, R., & Fedden, S. (2006). *The Ok-Oksapmin language family*. Paper presented at the School of Languages & Linguistics Postgraduate Conference.
Luria, A. R. (1973). *The working brain: An introduction to neuropsychology*. New York: Basic Books.
Luria, A. R. (1976). *Cognitive development: Its cultural and social foundations*. Cambridge, MA: Harvard University Press.
Maynard, A. E. (2009). Context and structure in social interaction and cognitive development. *Human Development*, *52*(5), 313–319.
Maynard, A. E., & Greenfield, P. M. (2006). *Cultural teaching and learning: Processes, effects, and development of apprenticeship skills*. Bekerman, Zvi.
Maynard, A. E., Greenfield, P. M., & Childs, C. (1999). Culture, history, biology, and body: Native and non-native acquisition of technological skill. *Ethos*, *27*(3), 379–402.

McComb, K., Packer, C., & Pusey, A. (1994). Roaring and numerical assessment in contests between groups of female lions, Panthera leo. *Animal Behaviour*, *47*(2), 379–387.

Mehler, J., & Bever, T. (1967). Cognitive capacity of very young children. *Science*, *158*(3797), 141–142.

Menninger, K. (1969). *Number words and number symbols: A cultural history of numbers*. Cambridge, MA: MIT Press.

Milroy, L. (1987). *Language and social networks* (2nd edition). Oxford: Blackwell.

Mimica, J. (1988). *Intimations of infinity: The cultural meanings of the Iqwaye counting and number system*. Oxford: BERG.

Moylan, T. (1981). Variation, change, and lexical diffusion in Oksapmin [microform]. S.l.: s.n.

Moylan, T. (1982). *The Oksapmin counting system*. Paper presented at the 1982 Meeting of the Northeastern Anthropological Association.

Nardi, B. A. (1996). Activity theory and human-computer interaction. In B. A. Nardi (Ed.), *Context and consciousness: Activity theory and human-computer interaction* (pp. 7–16). Cambridge, MA: MIT Press.

Nasir, N. S. (2000a). "All money ain't good money": The development of emergent goals and strategic approaches in the game of dominoes. *Dissertation Abstracts International Section A: Humanities and Social Sciences*, *61*(6-A), 2186.

Nasir, N. S. (2000b). "Points ain't everything": Emergent goals and average and percent understandings in the play of basketball among African American students. *Anthropology & Education Quarterly*, *31*(3), 283–305.

Nasir, N. S. (2008). Everyday pedagogy: Lessons from basketball, track, and dominoes. *Phi Delta Kappan*, *89*(7), 529–532.

Nasir, N. S., & Saxe, G. B. (2003). Ethnic and academic identities: A cultural practice perspective on emerging tensions and their management in the lives of minority students. *Educational Researcher*, *32*(5), 14–18.

Nelson, K. (1981). Cognition in a script framework. In J. H. Flavell & L. Ross (Eds.), *Social cognitive development*. Cambridge: Cambridge University Press.

Nelson, K. (1986). *Event knowledge*. Hillsdale, NJ: Erlbaum.

Nieder, A., Diester, I., & Tudusciuc, O. (2006). Temporal and spatial enumeration processes in the primate parietal cortex. *Science*, *313*(5792), 1431–1435.

Nieder, A., & Merten, K. (2007). A labeled-line code for small and large numerosities in the monkey prefrontal cortex. *Journal of Neuroscience*, *27*(22), 5986–5993.

Nissen, H. J., Damerow, P., & Englund, R. K. (1993). *Archaic bookkeeping: Early writing and techniques of the economic administration in the ancient Near East*. Chicago: University of Chicago Press.

Nunes, T., Schliemann, A. D., & Carraher, D. W. (1993). *Street mathematics and school mathematics*. Cambridge: Cambridge University Press.

Olson, D. R. (1976). Culture, technology, and intellect. In L. B. Resnick (Ed.), *The nature of intelligence* (pp. 189–202). Hillsdale, NJ: Erlbaum.

Pea, R. D. (1985). Beyond amplification: Using the computer to reorganize mental functioning. *Educational Psychologist*, *20*(4), 167–182.

Pea, R. D. (1993). Distributed intelligence and designs for education. In G. Salomon (Ed.), *Distributed cognitions*. Cambridge: Cambridge University Press.

Piaget, J. (1952). *The child's conception of number*. New York: Norton.

Piaget, J. (1962). *Play, Dreams, and Imitation in Childhood* (C. G. a. F. M. Hodgson, Trans.). New York: Norton.

Piaget, J. (1963). *The Origins of Intelligence in Children* (M. Cook, Trans.). New York: Norton.

Piaget, J. (1968). Quantification, conservation, and nativism. *Science, 162*(3857), 976–979.

Piaget, J. (1970). Piaget's theory. In P. H. Mussen (Ed.), *Carmichael's manual of child psychology* (pp. 703–731). New York: Wiley.

Piaget, J. (1977). *The development of thought: The equilibration of cognitive structures.* New York: Viking Penguin.

Pica, P., Lemer, C., Izard, V. r., & Dehaene, S. (2004). Exact and approximate arithmetic in an Amazonian indigene group. *Science. Special Cognition and Behavior, 306*(5695), 499–503.

Posner, J. K., & Baroody, A. J. (1979). Number conservation in two West African societies. *Journal of Cross Cultural Psychology, 10*(4), 479–496.

Pospisil, L. J. (1963). *Kapauku Papuan economy* (Vol. 67). New Haven: Yale University Publications in Anthropology.

Quinn, N., & Holland, D. (1987). Culture and cognition. In D. Holland & N. Quinn (Eds.), *Cultural models in language and thought.* New York: Cambridge University Press.

Robbins, J. (1999). This is our money: Modernism, regionalism, and dual currencies in Urapmin. In D. Akin & J. Robbins (Eds.), *Money and modernity: State and local currencies in Melanesia* (pp. 82–102). Pittsburgh, PA: University of Pittsburgh Press.

Rogers, J. R. (1949). *Patrol report no. 2–Telefomin.* Telefomin, West Sepik: National archives & public records services of Papua New Guinea.

Rogoff, B. (1981). Schooling and the development of cognitive skills. In H. C. Triandis & A. Heron (Eds.), *Handbook of cross-cultural psychology* (Vol. 4). Boston: Allyn & Bacon.

Roth, W.-M. (1996). Knowledge diffusion in a grade 4–5 classroom during a unit on civil engineering: An analysis of a classroom community in terms of its changing resources and practices. *Cognition & Instruction, 14*(2), 179–220.

Roth, W.-M., & Radford, L. (2011). *A cultural historical perspective on mathematics teaching and learning.* Rotterdam: Sense Publishers.

Saxe, G. B. (1977). A developmental analysis of notational counting. *Child Development, 48*, 1512–1520.

Saxe, G. B. (1979). Developmental relations between notational counting and number conservation. *Child Development, 50*(1), 180–187.

Saxe, G. B. (1981a). Body parts as numerals: A developmental analysis of numeration among the Oksapmin in Papua New Guinea. *Child Development, 52*(1), 306–316.

Saxe, G. B. (1981b). Number symbols and number operations: Their development and interrelation. *Topics in Language Disorders, 2*(1), 67–76.

Saxe, G. B. (1981c). The changing form of numerical reasoning among the Oksapmin. Indigenous Mathematics Project. Working Paper 14. *Report: ED229268.* 12pp. Jan.

Saxe, G. B. (1981d). When fourth can precede second: A developmental analysis of an indigenous numeration system among Ponam islanders in Papua New Guinea. *Journal of Cross-Cultural Psychology, 12*(1), 37–50.

Saxe, G. B. (1982a). Culture and the development of numerical cognition: Studies among the Oksapmin of Papua New Guinea. In C. J. Brainerd (Ed.), *Children's logical and mathematical cognition* (pp. 157–176). New York: Springer-Verlag.

Saxe, G. B. (1982b). Developing forms of arithmetical thought among the Oksapmin of Papua New Guinea. *Developmental Psychology, 18*(4), 583–594.

Saxe, G. B. (1983). Culture, counting and number conservation. *International Journal of Psychology, 18*(3–4), 313–318.

Saxe, G. B. (1985). Effects of schooling on arithmetical understanding: Studies with Oksapmin children in Papua New Guinea. *Journal of Educational Psychology*, 77(5), 503–513.

Saxe, G. B. (1991). *Culture and cognitive development: Studies in mathematical understanding*. Hillsdale, NJ: Erlbaum.

Saxe, G. B., & Bermudez, T. (1996). Emergent mathematical environments in children's games. In L. Steffe, P. Nesher, G. Golding, & B. Greer (Eds.), *Theories of mathematical learning* (pp. 51–68). Hillsdale, NJ: Erlbaum.

Saxe, G. B., & Esmonde, I. (2004). *Unintended consequences: The shifting structure of the Oksapmin's indigenous number system*. Paper presented at the American Educational Research Association, San Diego, CA.

Saxe, G. B., & Esmonde, I. (2005). Studying cognition in flux: A historical treatment of fu in the shifting structure of Oksapmin mathematics. *Mind, Culture, and Activity* Special Issue: *Combining longitudinal, cross-historical, and cross-cultural methods to study culture and cognition, 12*(3–4), 171–225.

Saxe, G. B., Gearhart, M., Shaughnessy, M., Earnest, D., Cremer, S., Sitabkhan, Y., Platas, L., & Young, A. (2009). A methodological framework and empirical techniques for studying the travel of ideas in classroom communities. In B. Schwartz, T. Dreyfus, & R. Hershkowitz (Eds.), *Transformation of knowledge in classroom interaction* (pp. 203–222). London: Routledge.

Saxe, G. B., & Guberman, S. R. (1998). Emergent arithmetical environments in the context of distributed problem solving: Analyses of children playing an educational game. In J. Greeno & S. Goldman (Eds.), *Thinking practices* (pp. 237–256). Hillsdale, NJ: Erlbaum.

Saxe, G. B., Guberman, S. R., & Gearhart, M. (1987a). Social processes in early number development. *Monographs of the Society for Research in Child Development, 52*(2), iii–viii, 3–162.

Saxe, G. B., Guberman, S. R., & Gearhart, M. (1987b). Social processes in early number development. *Monographs of the Society for Research in Child Development*, 52, Serial No. 162.

Saxe, G. B., & Moylan, T. (1982). The development of measurement operations among the Oksapmin of Papua New Guinea. *Child Development, 53*(5), 1242–1248.

Saxe, G. B., & Posner, J. (1982). The development of numerical cognition: Cross-cultural perspectives. In H. P. Ginsburg (Ed.), *The development of mathematical thinking* (pp. 291–317). New York: Academic Press.

Schaeffer, B., Eggleston, V. H., & Scott, J. L. (1974). Number development in young children. *Cognitive Psychology, 6*, 357–379.

Schmandt-Besserat, D. (1992). *Before writing*. Austin: University of Texas Press.

Schmandt-Besserat, D. (2004). Accounting with tokens in the ancient Near East, from http://www.utexas.edu/cola/depts/lrc/numerals/dsb/dsb1.html.

Scribner, S. (1985). Vygotsky's uses of history. In J. V. Wertsch (Ed.), *Culture, communication, and cognition: Vygotskian perspectives*. New York: Cambridge University Press.

Scribner, S., & Cole, M. (1981). *The psychology of literacy*. Cambridge, MA: Harvard University Press.

Sfard, A. (2005). Discourse in flux. *Mind, Culture, & Activity*, *12*(3 & 4), 233–250.

Sfard, A. (2008). *Thinking as communicating*. Cambridge: Cambridge University Press.

Siegal, M. (1991). A clash of conversational worlds: Interpreting cognitive development through communication. In L. Resnick, B. J. M. Levine, & S. D. Teasley (Eds.), *Perspectives on socially shared cognition* (pp. 23–40). Washington, DC: American Psychological Association.

Siegler, R. S., & Crowley, K. (1991). The microgenetic method: A direct means for studying cognitive development. *American Psychologist*, *46*, 606–620.

Siegler, R. S., & Crowley, K. (1992). Microgenetic methods revisited. *American Psychologist*, *47*(10), 1241–1243.

Sitabkhan, Y. (2008). *Child vendors on the trains in Mumbai, India: A comparative case study*. University of California, Berkeley.

Smith, D. E. (1925). *History of mathematics*. New York: Ginn & Company.

Sperber, D., & Hirschfeld, L. (2001). Culture, cognition, and evolution. In R. A. Wilson & F. C. Keil (Eds.), *The MIT encyclopedia of the cognitive sciences (MITECS)* (pp. cxi–cxxxii). Cambridge, MA: MIT Press.

Starkey, P., Spelke, E. S., & Gelman, R. (1990). Numerical abstraction by human infants. *Cognition*, *36*(2), 97–127.

Steffe, L. P., von Glasersfeld, E., Richards, J., & Cobb, P. (1983). *Children's counting types: Philosophy, theory, and applications*. New York: Praeger.

Swadling, P., Mawe, T., & Tomo, W. (1990). Archaeology of Telefolip. In B. Craig & D. Hyndman (Eds.), *Children of Afek: Tradition and change among the Mountain-Ok of central New Guinea*. (Vol. 40, pp. 109–114). Sydney: University of Sydney.

Swetz, F. (1987). *Capitalism and arithmetic: The new math of the 15th century, including the full text of the Treviso arithmetic of 1478*, translated by David Eugene Smith. La Salle, IL: Open Court.

Taylor, E. V. (2009). The purchasing practice of low-income students: The relationship to mathematical development. *The Journal of the Learning Sciences*, *18*, 370–415.

Tomasello, M. (1999). *The cultural origins of human cognition*. Cambridge, MA: Harvard University Press.

Toulmin, S. E. (1972). *Human understanding*. Princeton, NJ: Princeton University Press.

Tylor, E. B. (1920). *Primitive culture*. New York: J. P. Putnam's Sons.

Uller, C., Jaeger, R., Guidry, G., & Martin, C. (2003). Salamanders (Plethodon cinereus) go for more: Rudiments of number in an amphibian. *Animal Cognition*, *6*(2), 105–112.

van Langernhove, L., & Harré, R. (1999). Introducing positioning theory. In R. Harré & L. v. Langenhove (Eds.), *Positioning theory: Moral contexts of intentional action* (pp. 14–31). Oxford: Blackwell.

Vygotsky, L. S. (1978). *Mind in society: The development of higher psychological processes*. Cambridge, MA: Harvard University Press.

Vygotsky, L. S. (1986). *Thought and language*. Cambridge, MA: MIT Press.

Vygotsky, L. S. (1987). Thinking and speech. In R. W. Riber & A. S. Carton (Eds.), *The collected works of L. S. Vygotsky, Volume 1: Problems of general psychology*. New York: Plenum.

Wakeley, A., Rivera, S., & Langer, J. (2000a). Can young infants add and subtract? *Child Development*, *71*(6), 1525–1534.

Wakeley, A., Rivera, S., & Langer, J. (2000b). Not proved: Reply to Wynn. *Child Development*, *71*(6), 1537–1539.

Wartofsky, M. (1973). Perception, representation, and the forms of action: Towards an historical epistemology. In M. Wartofsky (Ed.), *Models: Representation and the scientific understanding*. Dordrecht: Reidel.

Wassmann, J., & Dasen, P. R. (1994). Yupno number system and counting. *Journal of Cross-Cultural Psychology*, *25*(1), 78–94.

Weeks, S. G. (1981a). Introduction. In S. G. Weeks (Ed.), *Oksapmin: Development and change* (pp. 13–33). Port Moresby: University of Papua New Guinea.

Weeks, S. G. (1981b). Education. In S. G. Weeks (Ed.), *Oksapmin: Development and change* (pp. 95–134). Port Moresby: University of Papua New Guinea.

Weeks, S. G. (1981c). Oksapmin chronology. In S. G. Weeks (Ed.), *Oksapmin: Development and change* (pp. 218–225). Port Moresby: University of Papua New Guinea.

Weeks, S. G. (1981d). *Oksapmin: Development and change*. Port Moresby: University of Papua New Guinea, Educational Research Unit.

Werner, H. (1948). *Comparative psychology of mental development*. New York: International Universities Press.

Werner, H., & Kaplan, B. (1963). *Symbol formation: An organismic-developmental approach to language and the expression of thought*. New York: Wiley.

Wertsch, J. V. (1998). *Mind as action*. New York: Oxford University Press.

Wesch, M. (2004). My first trip to Tumolbil: Number (Vol. Nekalimin.net Version 3.0; http://www.people.virginia.edu/~mlw5k).

Whorf, B. L. (1956). *Language, thought, and reality*. Cambridge, MA: MIT Press.

Wurm, S. (1983). Linguistic prehistory in the New Guinea area. *Journal of Human Evolution*, *12*, 25–35.

Wynn, K. (1992). Addition and subtraction by human infants. *Nature*, *358*(6389), 749–750.

Wynn, K. (2000). Addition and subtraction by human infants. *Muir, Darwin*.

Xu, F., & Arriaga, R. I. (2007). Number discrimination in 10-month-old infants. *British Journal of Developmental Psychology*, *25*(1), 103–108.

Xu, F., & Spelke, E. S. (2000). Large number discrimination in 6-month-old infants. *Cognition*, *74*(1), B1–B11.

Xu, F., Spelke, E. S., & Goddard, S. (2005). Number sense in human infants. *Developmental Science*, *8*(1), 88–101.

Zaslavsky, C. (1973). *Africa counts*. Boston: Prindle, Weber, & Schmidt.

Zebian, S. (2004). Influences of cultural artifacts and social practices on number conceptualization in adults: Experimental and ethnographic approaches to everyday numeric cognition. In B. N. Setiadi, A. Supratiknya, W. J. Lonner, & Y. H. Poortinga (Eds.), *Ongoing themes in psychology and culture (Online Ed.)*. Melbourne, FL: International Association for Cross-Cultural Psychology. Retrieved from http://www.iaccp.org.

Zimansky, P. (1993). Review of Before writing. Volume I: From counting to cuneiform; Before writing. Volume II: A catalogue of Near Eastern tokens. *Journal of Field Archaeology*, *20*(4), 513–517.

Index

BOOKS IN THE SERIES *(continued from page iii)*

Computation and Human Experience
Philip E. Agre

Situated Cognition: On Human Knowledge and Computer Representations
William J. Clancey

Communities of Practice: Learning, Meaning, and Identity
Etienne Wenger

Learning in Likely Places: Varieties of Apprenticeship in Japan
John Singleton, Editor

Talking Mathematics in School: Studies of Teaching and Learning
Magdalene Lampert and Merrie L. Blunk, Editors

Perspectives on Activity Theory
Yrjö Engeström, Reijo Miettinen, and Raija-Leena Punamäki, Editors

Dialogic Inquiry: Towards a Socio-cultural Practice and Theory of Education
Gordon Wells

Vygotskian Perspectives on Literacy Research: Constructing Meaning Through Collaborative Inquiry
Carol D. Lee and Peter Smagorinsky, Editors

Technology in Action
Christian Heath and Paul Luff

Changing Classes: School Reform and the New Economy
Martin Packer

Building Virtual Communities: Learning and Change in Cyberspace
K. Ann Renninger and Wesley Shumar, Editors

Adult Learning and Technology in Working-Class Life
Peter Sawchuk

Vygotsky's Educational Theory in Cultural Context
Alex Kozulin, Boris Gindis, Vladimir S. Ageyev, and Suzanne M. Miller, Editors

Designing for Virtual Communities in the Service of Learning
Sasha A. Barab, Rob Kling, and James H. Gray, Editors

Bakhtinian Perspectives on Language, Literacy, and Learning
Arnetha F. Ball and Sarah Warshauer Freedman, Editors

Beyond Communities of Practice: Language, Power and Social Context
David Barton and Karin Tusting, Editors

Human-Machine Reconfigurations: Plans and Situated Actions, 2nd Edition
Lucy Suchman

Psychotherapy in Everyday Life
Ole Dreier

Thinking as Communicating: Human Development, the Growth of Discourses, and Mathematizing
Anna Sfard

Assessment, Equity, and Opportunity to Learn
Pamela A. Moss et al., Editors

From Teams to Knots: Studies of Collaboration and Learning at Work
Yrjö Engeström

Acting in Anaesthesia: Ethnographic Encounters with Patients, Practitioners and Medical Technologies
Dawn Goodwin

The Materiality of Learning: Technology and Knowledge in Educational Practice
Estrid Sørensen

How Students Come to Be, Know, and Do: A Case for a Broad View of Learning
Leslie Rupert Herrenkohl and Véronique Mertl

Making Work Visible: Ethnographically Grounded Case Studies of Work Practice
Margaret H. Szymanski and Jack Whalen, Editors

Embodied Interaction: Language and Body in the Material World
Jürgen Streeck, Charles Goodwin, and Curtis LeBaron, Editors

The Learning in Doing series was founded in 1987 by Roy Pea and John Seely Brown.